Thermal Engineering Volume 1

Shiv Kumar

Thermal Engineering
Volume 1

Ane Books
Pvt. Ltd.

Springer

Shiv Kumar
Department of Mechanical and Automation
Engineering
Guru Gobind Singh Indraprastha University
Delhi, India

ISBN 978-3-030-67276-8   ISBN 978-3-030-67274-4   (eBook)
https://doi.org/10.1007/978-3-030-67274-4

Jointly published with ANE Books Pvt. Ltd.
In addition to this printed edition, there is a local printed edition of this work available via Ane Books in
South Asia (India, Pakistan, Sri Lanka, Bangladesh, Nepal and Bhutan) and Africa (all countries in the
African subcontinent).
ISBN of the Co-Publisher's edition: 978-9-3854-6232-0

This Springer imprint is published by the registered company Springer Nature Switzerland AG
The registered company address is: Gewerbestrasse 11, 6330 Cham, Switzerland

Dedicated to

My Parents

My Wife Dr. Kusum Lata and My Son Tanishq

Preface

Thermodynamics deals with the storage, transformation and transfer of energy. As an engineering discipline, Thermal Engineering deals with the innovative use of the laws of thermodynamics in solving relevant technological problems. This introductory textbook aims to provide undergraduate engineering students with the knowledge (basics principles and thermodynamics laws) they need to have to know, understand and analyze the thermodynamic problems they are likely to encounter in practice.

The book is developed in such a way that the most complex things are explained in simplest of manners to ensure thorough understanding for the reader. The book is concise and every concept is presented from an elementary and tangible perspective.

The subject matter is well illustrated with a innumerable examples. A great deal of attention is given to select the numerical problems and solving them. The theory and numerical problems at the end of each chapter also aim to enhance the creative capabilities of students.

Text books are dynamic and need to change with time. Suggestions from the teachers and students for the further improvement of the text are welcome and will be implemented in the next edition. The readers are requested to bring out the error to the notice, which will be suitably acknowledged.

Shiv Kumar

Acknowledgements

First of all, I would like to express my deep gratitude to God for giving me the strength and health for completing this book. I am very thankful to my colleagues in the mechanical engineering department for their highly appreciable help and my students for their valuable suggestions.

I am also thankful to my publishers Shri. Sunil Saxena and Shri. Jai Raj Kapoor of Ane Books Pvt. Ltd. and the editorial group for their help and assistance.

A special thanks goes to my wife Dr. Kusum for her help, support, and strength to complete the book.

Shiv Kumar

Contents

About the Author

Shiv Kumar is former Head of Department of Mechanical and Automation Engineering and Associate Professor at Guru Premsukh Memorial College of Engineering affiliated to Guru Gobind Singh Indraprastha University, Delhi, India. He obtained his Bachelor's degree in Mechanical Machine Design and Automation Engineering from R. E. C. (presently NIT), Jalandhar, and Master's degree in Thermal Engineering from Delhi College of Engineering, India. He has over 18 years of teaching experience. He has published several books in the field of Mechanical Engineering.

Chapter 1
Basic of Thermodynamics

Nomenclature

The following is the nomenclature introduced in this chapter:

STP	–	Standard Temperature and Pressure
NTP	–	Normal Temperature and Pressure
CV	m^3	Control volume
CS	m^2	Control surface
m	kg	Mass
V	m^3	Volume
v	m^3/kg	Specific volume
H	kJ	Enthalpy
h	kJ/kg	Specific enthalpy
E	kJ	Energy
e	kJ/kg	Specific energy
T	K	Temperature
p	kPa	Pressure
ρ	kg/m^3	Density
S	kJ/K	Entropy
s	kJ/kgK	Specific entropy
R	kJ/kgK	Gas constant
$\bar{R}$	kJ/kmol K	Universal gas constant
M	kg/kmol	Molecular weight
n	kmol	Number of moles

(continued)

© The Author(s) 2022
S. Kumar, *Thermal Engineering Volume 1*,
https://doi.org/10.1007/978-3-030-67274-4_1

(continued)

γ	—	Adiabatic index
c_p	kJ/kgK	Specific heat at constant pressure
c_v	kJ/kgK	Specific heat at constant volume
n	—	Polytropic index
W	N	Weight
w	N/m^3	Specific weight
g	m/s^2	Acceleration due to gravity
S	—	Specific gravity
h	m	Pressure head
Q	kJ	Heat transfer
W	kJ	Work
KE	J	Kinetic energy
PE	J	Potential energy
z	m	Datum head
U	kJ	Internal energy
u	kJ/kg	Specific internal energy
Z	—	Compressibility factor

1.1 Introduction

Thermodynamics is a science that deals with the storage, transformation, and transfer of energy. The energy is stored in the system in the form of internal energy (due to temperature), kinetic energy (due to motion), potential energy (due to elevation), and chemical energy (due to composition); it is transformed from one form of energy to another form. By transfer of energy, we mean that it can cross a boundary of the system either as heat[1] or work.

   OR

   Thermodynamics is a science that deals with the relation among the heat, work, and properties of the system in equilibrium.

---

[1] Heat (or heat transfer) is the type of transfer of energy, and it exists only due to difference in temperature between the system and the surrounding of the system. Heat cannot be stored in the system because it is not a property of the system.

1.2 Analysis of Matter

There are two points of view from which the behaviour of matter can be studied:

(i)   Microscopic view and
(ii) Macroscopic view.

(i)   **Microscopic View**. When the study is performed at the molecular level of the matter, it is called microscopic view. A certain quantity of matter is chosen to study the behaviour of the individual molecule. Then by adopting a statistical approach, collective molecular activity is analysed.
(ii) **Macroscopic View**. When the study is performed on the matter or whole system, it is called macroscopic view. A certain quantity of matter is chosen to study the behaviour of all molecules (i.e., whole system). It is also called the classical approach because all the molecules are considered under study.

1.3 Continuum

Materials, such as solids, liquids, and gases, consist of discrete molecules separated by empty space. However, certain physical phenomena, such as analysis of the fluid flow problem, are made by assuming the concept that matter exists as a continuum, i.e., matter is viewed as a continuum, homogeneous matter with no holes (empty space or void) in between the molecules, for example, between fluid particles in the case of fluids. The continuum idealization allows us to treat properties of the fluids continuous from point to point and vary continuously with no jump discontinuities. So a continuous and homogeneous fluid medium is called continuum. A continuum idealization is obvious in the statement we make, such as the density of water in a glass is the same at any point.

1.4 Standard Temperature and Pressure (STP) and Normal Temperature and Pressure (NTP)

1.4.1 Standard Temperature and Pressure (STP)

It refers to the condition of the standard atmospheric pressure of 760 mm of Hg (1.01325 bar) and a temperature of 15 °C (or 288 K), i.e., the values of temperature and pressure at STP are 15 °C and 760 mm of Hg, respectively.

1.4.2 Normal Temperature and Pressure (NTP)

It refers to the condition of the standard atmospheric pressure of 760 mm of mercury (1.01325 bar) and a temperature of 0 °C (or 273 K), i.e., the values of temperature and pressure at NTP are 0 °C and 760 mm of Hg, respectively.

1.5 Thermodynamic System

A system is defined as the quantity of matter (or region in space) upon which we have to make the study. In the case of a thermodynamic system, we study the thermodynamic behaviour of the system.

The region outside the system is called the **surroundings** (i.e., everything external to the system is called the surroundings). The real or imaginary surface that separates the system from its surroundings is called the **boundary**, as shown in Figs. 1.1 and 1.2. The boundary of a system can be fixed or movable. Note that the boundary is the contact surface shared by both the system and the surroundings. Mathematically, the boundary has negligible thickness and thus it can neither contain any mass nor occupy any volume in space (Fig. 1.3).

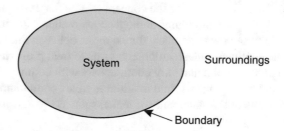

Fig. 1.1 System, surroundings and boundary

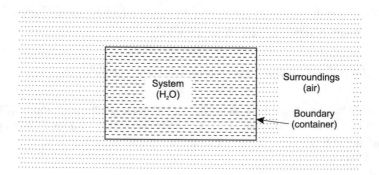

Fig. 1.2 System (water), surroundings (air), and boundary (container)

Fig. 1.3 A closed system
with a moving boundary

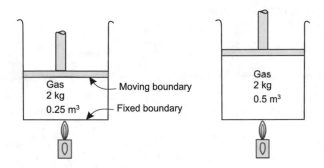

A system and its surroundings when put together is called a **universe**:

$$\boxed{Universe = System + Surroundings}$$

1.6 Types of Thermodynamic Systems

The thermodynamic system can be classified into three types:

1. Open system,
2. Closed system, and
3. Isolated system.

1.6.1 Open System

In this type of system, mass of fluid flow can cross the boundary of the system. If mass of fluid flow can cross the boundary of the system, then energy also transfers with mass of fluid flow because energy is an extensive property. In a simple way, the open system is defined as the system in which both mass and energy can cross the boundary of the system (Fig. 1.4).

The region that usually encloses the open system is called **control volume**. The surface which surrounds a control volume is referred to as the **control surface**.

An IC engine, a pump, a turbine, a nozzle, a diffuser, an inflating or deflating balloon, and hot coffee in an open flask are examples of the open system or control volume (Fig. 1.5).

Fig. 1.4 Open system

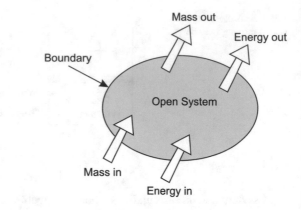

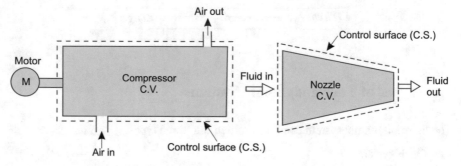

Fig. 1.5 Control volume (CV) and control surface (CS)

1.6.2 Closed System

A closed system is also known as a control mass; it consists of a fixed amount of mass and no mass can cross its boundary. But energy, in the form of heat or work, can cross the boundary, and the volume of the closed system is either fixed or changed. Examples of a closed system are (Fig. 1.6).

Fig. 1.6 Closed system

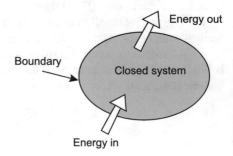

- Hot coffee in a closed stainless steel flask;
- Boiling of water in a closed pan;
- Expansion of gas in an internal combustion engine.

1.6.3 Isolated System

The isolated system is one in which both mass and energy cannot cross the boundary of the system. In this type of system, even chemical reactions don't take place, for example, hot coffee in a thermos flask (Fig. 1.7, Table 1.1).

Fig. 1.7 Isolated system

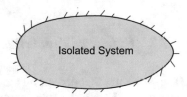

Table 1.1 Comparison of Open System, Closed System, and Isolated System

Open System	Closed System	Isolated System
Both mass and energy can cross the boundary of the system (or mass of fluid flow can cross the boundary of the system). It means if mass is transferred, then energy is also transferred. An open system is also called control volume Examples: – Hot coffee in an open flask – IC engine – Compressor – Turbine – Nozzle and diffuser, etc	Only energy can cross the boundary of the system, and a fixed amount of mass in the system. A closed system is also known as a control mass Examples: – Hot coffee in a closed stainless steel flask – Boiling of water in a closed pan – Expansion of gas in an internal combustion engine	Both mass and energy cannot cross the boundary of the system (or energy cannot cross the boundary of the system. It means if energy is not transferred, then mass transferred is impossible). Even there is no chemical reaction that takes place within the isolated system Example: – Hot coffee in a thermos flask

1.7 Properties of System

The characteristics used to describe the condition (or state) of the system are called **properties of the system**. Temperature—T, pressure—p, volume—V, energy—E, mass—m, etc., are properties of the system. These properties of a system can be classified into two types:

(i)   Intensive properties, and
(ii) Extensive properties.

1.7.1 Intensive Properties

The properties that are independent of the mass (or size) of the system are called **intensive properties**. Temperature, pressure, density, viscosity, specific heat, and thermal conductivity are examples of intensive properties.

1.7.2 Extensive Properties

The properties that depend on the mass (or size) of the system are called **extensive properties**. Energy, enthalpy, and entropy are examples of extensive properties.

Extensive properties per unit mass such as specific volume, specific energy, specific entropy, and specific enthalpy are called intensive properties, which are also known as specific properties.

An easy way to determine whether a property is intensive or extensive is to divide the system among four equal parts with an imaginary partition, as shown in Fig. 1.8.

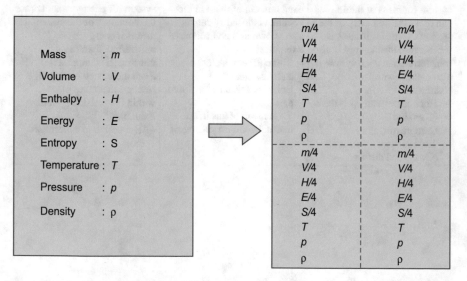

Fig. 1.8 Easy way to differentiate intensive and extensive properties

Each part will have the same value of intensive properties as the original system but 1/4th the value of the extensive properties.

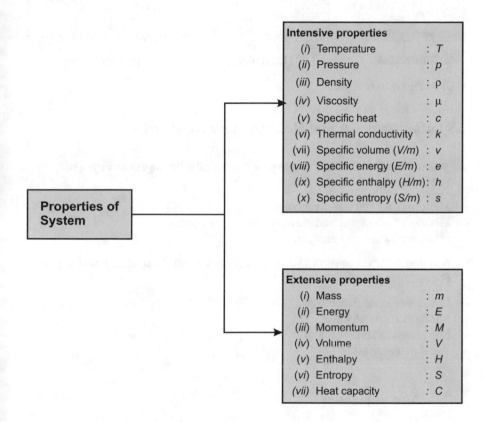

1.8 Phase

A phase is a quantity of matter which is uniform throughout, both in chemical composition and physical structure. Uniform chemical composition means that the chemical composition does not vary at any part of the system, and uniform physical structure means matter in all gas, or all liquid, or all solid.

For example, ice has a solid phase, water has a liquid phase, and steam has a vapour or gaseous phase (Fig. 1.9).

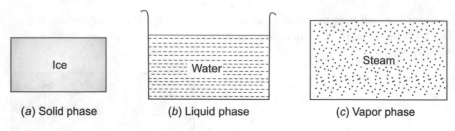

Fig. 1.9 Three phases of water

1.9 Homogeneous and Heterogeneous Systems

A system which consists of a single phase is called a **homogeneous system**. For example,

- Mixture of air and water vapour;
- Mixture of iso-octane and normal heptane (i.e., petrol);
- Mixture of water and milk, etc.

A system which consists of two or more phases is called a **heterogeneous system**. For example,

- Mixture of ice and mercury,
- Mixture of ice and water,
- Iron pieces in kerosene, and
- Wood pieces in oil.

1.10 Pure Substance

A pure substance is a single substance which has an unvarying molecular structure during the process of energy transfer. It may exist in more than one phase, but the chemical composition is the same in all phases. Thus, a mixture of liquid water and water vapour (steam), or a mixture of ice and liquid water, is all pure substance, for every phase has the same chemical composition. On the other hand, a mixture of liquid air and gaseous air is not a pure substance, since the composition of the liquid phase is different from that of the gaseous phase.

1.11 State

The state of a system is the specific condition of a system at any instant of time. The state of a system is described or measured by its properties such as temperature: T; pressure: p; and volume: V. Minimum two parameters are required to describe the state of a system in p–v and T-s diagrams. It is denoted by a point in p-v and T-s diagrams as shown in Fig. 1.10.

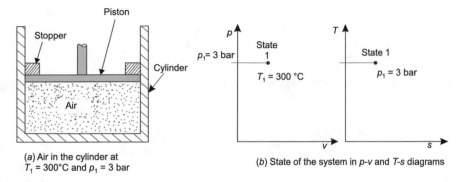

(a) Air in the cylinder at
$T_1 = 300°C$ and $p_1 = 3$ bar

(b) State of the system in p-v and T-s diagrams

Fig. 1.10 State

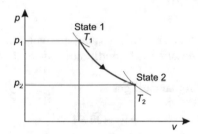

Fig. 1.11 One process, two states

1.12 Process

A process occurs when the system undergoes a change or series of changes in the state of the system. If the stopper of the piston is removed in Fig. 1.10a, the air inside the cylinder allows expanding from state 1 (i.e., p_1 and T_1) to state 2 (i.e., p_2 and T_2). So this change in state from p_1, T_1 to p_2, T_2 is called a process, as shown in Fig. 1.11.

1.13 Cycle

When the initial and final states of a series of processes of the system are identical, it is called a **cycle**. The thermodynamic cycles are shown in Fig. 1.12, with two processes and two states and with four processes and four states.

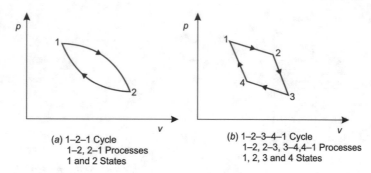

(a) 1–2–1 Cycle
1–2, 2–1 Processes
1 and 2 States

(b) 1–2–3–4–1 Cycle
1–2, 2–3, 3–4,4–1 Processes
1, 2, 3 and 4 States

Fig. 1.12 Cycle

1.14 Equation of State

The equation involving temperature, pressure, and volume (or specific volume or density) which used to describe the condition or state of the system is called an **equation of state**.

The equation of state for an ideal gas is

$$pV = mRT \tag{1.1}$$

where

p = absolute pressure of a perfect gas is in kPa,
V = volume of a perfect gas is in m³,
m = mass of a perfect gas is in kg,
R = gas constant is in kJ/kgK,
       = 0.287 kJ/kgK for air,
T = absolute temperature, i.e., temperature in K (kelvin).

or

$$\frac{pV}{m} = RT$$

or

$$pv = RT \tag{1.2}$$

where
$v = \frac{V}{m}$, specific volume is in m³/kg,

also
$v = \frac{1}{\rho}$, specific volume is the reciprocal of the density, and its SI unit is m³/kg,

∴

$$\frac{p}{\rho} = RT$$

$$p = \rho RT \tag{1.3}$$

Gas Constant: R. It is defined as the ratio of the universal gas constant $(\overline{R})$ to the molecular weight of a perfect gas.

Mathematically,

$$\text{Gas constant: } R = \frac{\text{Universal gas constant: } \overline{R}}{\text{Molecular weight: } M}$$

$$R = \frac{\overline{R}}{M}$$

where

$$\overline{R} = 8.314 \text{ kJ/kmol K}$$

S. no.	Ideal gas	Molecular weight: M kg/kmol	Universal gas constant: $\overline{R}$ kJ/kmol K	Gas constant: R kJ/kgK
1	Air	28.92	8.314	$\frac{8.314}{44} = 0.287$
2	CO$_2$	44	8.314	$\frac{8.314}{44} = 0.189$
3	O$_2$	32	8.314	$\frac{8.314}{32} = 0.259$
4	N$_2$	28	8.314	$\frac{8.314}{28} = 0.297$
5	H$_2$O(V) (Superheated steam)	18	8.314	$\frac{8.314}{18} = 0.462$

Substituting

$$R = \frac{\overline{R}}{M} \text{ in Eq. (1.1), we get}$$
$$pV = \frac{m\overline{R}T}{M}$$

$$pV = n\overline{R}T \tag{1.4}$$

where

$n = \frac{m}{M}$, number of moles. It is defined as the ratio of the mass of a gas to the molecular weight of a gas.

Equations (1.1), (1.2), (1.3), and (1.4) are all forms of the equations of states. These equations are applicable at any state of the system for an ideal gas; for example, if we know V, m, R, and T, then p can be found out by using Eq. (1.1).

1.15   Thermodynamics Processes

Thermodynamics processes are used to study the thermodynamics behaviour of a series of changes in the state of the system. The following are the main thermodynamic process:

(i)   Isothermal process,
(ii)   Isobaric process,
(iii)   Isochoric process,
(iv)   Adiabatic process, and
(v)   Polytropic process.

1.15.1   Isothermal Process [T = C]

The isothermal process takes place at constant temperature. The volume (or specific volume) of the gas varies inversely with pressure at constant temperature in this process.

From the equation of state,

$$pv = RT$$

For any process, R is constant and T is constant for the isothermal process (Fig. 1.13):

$$\therefore$$

$$pv = C$$

or

$$\frac{pV}{m} = C \quad | \because \text{ Specific volume: } v = \frac{V}{m}$$

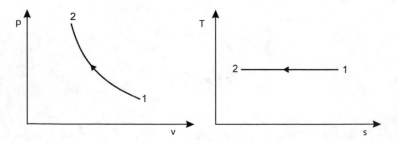

Fig. 1.13   Isothermal compression process in *p-v* and *T-s* diagrams

or

$$pV = C$$

Hence, the product of absolute pressure and volume (or specific volume) is constant.

For isothermal process 1–2,

$$p_1 v_1 = p_2 v_2$$

or

$$\frac{v_2}{v_1} = \frac{p_1}{p_2}$$

The isothermal process follows **Boyle's law** which states that the specific volume of a perfect gas is inversely proportional to the absolute pressure when the temperature is kept constant.

Mathematically,

$$v \propto \frac{1}{p} \quad \text{at } T = C$$

or

$$pv = C$$

1.15.2 *Isobaric Process [P = C]*

The isobaric process takes place at constant pressure. In this type of process, the volume (or specific volume) of the gas varies directly with the temperature at constant pressure.

From the equation of state (Fig. 1.14),

$$pv = RT$$

or

$$\frac{v}{T} = \frac{R}{p}$$

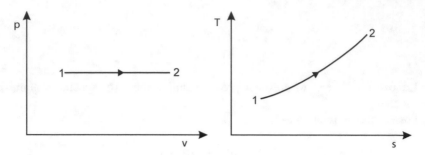

Fig. 1.14 Isobaric process in *p-v* and *T-s* diagrams

or

$$\frac{v}{T} = C$$

For isothermal process 1–2,

$$\frac{v_1}{T_1} = \frac{v_2}{T_2}$$

or

$$\frac{\boldsymbol{T}_2}{\boldsymbol{T}_1} = \frac{\boldsymbol{v}_2}{\boldsymbol{v}_1}$$

The isobaric process follows **Charles' law** which states that the specific volume of a perfect gas is directly proportional to the absolute temperature when the pressure is kept constant.

Mathematically,

$$v \propto T \quad \text{at } p = C$$
$$\frac{v}{T} = C$$

1.15.3   *Isochoric Process (or Isometric Process) [V = C]*

The isochoric process takes place at constant volume. In this type of process, the pressure of the gas varies directly with the temperature at constant volume (Fig. 1.15).

From the equation of state,

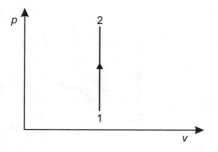
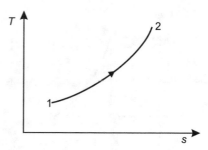

Fig. 1.15 Isochoric process in *p-v* and *T-s* diagrams

$$pv = RT$$
$$\frac{p}{T} = \frac{R}{v}$$
$$\frac{p}{T} = C$$

For isochoric process 1–2,

$$\frac{p_1}{T_1} = \frac{p_2}{T_2}$$

or

$$\frac{T_2}{T_1} = \frac{p_2}{p_1}$$

The isochoric process follows **Gay-Lussac's law** (or **Amonton's law**) which states the absolute pressure of a perfect gas is directly proportional to the absolute temperature when the volume is kept constant.

Mathematically,

$$p \propto T \quad \text{at } v = C$$

or

$$\frac{p}{T} = C$$

1.15.4 Adiabatic Process [$pv^{\gamma} = C$]

When there is no heat transfer between the system and surroundings during a process, it is known as an adiabatic process. The properties like pressure, temperature, and

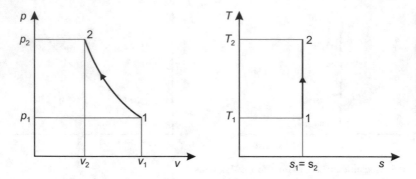

Fig. 1.16 Adiabatic process (i.e., adiabatic compression) in p-v and T-s diagrams

volume of the system will vary during this process. Normally, this process is considered as an ideal process and also called reversible adiabatic (or adiabatic isoentropic). In this process, the entropy of the system will remain constant without the transfer of heat to or from the surroundings.

This process can be represented by the equation (Fig. 1.16)

$$pv^\gamma = C$$

$$\frac{RT}{v}v^\gamma = C$$

$$[pv = RT \text{ or } p = \frac{RT}{v}]$$

$$Tv^{\gamma-1} = C$$

where

$$\gamma = \frac{c_p}{c_v}, \text{ adiabatic index. It is property of the gas.}$$

$$= 1.4 \text{ for air.}$$

Adiabatic index (γ) is defined as the ratio of specific heat at constant pressure (c_p) and the specific heat at constant volume (c_v).

For adiabatic process 1–2,

$$p_1 v_1^\gamma = p_2 v_2^\gamma$$

$$\frac{p_2}{p_1} = \left(\frac{v_1}{v_2}\right)^\gamma = \left(\frac{v_2}{v_1}\right)^{-\gamma} \tag{1.5}$$

From the equation of state,

$$pv = RT$$

$$\frac{pv}{T} = R = C$$

For process 1–2,

$$\frac{p_1 v_1}{T_1} = \frac{p_2 v_2}{T_2}$$

or,

$$\frac{v_1}{v_2} = \frac{p_2}{p_1} \times \frac{T_1}{T_2}$$

Substituting the value of $\frac{v_1}{v_2}$ in Eq. (1.5), we get

$$\frac{p_2}{p_1} = \left(\frac{p_2}{p_1} \times \frac{T_1}{T_2}\right)^{\gamma} = \left(\frac{p_2}{p_1}\right)^{\gamma} \times \left(\frac{T_1}{T_2}\right)^{\gamma}$$

or,

$$\left(\frac{T_2}{T_1}\right)^{\gamma} = \left(\frac{p_2}{p_1}\right)^{\gamma-1}$$

or,

$$\boldsymbol{\frac{T_2}{T_1} = \left(\frac{p_2}{p_1}\right)^{\frac{\gamma-1}{\gamma}}} \tag{1.6}$$

Substituting the value of $\frac{p_2}{p_1}$ from Eq. (1.5) in Eq. (1.6), we get

$$\frac{T_2}{T_1} = \left[\left(\frac{v_1}{v_2}\right)^{\gamma}\right]^{\frac{\gamma-1}{\gamma}}$$

$$\frac{T_2}{T_1} = \left(\frac{v_1}{v_2}\right)^{\gamma-1} \tag{1.7}$$

From Eqs. (1.15) and (1.16), we get

$$\boldsymbol{\frac{T_2}{T_1} = \left(\frac{p_2}{p_1}\right)^{\frac{\gamma-1}{\gamma}} = \left(\frac{v_1}{v_2}\right)^{\gamma-1}}$$

also

$$\boldsymbol{\frac{T_2}{T_1} = \left(\frac{p_2}{p_1}\right)^{\frac{\gamma-1}{\gamma}} = \left(\frac{v_1}{v_2}\right)^{\gamma-1} = \left(\frac{\rho_2}{\rho_1}\right)^{\gamma-1}}$$

For adiabatic process:

(a)   **Relation between p and T**

$$\frac{T_2}{T_1} = \left(\frac{p_2}{p_1}\right)^{\frac{\gamma-1}{\gamma}}$$

(b)   **Relation between p and v**

$$\left(\frac{p_2}{p_1}\right)^{\frac{\gamma-1}{\gamma}} = \left(\frac{v_1}{v_2}\right)^{\gamma-1}$$

or

$$\frac{p_2}{p_1} = \left(\frac{v_1}{v_2}\right)^{\gamma}$$

(c)   **Relation between T and v**

$$\frac{T_2}{T_1} = \left(\frac{v_1}{v_2}\right)^{\gamma-1} = \left(\frac{v_2}{v_1}\right)^{1-\gamma}$$

(d)   **Relation between T and ρ**

$$\frac{T_2}{T_1} = \left(\frac{\rho_2}{\rho_1}\right)^{\gamma-1}$$

1.15.5   *Polytropic Process [$pv^n = C$]*

The polytropic process is a reversible process like the adiabatic process but the difference is that in the case of the polytropic process, the entropy change occurs due to reversible heat transfer. This process can be represented by the equation (Fig. 1.17):

$$pv^n = C$$

where

$n =$ polytropic index. It is not a property of the gas. The value of n depends upon the process

   Characteristics of the polytropic process:

(a)   Entropy of the process changes.
(b)   Both heat and work transfer take place.

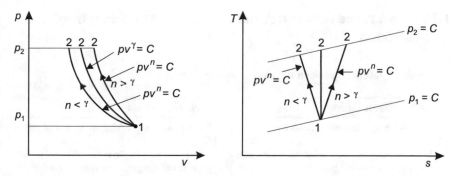

Fig. 1.17 Polytropic process in p-v and T-s diagrams

The relations among T, p, and v are obtained by replacing n instead of γ in relations for the adiabatic process:

$$\frac{T_2}{T_1} = \left(\frac{p_2}{p_1}\right)^{\frac{n-1}{n}} = \left(\frac{v_1}{v_2}\right)^{n-1}$$

For polytropic process:

(a) **Relation between p and T**

$$\frac{T_2}{T_1} = \left(\frac{p_2}{p_1}\right)^{\frac{n-1}{n}}$$

(b) **Relation between p and v**

$$\frac{p_2}{p_1} = \left(\frac{v_1}{v_2}\right)^{n}$$

(c) **Relation between T and v**

$$\frac{T_2}{T_1} = \left(\frac{v_1}{v_2}\right)^{n-1} = \left(\frac{v_2}{v_1}\right)^{1-n}$$

(d) **Relation between T and ρ**

$$\frac{T_2}{T_1} = \left(\frac{\rho_2}{\rho_1}\right)^{n-1}$$

1.16   Difference Between Adiabatic Process and Polytropic Process

The main difference between the two processes are as follows:

S. no.	Adiabatic process	Polytropic process
1	Adiabatic process follows the law of $pv^{\gamma} = C$, where $\gamma = c_p/c_v$, adiabatic index. It is a property of the gas	Polytropic process follows the law of $pv^n = C$, where $n =$ polytropic index. It is not a property of the gas. The values of n depend upon the process. The value of n is found out by $$p_1 v_1^n = p_2 v_2^n$$ or $$n = \frac{\log_e \frac{p_1}{p_2}}{\log_e \frac{v_2}{v_1}}$$
2	Only work transfer takes a plate and no heat transfer occurs	Both heat and work transfer take place
3	Entropy remains constant in this process which shows the process is reversible	Polytropic process is a reversible process, but entropy change takes place (an increase of entropy may result from reversible heat transfer to the system from the surroundings and decrease of entropy when reversible heat transfer to the surroundings from the system)

1.17   All Processes are Defined on the Basis of Polytropic Law $[PV^n = C]$ by Varying the Values of Polytropic Index n

The following processes are defined on the basis of the polytropic law $pV^n = C$ by varying the values of n.

(i)   Isothermal process,
(ii)   Isobaric process,
(iii)   Isochoric process (or Isometric process),
(iv)   Adiabatic process, and
(v)   Polytropic process.

(i)   **Isothermal Process**. For the isothermal process, the value of polytropic index $n = 1$.

From polytropic law,

$$pV^n = C$$

For $n = 1$,

$$pV = C$$

which represents constant temperature or isothermal process. Both the compression and expansion processes are shown in the upper left quadrant and the lower right quadrant, respectively, in Fig. 1.18.

(ii) **Isobaric Process.** For the isobaric process, the value of polytropic index $n = 0$.

From polytropic law,

$$pV^n = C$$

For $n = 0$,

$$pV^0 = C$$
$$p = C \quad [\because V^0 = 1]$$

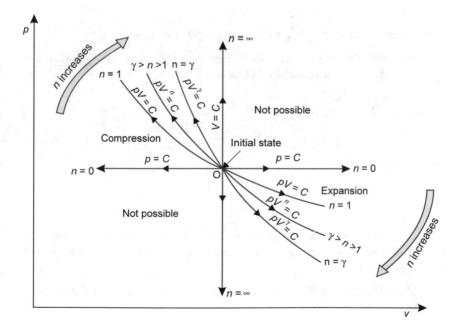

Fig. 1.18 Process for various values n

which shows a constant pressure or isobaric process. Both the compression and expansion processes are shown in Fig. 1.18.

(iii) **Isochoric Process**. For the isochoric process, the value of polytropic index $n = \infty$.

From polytropic law,

$$pV^n = C$$
$$p^{1/n}V = C^{1/n}$$
$$p^{1/n}V = C'$$

where C' is another constant.

For $n = \infty$,

$$p^{1/\infty}V = C'$$
$$p^0 V = C' \quad [\because \frac{1}{\infty} = 0]$$

or

$$V = C'$$

which indicates a constant volume or isochoric process.

(iv) **Adiabatic Process**. For the adiabatic process, the value of polytropic index $n = \gamma$, where γ is the adiabatic index.

From polytropic law,

$$pV^n = C$$

For $n = \gamma$,

$$pV^\gamma = C$$

which represents the adiabatic process. Both the compression and expansion processes are shown in the upper left quadrant and the lower right quadrant, respectively, in Fig. 1.18.

(v) **Polytropic Process**. For the polytropic process, the value of n lies between 1 and n.

That is,

$$pV^n = C$$

Taking $\log_e$ on both sides, we get

$$\log_e p + n \log_e V = \log_e C$$

On differentiation, we get

$$\frac{dp}{p} + n\frac{dV}{V} = 0$$

or

$$\frac{dp}{p} = -n\frac{dV}{V}$$

or

$$\frac{dp}{dV} = -n\frac{p}{V}$$

The slope of the process thus increases in the negative direction with an increase in the value of n (Table 1.2).

Table 1.2 Summary of processes for perfect gas

S. no.	Process	Law	p-v-T relation	Polytropic Index: n
1	Isothermal, i.e., $T = C$	Boyle's law	$p_1 v_1 = p_2 v_2$	$n = 1$
2	Isobaric, i.e., $p = C$	Charles' law	$\frac{v_1}{T_1} = \frac{v_2}{T_2}$	$n = 0$
3	Isochoric OR Isometric, i.e., $V = C$	Gay-Lussac's law	$\frac{p_1}{T_1} = \frac{p_2}{T_2}$	$n = \infty$
4	Adiabatic	$pv^\gamma = C$	$\frac{T_2}{T_1} = \left(\frac{p_2}{p_1}\right)^{\frac{\gamma-1}{\gamma}} = \left(\frac{v_1}{v_2}\right)^{\gamma-1}$	$n = \gamma$
5	Polytropic	$pv^n = C$	$\frac{T_2}{T_1} = \left(\frac{p_2}{p_1}\right)^{\frac{n-1}{n}} = \left(\frac{v_1}{v_2}\right)^{n-1}$	$n = n$

1.18   Avogadro's Law

Avogadro's law (also called Avogadro's hypothesis) is a principle stated in 1811 by the Italian chemist Amedeo Avogadro (1776–1856) that 'equal volumes of gases at the same temperature and pressure contain the same number of molecules regardless of their chemical nature and physical properties'. From Avogadro's law, the converse follows that equal numbers of molecules of any gases [i.e., the lightest gas (hydrogen) as for a heavy gas such as carbon dioxide or bromine] under identical conditions occupy equal volumes.

Mathematically,

$$pV = mRT$$
$$pV = n\overline{R}T$$

where

p is the pressure of the gas,
V is volume of the gas,
m is mass of the gas,
T is temperature of the gas,
R is gas constant,
n is number of moles, and
$\overline{R}$ is universal gas constant.

$$\frac{nT}{pV} = \frac{1}{\overline{R}} = \text{Constant}$$
$$\frac{n_1 T_1}{p_1 V_1} = \frac{n_2 T_2}{p_2 V_2} = \frac{n_3 T_3}{p_3 V_3}$$

As T, p and V are constant

$$n_1 = n_2 = n_3$$

The density of oxygen at NTP (i.e., at 0 °C and 101.325 kPa) is 1.429 kg/m³, i.e.,

$$\rho = 1.429 \text{ kg/m}^3$$

$\therefore$ Specific volume at O_2 at NTP:

$$v = \frac{1}{\rho} = \frac{1}{1.429} \text{ m}^3/\text{kg}$$

and volume of 32 kg (or kg molecule briefly written as 1 kg mol):

$$V = vM = \frac{1}{1.429} \times 32 = 22.4 \text{ m}^3$$

Similarly, it can be proved that the volume of 1 kg mol of any gas at NTP is 22.4 m^3.

Note that 1 g-mole of all gases occupies a volume of 22.4 L at NTP.

1.19   Characteristic Equation of a Gas

Boyle's law and Charles' law are applicable only if one parameter is constant and two parameters are variable. In engineering practice, all the three parameters, namely the pressure, volume, and temperature, vary simultaneously and therefore the Boyle's law and Charles' law are not applicable. By combining the two gas laws, a general equation for a given mass of a gas undergoing changes in temperature, pressure, and volume can be obtained as follows:

From Boyle's law

$$v \propto \frac{1}{p} \quad \text{at} \quad T = C$$

From Charles' law

$$v \propto T \quad \text{at} \quad p = C$$

Therefore,

$$v \propto \frac{T}{p} \quad \text{when } T \text{ and } p \text{ both vary}$$

or

$$v = \frac{RT}{p}$$
$$pv = RT$$

where

R   is called characteristic gas constant

$$\frac{pV}{m} = RT \quad \left[\because \text{ Specific volume} : \ v = \frac{V}{m} \right]$$
$$pV = mRT$$
$$\frac{pV}{T} = mR$$

Since mass of the gas is constant,

$\therefore$

$$\frac{pV}{T} = \text{constant}$$

$$\frac{p_1 V_1}{T_1} = \frac{p_2 V_2}{T_2} = \text{constant}$$

The equation $pV = mRT$ is called the characteristic equation of the gas. The unit of the characteristic gas constant R is found as follows:

$p = $ absolute pressure of the gas is in kPa
$V = $ volume of the gas is in m^3
$m = $ mass of the gas is in kg
$T = $ temperature of the gas is in K

$\therefore$

$$\text{kPa} \cdot \text{m}^3 = \text{kg} \times R \times \text{K}$$

$$k\frac{N}{m^2} \cdot m^3 = \text{kg K} \times R \quad [\because \; \text{Pa} = \frac{N}{m^2}]$$

$$\frac{\text{kNm}}{\text{kgK}} = R$$

or

$$R = \text{kJ/kgK} \quad [\because \; \text{J} = \text{Nm}]$$

1.20   Dalton's Law of Partial Pressure

It states that the pressure of a mixture of ideal gases is equal to the sum of the partial pressure of constituent gases at constant temperature and volume (Fig. 1.19).

For a mixture of gas A and gas B,

$$pV = mRT$$
$$pV = n\overline{R}T$$

For gas A,

$$p_A V = n_A \overline{R} T$$

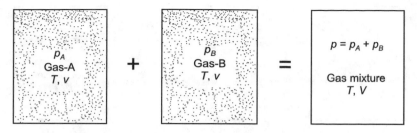

Fig. 1.19 Dalton's law of partial pressure for mixture of two ideal gases

For gas B,

$$p_B V = n_B \overline{R} T$$

By adding of gas A and gas B, we have

$$p_A V + p_B V = n_A \overline{R} T + n_B \overline{R} T$$
$$(p_A + p_B) V = (n_A + n_B) \overline{R} T$$
$$(p_A + p_B) V = n \overline{R} T \quad [\because n = n_A + n_B]$$
$$(p_A + p_B) V = p V \quad [\text{From mixture of gases}; pv = n \overline{R} T]$$

or

$$p = p_A + p_B$$

where

$p =$ pressure of the mixture of gases,
$p_A =$ partial pressure of gas A in the gas mixture, and
$p_B =$ partial pressure of gas B in the gas mixture.

Partial pressure is defined as the pressure of each individual gas in the mixture of gases.

1.21 Amagat's Law of Partial Volume

It states that the volume of a mixture of ideal gases is equal to the sum of the volumes of each gas that would occupy the mixture at constant pressure and temperature (Fig. 1.20).

For a mixture of gas A and gas B,

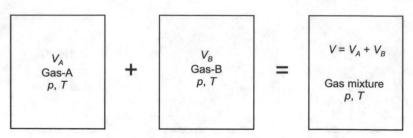

Fig. 1.20 Amagat's law of partial volume

$$pV = mRT$$
$$pV = n\overline{R}T$$

For gas A,

$$pV_A = n_A\overline{R}T$$

For gas B,

$$pV_B = n_B\overline{R}T$$

By adding of gas A and gas B, we have

$$pV_A + pV_B = n_A\overline{R}T + n_B\overline{R}T$$
$$p(V_A + V_B) = (n_A + n_B)\overline{R}T$$
$$p(V_A + V_B) = n\overline{R}T$$
$$p(V_A + V_B) = pV$$

or

$$V = V_A + V_B$$

where

$V =$ volume of the mixture of gases,
$V_A =$ partial volume of gas A in the gas mixture and
$V_B =$ partial volume of gas B in the gas mixture.

Problem 1.1 A quantity of a gas occupies 0.125 m^3 at a pressure of 4.5 bar. It is expanded to a pressure of 1.2 bar at constant temperature. Find the volume of gas after expansion.

    Solution: Given data:

$$V_1 = 0.125 \text{ m}^3$$

$$p_1 = 4.5 \text{ bar}$$
$$p_2 = 1.2 \text{ bar}$$
$$V_2 = ?$$
$$T_1 = T_2$$

We know that

$$\frac{p_1 V_1}{T_1} = \frac{p_2 V_2}{T_2}$$
$$p_1 V_1 = p_2 V_2 \quad [\because T_1 = T_2]$$
$$4.5 \times 0.125 = 1.2 \times V_2$$

or

$$V_2 = 0.4687 \text{ m}^3$$

Problem 1.2 A certain volume of a gas at NTP is heated until its temperature becomes 300 °C and its volume is doubled. Find the final pressure.

Solution: Given data:
   At NTP,

$$T_1 = 0\,^{\circ}\text{C} = 273 \text{ K}$$
$$p_1 = 101.325 \text{ kPa}$$
$$T_2 = 300\,^{\circ}\text{C} = (300 + 273) \text{ K} = 573 \text{ K}$$
$$V_2 = 2V_1$$
$$p_2 = ?$$

We know that

$$\frac{p_1 V_1}{T_1} = \frac{p_2 V_2}{T_2}$$
$$\frac{101.325 \times V_1}{273} = \frac{p_2 \times 2V_1}{573}$$

or

$$p_2 = 106.335 \text{ kPa}$$

Problem 1.3 10 L of air at a pressure of 10 bar and a temperature of 80 °C is expanded to a pressure of 3 bar. The temperature after the expansion is 30 °C. Determine the final volume of air.

Solution: Given data:

$$V_1 = 10 \text{ L}$$
$$p_1 = 10 \text{ bar}$$
$$T_1 = 80\,^{\circ}\text{C} = (80 + 273)\text{K} = 353 \text{ K}$$
$$p_2 = 3 \text{ bar}$$
$$T_2 = 30\,^{\circ}\text{C} = (30 + 273)\text{K} = 303 \text{ K}$$
$$V_2 = ?$$

We know that

$$\frac{p_1 V_1}{T_1} = \frac{p_2 V_2}{T_2}$$
$$\frac{10 \times 10}{353} = \frac{3 \times V_2}{303}$$

or

$$V_2 = \mathbf{28.61 \text{ L}}$$

Problem 1.4 If 1 kg of air occupies a volume of 0.773 m^3 at normal temperature and pressure, find the value of the gas constant for air in kJ/kgK.

Solution: Given data:

$$m = 1 \text{ kg}$$
$$V = 0.773 \text{ m}^3$$

At NTP,

$$T = 0\,^{\circ}\text{C} = 273 \text{ K}$$
$$p = 101.325 \text{ kPa}$$

According to the equation of state,

$$pV = mRT$$
$$101.325 \times 0.773 = 1 \times R \times 273$$

or Gas constant:

$$R = \mathbf{0.2869 \text{ kJ/kgK}}$$

Problem 1.5 A gas of molecular weight of 27.62 occupies 150 L at a pressure of 1.35 bar and a temperature of 20 °C. Find the mass of the gas.

Solution: Given data:
  Molecular weight:

$$M = 27.62$$

  ∴ Gas constant:

$$R = \frac{\overline{R}}{M} = \frac{8.314}{27.62} = 0.301 \text{ kJ/kgK}$$

  Volume:

$$V = 150 \text{ L}$$
$$= \frac{150}{1000} \text{ m}^3 = 0.15 \text{ m}^3$$

  Pressure:

$$p = 1.35 \text{ bar}$$
$$= 1.35 \times 10^2 \text{ kPa}$$

  Temperature:

$$T = 20\,^{\circ}\text{C} = (20 + 273)\text{K} = 293 \text{ K}$$

  Now,

$$pV = mRT$$
$$1.35 \times 10^2 \times 0.15 = m \times 0.301 \times 293$$

or

$$m = \mathbf{0.229 \text{ kg}}$$

1.22   Thermodynamic Properties of Fluids

In order to study the thermodynamic behaviour of a fluid, some important properties of fluids are listed as follows:

1. Density,
2. Specific volume,
3. Specific weight,
4. Specific gravity,
5. Temperature, and
6. Pressure.

1.23 Density

The density of a fluid is defined as the ratio of the mass of a fluid to its volume. It is denoted by ρ (rho).

Mathematically,

Density:

$$\rho = \frac{\text{Mass of fluid:} m}{\text{Volume of fluid:} V}$$
$$\rho = \frac{m}{V}$$

The SI unit of density ρ is kg/m^3.

The density (ρ) is also known as specific mass or mass density.

Remember:

Density of water : $\rho = 1000$ kg/m^3 at temperature 4°C

Density of mercury : $\rho = 13600$ kg/m^3 at NTP

Densityofair : $\rho = 1.29$ kg/m^3 at NTP

1.24 Specific Volume

It is defined as the volume per unit mass of a fluid. It is denoted by v.

Mathematically,

$$\text{Specific volume} : v = \frac{\text{volume of fluid} : V}{\text{Mass of fluid} : m}$$
$$v = \frac{V}{m} = \frac{1}{m/V} = \frac{1}{\rho}$$

It is the reciprocal of density. The SI unit of specific volume: $v = \frac{1}{\rho} = \frac{1}{\text{kg/m}^3} = \frac{\text{m}^3}{\text{kg}}$.

It is commonly applied in gases.

1.25 Specific Weight

It is defined as weight per unit volume of a fluid. It is denoted by w.
Mathematically,
Specific weight:

$$w = \frac{\text{Weight of fluid: } W}{\text{Volume of fluid: } V}$$
$$w = \frac{W}{V} = \frac{mg}{V} = \rho \mathbf{g} \quad [\because W = mg]$$

It is also defined as the product of the density of a fluid (ρ) and acceleration due to gravity (g).
The SI units of specific weight:

$$w = \rho g$$
$$= \frac{kg}{m^3} \times \frac{m}{s^2} = \frac{N}{m^3} \quad \left[\because \frac{kg\ m}{s^2} = N\right]$$

Specific weight is also known as weight density.

1.26 Specific Gravity

It is defined as the ratio of the specific weight of a given fluid to the specific weight of a standard fluid. It is denoted by S.
We know that fluids are classified into two groups:

(i) Liquids and
(ii) Gases.

1.26.1 *Specific Gravity for Liquids*

It is defined as the ratio of the specific weight of a given liquid to the specific weight of water. It is denoted by S_l.
Mathematically,
Specific gravity for liquid:

$$S_l = \frac{\text{Specific weight of given liquid}}{\text{Specific weight of water at } 4°C \text{ temperature}}$$

$$S_l = \frac{w}{w_{\text{water}}} = \frac{\rho g}{\rho_{\text{water}} \cdot g} = \frac{\rho}{\rho_{\text{water}}}$$

Specific gravity for liquids is also defined as the ratio of the density of a given liquid to the density of the water.

Density of water at 4 °C temperature is 1000 kg/m³,

i.e.,

$$\rho_{\text{water}} = 1000 \text{ kg/m}^3$$

∴ The density of given liquid:

$$\rho = S_l \times \rho_{\text{water}} = 1000 \; S_l \;\; \text{kg/m}^3$$

1.26.2   Specific Gravity for Gases

It is defined as the ratio of the specific weight of a given gas to the specific weight of air at NTP. It is denoted by S_g.

Mathematically,

Specific gravity for gases:

$$S_g = \frac{\text{Specific weight of given gas}}{\text{Specific weight of air at NTP}}$$

$$S_g = \frac{w}{w_{\text{air}}} = \frac{\rho g}{\rho_{\text{air}} \cdot g} = \frac{\rho}{\rho_{\text{air}}}$$

Specific gravity for gases is also defined as the ratio of the density of a given gas to the density of air at NTP.

The density of air at NTP is 1.29 kg/m³.

That is,

$$\rho_{\text{air}} = 1.29 \text{ kg/m}^3$$

∴ The density of given gas:

$$\rho = S_g \times \rho_{\text{air}}$$
$$\rho = \mathbf{1.29 \; S_g \; kg/m^3}$$

Remember:

Specific gravity of water : $S = 1$
Specific gravity of mercury : $S = 13.6$
Specific gravity of air : $S = 1$

1.27 Temperature

It is defined as a measure of the intensity of hotness or coldness of a body. A body is said to be at a high temperature (or hot) when it shows a high level of thermal energy in it. Similarly, a body is said to be at a low temperature (or cold) when it shows a low level of thermal energy in it. Temperatures are measured by a thermometer in degree centigrade (°C) and degree Fahrenheit (°F) scales.

Relation between degree Centigrade (°C) and degree Fahrenheit (°F):

$$°C = \frac{5}{9} \left(°F - 32 \right)$$

Relation between absolute temperature (K) and degree centigrade (°C):

$$T\,K = T_1\,°C + 273$$

where

$T =$ absolute temperature and
$T_1 =$ degree celsius temperature.

1.28 Pressure

Pressure is defined as the normal force exerted by a fluid per unit area. It is also known as the intensity of pressure. It is usually more convenient to use pressure rather than force to describe the influences upon fluid behaviour. The counterpart of pressure in solids is normal stress, since pressure is defined as force per unit area. That is,

Pressure:

$$p = \frac{\text{Force} : F}{\text{Area} : A} = \frac{F}{A}$$

If 'F' is expressed in Newton (N) and 'A' expressed in square metre (m²), then the unit of 'p' will be N/m². The standard unit of pressure is pascal (Pa) or N/m².

Pressure is also used for solids as synonymous to normal stress, which is force acting perpendicular to the surface per unit area. For example, for an object sitting on a surface, the force pressing on the surface is the weight of the object, but in different orientations it might have a different area in contact with the surface and therefore exert different pressure as shown in Fig. 1.21.

There are many physical situations where pressure is the most important variable. If you are peeling an apple, then pressure is the key variable. If the knife is sharp, then the area of contact is small and you can peel with less force exerted on the knife. If you get an injection, then pressure is the most important variable in getting the needle through your skin. It is better to have a sharp needle than a dull one since the small area of contact implies that less force is required to push the needle through the skin.

Some important pressure conversion:

$$1\ \text{Pa} = 1\ \text{N/m}^2$$
$$1\ \text{bar} = 10^5\ \text{Pa} = 0.1\ \text{MPa} = 100\ \text{kPa}$$
$$1\ \text{atm} = 101325\ \text{Pa} = 101.325\ \text{kPa} = 1.01325\ \text{bar}$$
$$1\ \text{kgf/cm}^2 = 9.81\ \text{N/cm}^2 = 9.81 \times 10^4\ \text{N/m}^2\ \text{or Pa}$$
$$1\ \text{kgf/cm}^2 = 0.981\ \text{bar} = \frac{0.981}{1.01325}\ \text{atm} = 0.968\ \text{atm}\ \left[\because\ 1\ \text{bar} = \frac{1}{1.01325}\ \text{atm}\right]$$

Note that the pressure units **kgf/cm²**, **bar,** and **atm** are almost equivalent to each other.

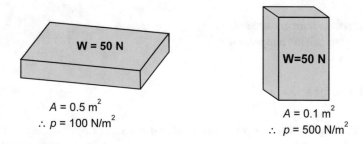

Fig. 1.21 Pressure varies with the area of contact with the surface

1.29 Pressure Head

Consider a vessel containing some liquid. We know that the liquid will exert pressure on all sides as well as the bottom of the vessel. Now consider a bottomless cylinder being made to stand in the liquid as shown in Fig. 1.22.

Let

$$w = \rho g, \text{ specific weight of the liquid.}$$
$$h = \text{height of liquid in the cylinder.}$$
$$A = \text{cross-sectional area of the cylinder.}$$

Then, weight of liquid contained in the cylinder

$$= \text{mass of liquid} \times \text{acceleration due to gravity}$$
$$= m \times g = \rho V g$$
$$= \rho A h g \quad (\because \text{ Mass} : m = \rho V, \text{ Volume} : V = Ah)$$

Therefore, intensity of pressure at the base of the cylinder is given by

$$p = \frac{\text{weight of liquid in the cylinder}}{\text{cross-sectional area of the cylinder base}}$$
$$p = \frac{\rho A h g}{A} = \rho g h \tag{1.8}$$

where

$$\rho g = w = \text{constant for a given liquid.}$$
$$p \propto h$$

Thus, it is proved that the intensity of pressure at any point in a liquid is directly proportional to the depth of the point measured from the free surface of the liquid as shown in Fig. 1.22b.

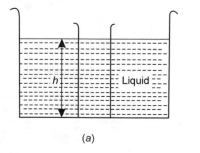
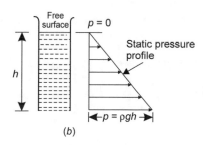

(a)   (b)

Fig. 1.22 Pressure head

From Eq. (1.8), we get

$$h = \frac{p}{\rho g}$$

Pressure head:

$$\boldsymbol{h = \frac{p}{\rho g} = \frac{p}{w}} \quad \text{(Specific weight : } w = \rho g)$$

Here, p is called intensity of static pressure and h is called pressure head of the liquid.

If p is expressed in N/m^2 and
w is expressed in N/m^3,

then

$$h = \frac{\text{N/m}^2}{\text{N/m}^3} = \text{m}$$

1.30   Laws of Liquid Pressure, Hydrostatic Equation, and Its Application

(a)   **Laws of liquid pressure**:

    (i)   The surface of a liquid which is subjected to atmospheric pressure is called the free surface of the liquid.

    (ii)   The free surface of a liquid is always horizontal.

    (iii)   Intensity of pressure at any point in a liquid is the same in all directions.

    (iv)   Intensity of pressure at any point in a liquid is directly proportional to the depth of the point from the free surface of the liquid.

    (v)   Pressure on a surface submerged in a liquid acts normal to the surface (Fig. 1.23).

(b)   **Hydrostatic equation**:

Fig. 1.23 Free surface

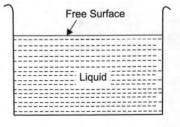

$$p = \rho g h \qquad (1.9)$$

where

$p =$ intensity of pressure of a static liquid at any point in it,
$\rho =$ density of liquid,
$g =$ acceleration due to gravity $= 9.81$ m/s^2, and
$h =$ depth of the point from the free surface of liquid.

Equation (1.9) is called **hydrostatic equation**.

(c)   **Application**: Hydrostatic equation is used to find out 'total pressure' on the immersed surface in a liquid and the position of the 'centre of pressure'.

1.31 Pascal's Law

Pascal's law states that **the intensity of pressure at any point in a fluid at rest is the same in all directions**.

Proof

Let us consider an arbitrarily small fluid element of wedge shape ABC as shown in Fig. 1.24. Let the width of the fluid element be unity and p_x, p_y, and p_z be the intensity of pressure acting on the face *AC*, *BC*, and *BA*, respectively.

$$\theta = \text{angle of the element of the fluid.}$$

As the element of liquid is at rest, therefore sum of the horizontal and vertical components of the liquid forces must be equal to zero.

Resolving the forces horizontally:

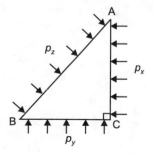

(a) Pressure on a fluid element   (b) Force on a fluid element

Fig. 1.24 Pascal's law

$$F_z \cdot \sin\theta - F_x = 0$$
$$p_z BA \sin\theta - p_x AC = 0$$
$$p_z AC - p_x AC = 0$$

or

$$p_z = p_x \tag{1.10}$$

$$\because F_z = p_z.BA$$
$$F_x = p_x.AC$$
$$\angle ABC; \sin\theta = \frac{AC}{BA}$$
$$\text{or } AC = BA.\sin\theta$$

Now resolving the forces vertically

$$F_z \cos\theta + W - F_y = 0$$

where $W =$ weight of the liquid element acts vertically in downward direction

$$F_z = p_z \cdot BA$$

and

$$F_y = p_y \cdot BC$$

$$\therefore p_z \cdot BA \cos\theta + W - p_y \cdot BC = 0$$
Since the element is very small, we are neglecting the weight of the liquid,

i.e.,

$$W \approx 0$$

$$\therefore$$

$$p_z BA \cos\theta - p_y BC = 0$$

or

$$p_z BC - p_y BC = 0$$

or

$$P_z = P_y \quad\quad\quad (1.11)$$

$$\therefore \angle ABC,$$

$$\cos\theta = \frac{AC}{BA}$$

or $BC = BA\cos\theta$

from Eqs. (1.10) and (1.11), we get

$$P_x = P_y = P_z$$

Hence, the intensity of pressure at any point in a fluid at rest is the same in all directions.

1.32 Atmospheric Pressure and Its Measurement

The atmospheric pressure is defined as the normal force exerted by atmospheric air per unit area. It is classified into two categories:

(i)   Standard atmospheric pressure and
(ii)   Local atmospheric pressure.

(i)   **Standard Atmospheric Pressure**: This pressure is a measurement at sea level and 15 °C temperature. The value of the standard atmospheric pressure is fixed:

i.e.,

$$760 \text{ mm of Hg} = 101.325 \text{ kP} = 1.01325 \text{ bar}$$

(ii)   **Local Atmospheric Pressure**: This pressure is measured at condition and place. The value of the local atmospheric pressure is not fixed and varies at a given place with a change in weather.

1.32.1 Measuring Atmospheric Pressure

The standard instrument used to measure the atmospheric pressure is called **Barometer**. It is also known as Torricelli Barometer, because the first measurement of atmospheric pressure began with a simple experiment performed by Evangelista Torricelli in 1643. In this experiment, Torricelli immersed a tube, sealed at one end, in a container of mercury as shown in Fig. 1.25. Atmospheric pressure then forced mercury up into the tube that was considerably higher than the mercury in a container.

Writing a force balance in the vertical direction gives

$$p_{atm} A = W = \rho g h A$$
$$\boldsymbol{p_{atm} = \rho g h}$$

where

$\rho =$ density of mercury,
$g =$ acceleration due to gravity, and
$h =$ height of the mercury column above the free surface.

Note that the length and the cross-sectional area of the tube have no effect on the height of the fluid column of a barometer as shown in Fig. 1.26.

If mercury is used in a barometer, it is called a mercury barometer and if water is used in a barometer, it is called a water barometer. Both types of barometers are used to measure standard atmospheric pressure, as shown in Fig. 1.27.

The standard atmospheric pressure provides the force necessary to push the mercury up (in a mercury barometer) the evacuated tube which is 760 mm, and to push the water up (in a water barometer) the evacuated tube is 10.33 m.

Mathematically,

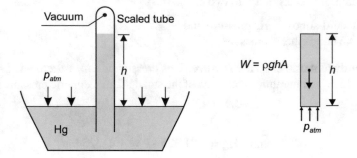

Fig. 1.25 Mercury barometer (Torricelli's barometer)

Fig. 1.26 Fluid column of a barometer is independent of the cross-sectional area of the tube

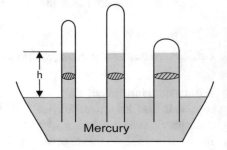

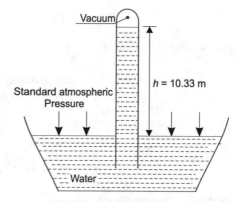

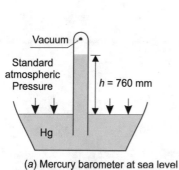

(a) Mercury barometer at sea level and 15°C atmospheric temperature

(b) Water barometer at sea level and 15°C atmospheric temperature

Fig. 1.27 Barometer

For mercury barometer

Standard atmospheric pressure,

$$p_{atm} = \text{density} \times g \times \text{height}$$
$$= \rho g h$$

Substituting the value of

$$\rho = 13600 \text{ kg/m}^3$$
$$g = 9.81 \text{ m/s}^2$$

and

$$h = 760 \text{ mm} = 0.76 \text{ m}$$

In above equation, we get

$$= 13600 \times 9.81 \times 0.76$$
$$= 101396.16 \text{ N/m}^2$$
$$= 101.39 \text{ kPa}$$
$$= 1.013 \text{ bar.}$$

For water barometer

Standard atmospheric pressure,

$$p_{atm} = \text{density} \times g \times \text{height}$$

$$= \rho g h$$

Substituting the value of

$$\rho = 1000 \text{ kg/m}^3$$
$$g = 9.81 \text{ m/s}^2$$

and

$$h = 10.33 \text{ m}$$

In above equation, we get

$$= 1000 \times 9.81 \times 10.33$$
$$= 101337 \text{ N/m}^2$$
$$= 101.33 \text{ kPa}$$
$$= 1.013 \text{ bar}$$

So, standard atmospheric pressure

$$= 760 \text{ mm of Hg} = 10.33 \text{ m of water} = 1.013 \text{ bar}$$

In practical, water barometer is not used, because of

(i)   Its big size, i.e., water rise in evacuated tube is 10.33 m; as compared to mercury, it is 0.76 m.
(ii) The high vapour pressure as compared to mercury; consequently, the water barometer does not give consistent and accurate readings.

1.32.2 Aneroid Barometer

The aneroid barometer consists of flexible short bellows called the aneroid cell, which is tightly sealed after removing the air at near zero pressure. The aneroid expands or squeezes due to variation in the atmospheric pressure. This motion causes the link L to move the pointer over the scale S. The deflections indicated by the pointer are relative to complete vacuum. The deflection given by the pointer indicate the absolute pressure.

The aneroid barometer is smaller and compact as compare to mercury barometer and much easier to read. It is the heart of the altimeter used in modern aviation (Fig. 1.28).

Problem 1.6 Find the depth of oil of specific gravity of 0.82 which produces an intensity of pressure equal to 2.5 kN/m². Also, find the pressure head in terms of water and mercury.

Fig. 1.28 Aneroid
barometer

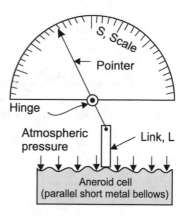

Solution: Given data (Fig. 1.29):
Specific gravity of oil:

$$S = 0.82$$

∴ Density of oil:

$$\rho_{oil} = \text{specific gravity of oil} \times \text{density of water}$$
$$= 0.82 \times 1000 = 820 \text{ kg/m}^3$$

Intensity of pressure:

$$p = 2.5 \text{ kN/m}^2 = 2500 \text{ N/m}^2$$

Now intensity of pressure:

$$p = \rho_{oil}gh_1$$

where

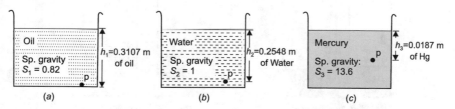

Fig. 1.29 Different values of pressure heads for oil, water, and mercury of the same intensity of pressure (p)

$h_1 =$ depth of oil and

$\therefore$

$$h_1 = \frac{p}{\rho_{\text{oil}}g} = \frac{2500}{820 \times 9.81} = \textbf{0.3107 m}$$

Set h_2 and h_3 pressure heads of water and mercury at the same intensity of pressure (p).

$\therefore$

$$h_2 = \frac{p}{\rho_{\text{water}}\,g} \text{ for water} = \frac{2500}{1000 \times 9.81} = \textbf{0.2548 m of water}$$

and

$$h_3 = \frac{p}{\rho_{\text{Hg}}g} \text{ for mercury}$$
$$= \frac{2500}{13600 \times 9.81} \quad \because \; \rho_{\text{Hg}} = 13600 \text{ kg/m}^3$$
$$= \textbf{0.0187 m of Hg}$$

Problem 1.7 Convert pressure head of mercury into equivalent pressure head of water (Fig. 1.30).

Solution:
For mercury barometer
Atmospheric pressure:

$$p = \rho_{\text{H}}gh_{\text{Hg}}$$

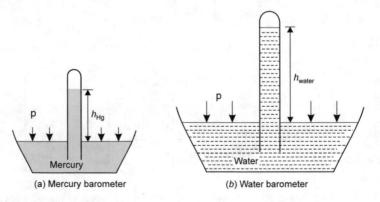

(a) Mercury barometer   (b) Water barometer

Fig. 1.30 Barometer

where

$$\rho_{Hg} = \text{density of mercury}$$
$$= 13600 \text{ kg/m}^3$$
$$h_{Hg} = \text{pressure hand of mercury in metre (m)}$$

$$p = 1360 \, gh_{Hg} \tag{1.12}$$

For water barometer
Atmospheric pressure:

$$p = \rho_{\text{water}} \, gh_{\text{water}}$$

where

$$\rho_{\text{water}} = \text{density of water}$$
$$= 1000 \text{ kg/m}^3$$
$$h_{\text{water}} = \text{pressure hand of mercury in metre (m)}$$

$$p = 1000 \, gh_{\text{water}} \tag{1.13}$$

Equating Eq. (1.12) with Eq. (1.13), we get

$$13600 \, gh_{Hg} = 1000 \, gh_{\text{water}}$$

or

$$h_{\text{water}} = 13.6 \, h_{Hg}$$

Problem 1.8 An open tank contains water up to a depth of 4 m and above it, an oil of specific gravity 0.9 for a depth of 1.5 m. Find the intensity of pressure.

(i)   at the interface of the two liquids and
(ii) at the bottom of the tank.

Solution: Given data:
   Height of water:

$$h_1 = 4 \text{ m}$$

   Height of oil:

Fig. 1.31 Schematic for
Problem 1.8

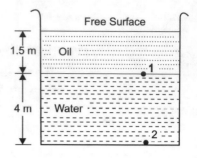

$$h_2 = 1.5 \text{ m}$$

Specific gravity of oil:

$$S_l = 0.9$$

Now

(i)   Pressure intensity at the interface of the two liquids, i.e., at point '1'

$$p_1 = \rho g h_2 = 900 \times 9.81 \times 1.5 \frac{\text{kg}}{\text{m}^3} \cdot \frac{\text{m}}{\text{s}^2} \cdot \text{m}$$
$$= 13243.5 \text{ N/m}^2 \text{ or Pa} = \mathbf{13.243 \text{ kPa}}$$

(ii) Pressure intensity at the bottom, i.e., at point '2' (Fig. 1.31)

$$p_2 = (\rho g h_2)_{\text{oil}} + (\rho g h_1)_{\text{water}}$$
$$= p_1 + (\rho g h_1)_{\text{water}}$$
$$= 13243.5 + 1000 \times 9.81 \times 4$$
$$= 52483.5 \text{ N/m}^2 \text{ or Pa} = \mathbf{52.48 \text{ kPa}}$$

Problem 1.9 An open tank contains mercury up to a depth of 3 m and above water
up to a depth of 2 m and above water an oil of specific gravity 0.92 for a depth of
1.2 m. Find the intensity of pressure

(i)   at the interface of the oil and water,
(ii)   at the interface of the water and mercury, and
(iii) at the bottom of the tank.

Solution: Given data:
   Height of mercury:

$$h_1 = 3 \text{ m}$$

Height of water:

Fig. 1.32 Schematic for Problem 1.9

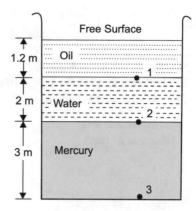

$$h_2 = 2 \text{ m}$$

Height of oil:

$$h_3 = 1.2 \text{ m}$$

Specific gravity of oil (Fig. 1.32):

$$S_{\text{oil}} = 0.92$$

Now,

(i)   Pressure intensity at the interface of the oil and water, i.e., at point '1'

$$
\begin{aligned}
p_1 &= (\rho g h_3)_{\text{oil}} \\
&= 920 \times 9.18 \times 1.2 \frac{\text{kg}}{\text{m}^3} \cdot \frac{\text{m}}{\text{s}^2} \cdot \text{m} \\
&= 10830.24 \text{ N/m}^2 \text{ or Pa} = \mathbf{10.83 \text{ kPa}}
\end{aligned}
$$

(ii)   Pressure intensity at the interface of the water and mercury, i.e., at point '2'

$$
\begin{aligned}
p_2 &= (\rho g h_3)_{\text{oil}} + (\rho g h_2)_{\text{water}} \\
&= p_1 + (\rho g h_2)_{\text{water}} \\
p_2 &= 10830.24 + 1000 \times 9.81 \times 2 \\
&= 30450.24 \text{ N/m}^2 \text{ or Pa} = \mathbf{30.45 \text{ kPa}}
\end{aligned}
$$

(iii)   Pressure intensity at the bottom of the tank, i.e., at point '3'

$$p_3 = (\rho g h_3)_{\text{oil}} + (\rho g h_2)_{\text{water}} + (\rho g h)_{\text{Hg}}$$

$$= p_2 + (\rho g h)_{\text{Hg}} = 30450.24 + 13600 \times 9.81 \times 3$$
$$= 430698.24 \text{ N/m}^2 = \textbf{430.69 kPa}$$

Problem 1.10 A hydraulic press has a ram of 400 mm diameter and a plunger of 50 mm diameter. Find the weight lifted by the hydraulic press when the force applied at the plunger is 800 N (Fig. 1.33).

Solution: Given data:
  Diameter of ram:

$$D = 400 \text{ mm} = 0.4 \text{ m}$$

Diameter of plunger:

$$d = 50 \text{ mm} = 0.05 \text{ m}$$

Force on plunger:

$$F = 800 \text{ N}$$

Now,
Weight lifted:

$$W = ?$$

Area of ram:

$$A = \frac{\pi}{4}D^2 = \frac{3.14}{4}(0.4)^2$$
$$= 0.1256 \text{ m}^2$$

Area of plunger:

Fig. 1.33 Hydranlic press

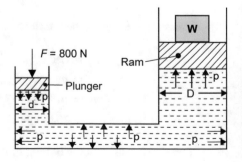

$$a = \frac{\pi}{4}d^2 = \frac{3.14}{4}(0.05)^2 = 1.962 \times 10^{-3}\,\text{m}^2$$

Pressure intensity created by 800 N on the water through plunger is given by

$$p = \frac{F}{a} = \frac{800\,\text{N}}{1.962 \times 10^{-3}\,\text{m}^2} = \mathbf{407747.19\,N/m^2}$$

According to Pascal's law, the intensity of pressure applied to a confined fluid at any point is transmitted undiminished throughout the fluid in all directions and acts upon every part of the confining vessel at right angles to its interior surfaces.

The pressure intensity at the ram due to plunger = pressure intensity at ram due to weight W.

$$407747.19\,\text{N/m}^2 = \frac{W}{A} \quad \left[\because \frac{F}{a} = \frac{W}{A} \right]$$

or Weight:

$$W = 407747.19 \times 0.1256\,\text{N/m}^2$$
$$= 51213.04\,\text{N} = \mathbf{51.213\,kN}$$

1.33 Absolute, Gauge, and Vacuum Pressure

1.33.1 Absolute Pressure

An absolute zero of pressure will occur when molecular momentum is zero. Such a situation can occur only when there is a perfect vacuum. The pressure measured with reference to absolute zero is called **absolute pressure**. It is abbreviated as p_{abs}.

1.33.2 Gauge Pressure

The pressure measured by a gauge (instrument) relative to the atmospheric pressure is called **gauge pressure**. The instrument by which gauge pressure is measured is called a pressure gauge in which the atmospheric pressure is marked as zero.

That is, the pressure gauges always indicate the pressure above the atmospheric pressure. It is abbreviated as p_g.

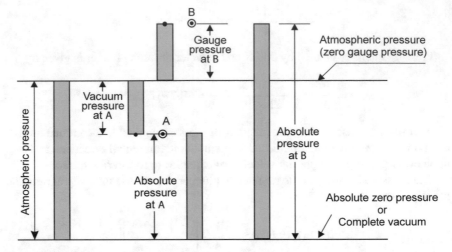

Fig. 1.34 Absolute, gauge, vacuum, and atmospheric pressure

1.33.3   Vacuum Pressure

The pressure of a fluid to be measured that is less than the atmospheric pressure is called **vacuum pressure**. It is also known as negative or suction pressure. The vacuum pressure of a liquid is measured by an instrument called a vacuum gauge. It is abbreviated as p_{vac}.

The relations among absolute, atmospheric, gauge, and vacuum pressure are shown in Fig. 1.34.

Mathematically,

(i)   For positive gauge pressure,

Absolute pressure (p_{abs}) = Atmospheric pressure (p_{atm}) + Gauge pressure (p_g)

$$p_{abs} = p_{atm} + p_g$$

(ii)   For vacuum pressure or negative gauge pressure,

Absolute pressure (p_{abs}) = Atmospheric pressure (p_{atm}) − Vacuum pressure (p_{vac})

$$p_{abs} = p_{atm} - p_{vac}$$

Problem 1.11 Determine the vacuum pressure in meters of water when the absolute pressure is 0.622001 bar. Assume atmospheric pressure as 10.33 m of water.

Solution: Given data:

Absolute pressure:

$$p_{abs} = 0.622001 \text{ bar} = 62200.1 \text{ N/m}^2$$

Atmospheric pressure:

$$p_{atm} = 10.33 \text{ m of water}$$

Now,
Absolute pressure:

$$p_{abs} = \rho g h$$

where

$$h = \text{absolute pressure head in metre of water}$$
$$\rho = 1000 \text{ kg/m}^3, \text{ density of water}$$

Absolute pressure head in water:

$$h - \frac{p_{abs}}{\rho g} = \frac{62200.1 \text{ N/m}^2}{1000 \times 9.81 \frac{kg}{m^3} \cdot \frac{m}{s^2}} = 6.340 \text{ m} \quad \left[\because 1 \text{ N} = \frac{kg \cdot m}{s^2} \right]$$

We know that

Absolute pressure head $=$ atmospheric pressure head $-$ vacuum pressure head
$$6.340 = 10.33 - \text{vacuum pressure head}$$

or

$$\text{Vacuum pressure head} = \textbf{3.99 m of water}$$

Problem 1.12 A point lies 7.5 m below the free surface of water. Determine the pressure of the point in

(i)   kPa
(ii)   mm of Hg.

If atmospheric pressure of 760 mm of Hg, determine the absolute pressure in

(iii)   m of water,
(iv)   bar, and
(v)   mm of Hg.

Solution:

(i)   The pressure of water at point A:

$$p = \rho g h$$
$$= 1000 \times 9.81 \times 7.5 \frac{\text{kg}}{\text{m}^3} \cdot \frac{\text{m}}{\text{s}^2} \cdot \text{m}$$
$$= 73575 \text{ N/m}^2 \text{ or Pa} = \textbf{73.575 kPa}$$

(ii)   Let h_{Hg} = pressure head of mercury in metre equivalent to given 7.5 m of water (Fig. 1.35)

$$(\rho g h)_{Hg} = (\rho g h)_{water}$$
$$\rho_{Hg} \times g \times h_{Hg} = \rho_{water} \times g \times h_{water}$$
$$h_{Hg} = \frac{\rho_{water}}{\rho_{Hg}} h_{water}$$
$$= \frac{1000}{13600} \times 7.5$$
$$= 0.55147 \text{ m}$$
$$= \textbf{551.47 mm.}$$

(iii)   Absolute pressure at point A:

$$p_{abs} = p_{atm} + (\rho g h)_{water}$$
$$= (\rho g h)_{Hg} + (\rho g h)_{water}$$
$$= 13600 \times 9.81 \times 0.76 + 1000 \times 9.81 \times 7.5$$
$$= 174971.16 \text{ N/m}^2$$

Let

h_w = absolute pressure head of water in metre,

Fig. 1.35 Schematic for Problem 1.12

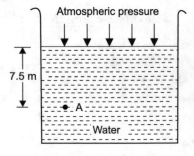

Atmospheric pressure

7.5 m

A

Water

$\therefore$

$$p_{abs} = \rho g h_w$$
$$174971.16 = 1000 \times 9.81 \times h_w$$

or

$$h_w = \textbf{17.836 m of water}$$

(iv)   Absolute pressure:

$$p_{abs} = 174971.66 \text{ N/m}^2$$
$$= \textbf{1.749 bar} \quad \left[\because \ 1 \text{ bar} = 10^5 \text{ N/m}^2 \right]$$

(v)   Let

$$h_{Hg} = \text{absolute pressure head of mercury in metre}$$

$\therefore$

$$p_{abs} = \rho g h_{Hg}$$
$$174971.16 = 13600 \times 9.81 \times h_{Hg}$$

or

$$h_{Hg} = 1.31147 \text{ m} = \textbf{1311.47 mm of Hg}$$

Or

$$h_{Hg} = \frac{h_w}{13.6} = \frac{17.836}{13.6}$$
$$= 1.31147 \text{ m} = \textbf{1311.47 mm of Hg}$$

Problem 1.13 The mercury barometer can be used to measure the height of a building. If the barometric readings at the bottom and at the top of a building are 750 mm and 730 mm of mercury, respectively, determine the height of the building. Take the densities of air and mercury to be 1.2 kg/m³ and 13,600 kg/m³, respectively.

Solution: Given data:

Barometer reading at the bottom of a building:

$$h_1 = 750 \text{ mm of Hg} = 0.75 \text{ m of Hg}$$

∴ Atmospheric pressure at the bottom of a building:

$$p_1 = \rho_{Hg}h_1g$$

Barometer reading at the top of a building:

$$h_2 = 730 \text{ mm of Hg} = 0.73 \text{ m of Hg}$$

∴ Atmospheric pressure at the top of a building:

$$p_2 = \rho_{Hg}h_2g$$

Pressure difference between the top and the bottom of a building:

$$\begin{aligned} \Delta p = p_1 - p_2 &= \rho_{Hg}h_1g - \rho_{Hg}h_2g \\ &= (h_1 - h_2)\rho_{Hg}g \\ &= (0.75 - 0.73) \times 13600 \times 9.8 = 2668.32 \text{ Pa} \end{aligned}$$

also

$$\Delta p = \rho_{air}gh$$

where h is column of air or height of a building.

∴

$$2668.32 = 1.2 \times 9.81 \times h$$

or

$$h = \mathbf{226.66 \text{ m}}$$

Problem 1.14 The water in a tank is pressurized by air, and the pressure is measured by a multi-fluid manometer as shown in Fig. 1.36. Find the gauge pressure of air in the tank if $h_1 = 0.2$ m, $h_2 = 0.3$ m, and $h_3 = 0.46$ m. Take the densities of water, oil, and mercury to be 1000 kg/m³, 850 kg/m³, and 13,600 kg/m³, respectively.

Solution:
We know that the gauge pressure on the datum line (2)–(2) at points D and E is the same.

∴

$$p_D = p_E = \rho_{Hg}gh_3$$
$$= 13600 \times 9.81 \times 0.46 = 61371.36 \text{ N/m}^2 \text{ or Pa}$$

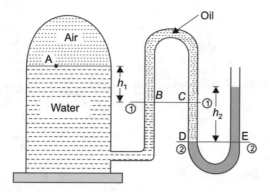

Fig. 1.36 Schematic for Problem 1.14

The gauge pressure below the datum line (1)–(1) at points B and C is the same.
∴

$$p_B = p_C = p_D - \rho_{\text{oil}}gh_2$$
$$= 61371.36 - 850 \times 9.81 \times 0.3$$
$$= 61371.36 - 2501.55 = 58869.81 \text{ Pa}$$

The gauge pressure of air at point A:

$$p_A = p_B - \rho_{\text{H2O}}gh_1$$
$$= 58869.81 - 1000 \times 9.81 \times 0.2$$
$$= 56907.81 \text{ Pa} = \textbf{56.907 kPa}$$

Problem 1.15 Repeat Problem 1.14. Determine the absolute pressure of air in the tank.

Solution:
The absolute pressure of air at point A:

$$p_{A,\text{abs}} = p_A + p_{\text{atm}} = 56.907 + 100$$
$$= \textbf{156.907 kPa}$$

Problem 1.16 The absolute pressure in water at a depth of 5 m is read to be 145 kPa. Determine

(a) the local atmospheric pressure and
(b) the absolute pressure at a depth of 5 m in a liquid whose specific gravity is 0.85 at the same location.

Solution: Given data (Fig. 1.37):

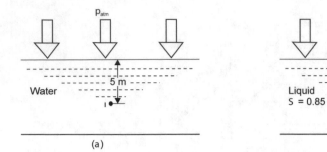

Fig. 1.37 Schematic forb Problem 1.16

(a)   For water:

$$h_1 = 5 \text{ m}$$

Absolute pressure:

$$p_1 = 145 \text{ kPa} = 145 \times 10^3 \text{ Pa}$$

also

$$p_1 = p_g + p_{atm}$$

$$145 \times 10^3 = \rho_{H2O}gh_1 + p_{atm}$$
$$145 \times 10^3 = 1000 \times 9.81 \times 5 + p_{atm}$$

or

$$p_{atm} = 95950 \text{ Pa} = \textbf{95.95 kPa}$$

(b)   For liquid whose specific gravity: $S = 0.85$

Density:

$$\rho = S \times 1000 = 0.85 \times 1000 = 850 \text{ kg/m}^3$$
$$h_1 = 5 \text{ m}$$

Absolute pressure:

$$p_1 = \rho g h_1 + p_{atm}$$
$$= 850 \times 9.81 \times 5 + 95950$$
$$= 137642.5 \text{ Pa} = \textbf{137.642 kPa}$$

Problem 1.17 The gauge pressure in a liquid of 3 m is read to be 28 kPa. Determine the gauge pressure in the same liquid at a depth of 9 m.

Solution: Given data:
   At point 1,

$$h_1 = 3 \text{ m}$$

Gauge pressure:

$$p_1 = 28 \text{ kPa} = 28 \times 10^3 \text{ Pa}$$

also

$$p_1 = \rho g h_1$$
$$28 \times 10^3 = \rho \times 9.81 \times 3$$

or

$$\rho = 951.41 \text{ kg/m}^3$$

At state 2 (Fig. 1.38),

$$h_2 = 9 \text{ m}$$

Gauge pressure:

$$p_2 = \rho g h_2$$
$$= 951.41 \times 9.81 \times 9 = 83999.98 \text{ Pa} \approx \textbf{84 kPa}$$

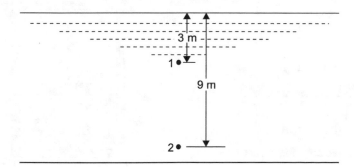

Fig. 1.38 Schematic for Problem 1.17

Problem 1.18 A gas is contained in a vertical, frictionless piston-cylinder arrangement. The piston has a mass of 4 kg and a cross-sectional area of 0.0035 m². A compressed spring above the piston exerts a force of 60 N on the piston. If the atmospheric pressure is 95 kPa, determine the pressure inside the cylinder.

Solution: Given data:

Mass of the piston:

$$m_p = 4 \text{ kg}$$

∴ Weight of the piston (Fig. 1.39):

$$W_p = m_p \, g = 4 \times 9.81 = 39.24 \text{ N}$$

Cross-sectional area:

$$A = 0.0035 \text{ m}^2$$

Force exerted by the spring on the piston:

$$F_s = 60 \text{ N}$$

Atmospheric pressure:

$$p_{\text{atm}} = 95 \text{ kPa} = 95 \times 10^3 \text{ Pa}$$

For equilibrium condition of the piston,

$$\text{Net downward force} = \text{Net upward force}$$
$$p_{\text{atm}} A + F_s + W_p = pA$$

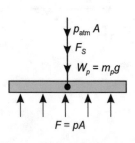

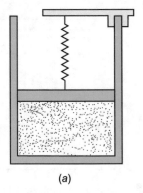

(b) Free body diagram of the piston

(a)

Fig. 1.39 Schematic for Problem 1.18

$$95 \times 10^3 \times 0.0035 + 60 + 39.24 = p \times 0.0035$$

or

$$p = 12334.28 \text{ Pa} = \mathbf{123.34 \text{ kPa}}$$

Problem 1.19 The air in a circular cylinder is heated until the spring is compressed to 50 mm. Find the work done by the air on the frictionless piston. The spring is initially unstretched as shown in Fig. 1.40.

Solution: Given data:
   Final spring compression:

$$x_2 = 50 \text{ mm} = 0.050 \text{ m}$$

Initial spring compression:

$$x_1 = 0$$

Stiffness of spring:

$$k = 10 \text{ kN/m}$$

Mass of piston:

$$m = 50 \text{ kg}$$

Diameter of piston:

$$d = 100 \text{ m} = 0.10 \text{ m}$$

Fig. 1.40 Schematic for
Problem 1.19

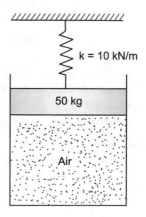

Fig. 1.41 Piston at initial condition

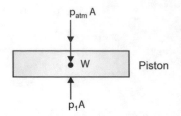

The pressure in the cylinder is initially found from a force balance (Fig. 1.41),

$$p_1 A = p_{atm} A + W$$

$$p_1 \frac{\pi}{4} d^2 = p_{atm} \times \frac{\pi}{4} d^2 \times mg$$

$$p_1 \times \frac{3.14 \times (0.1)^2}{4} = 101325 \times \frac{3.14}{4} \times (0.1)^2 + 50 \times 9.81 \quad \because \ p_{atm} = 101325 \ \text{Pa}$$

$$7.85 \times 10^{-3} p_1 = 795.40 + 490.5 = 1285.9$$

or

$$p_1 = 163.80 \times 103 \ \text{Pa}$$
$$= 163.80 \ \text{kPa}$$

Work done by the air on the piston,

W_{Net} = work required to move the piston + work required to compress the spring

$$= p \times A \times x + \frac{1}{2} k \left(x_2^2 - x_1^2 \right)$$

$$= 163.80 \times \frac{\pi}{4} d^2 \times 0.050 + \frac{1}{2} \times 10 \times \left[(0.050)^2 - (0)^2 \right]$$

$$= 163.80 \times \frac{3.14}{4} \times (0.1)^2 \times 0.050 + \frac{1}{2} \times 10 (0.050)^2$$

$$= 0.0643 + 0.0125 = \mathbf{0.0768 \ kJ}$$

Problem 1.20 A 100 kg mass drops 3 m as shown in Fig. 1.42, resulting in an increased volume in the cylinder of $0.002 \ \text{m}^3$. The weight W and the piston maintain a constant gauge pressure of 100 kPa. Determine the net work done by the gas on the surroundings. Neglect all friction.

Solution: Given data:
   Mass on the pulley:

$$m = 100 \ \text{kg}$$

Fig. 1.42 Schematic for
Problem 1.20

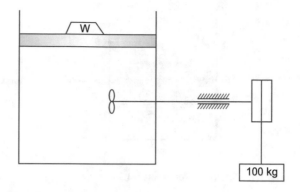

Mass drop on the pulley:

$$x = 3 \text{ m}$$

Volume increased in cylinder:

$$dV = 0.002 \text{ m}^3$$

Gauge pressure in cylinder:

$$p_G = 100 \text{ kPa}$$

Absolute pressure in cylinder:

$$p = p_G + p_{\text{atm}} = 100 + 101.325 = 201.325 \text{ kPa}$$

The paddle wheel does work on the system due to the 100 kg mass dropping 3 m:

$$W_1 = -mg\,x = -100 \times 9.81 \times 3 = -2943 \text{ kJ}$$

The work done by the frictionless piston:

$$W_2 = pdV = 201.325 \times 0.002 = 0.40026 \text{ kJ}$$

The net work done:

$$W_{\text{Net}} = W_1 - W_2 = -2.943 + 0.4026 = \mathbf{-2.54 \text{ kJ}}$$

Problem 1.21 A vertical piston-cylinder arrangement contains a gas at a pressure
of 100 kPa. The piston has a mass of 5 kg and a diameter of 120 mm. The pressure
of the gas is to be increased by placing some weights on the piston. Determine the

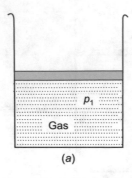

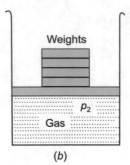

Fig. 1.43 Schematic for Problem 1.21

local atmospheric pressure and the mass of the weights that will double the pressure of the gas inside the cylinder (Fig. 1.43).

Solution: Given data:
   Pressure at initial condition:

$$p_1 = 100 \text{ kPa} = 100 \times 10^3 \text{ Pa}$$

Mass of piston:

$$m_p = 5 \text{ kg}$$

$\therefore$ Weight of piston:

$$W_p = m_p g = 5 \times 9.81 = 49.05 \text{ N}$$

Diameter of piston:

$$D = 120 \text{ mm} = 0.12 \text{ m}$$

$\therefore$ Cross-sectional area:

$$A = \frac{\pi}{4} D^2 = \frac{3.14}{4} \times (0.12)^2 = 0.011304 \text{ m}^2$$

For equilibrium condition, downward force equal to upward forces, we get

$$W_p + p_{\text{atm}} A = p_1 A \quad \text{for Fig. 1.43a}$$
$$49.05 + p_{\text{atm}} \times 0.011304 = 100 \times 10^3 \times 0.011304$$
$$49.05 + 0.011304 p_{\text{atm}} = 1130.4$$

or

$$0.011304\,p_{atm} = 1081.35$$
$$p_{atm} = 95660.82 \text{ Pa}$$
$$= \mathbf{95.66 \text{ kPa}}$$

Case II
   Let

$$W = \text{net weight placed on the piston}$$
$$p_2 = 2p_1 = 2 \times 100 \times 10^3 \text{ Pa} = 200 \times 10^3 \text{ Pa}$$

For equilibrium condition,

$$\text{downward force} = \text{upward force}$$
$$W_p + W + p_{atm}\,A = p_2 A \quad \text{for Fig. 1.43b}$$
$$49.05 + W + 95660.82 \times 0.011304 - 200 \times 10^3 \times 0.011304$$
$$49.05 + W + 1081.349 = 2260.8$$

or

$$W = 1130.40 \text{ N}$$

also

$$W = Mg$$

∴

$$1130.40 = M \times 9.81$$

or

$$M = \mathbf{115.23 \text{ kg}}$$

1.34   Thermodynamic Equilibrium

Thermodynamics deals with equilibrium states. The word equilibrium implies a state of balance. In an equilibrium state, there are no unbalanced potentials (or driving forces) within the system. The thermodynamic equilibrium is defined on the basis of the following three types of equilibria:

1.   Thermal equilibrium (equality of temperature, i.e., no temperature gradient),
2.   Mechanical equilibrium (equality of pressure, i.e., no unbalance force), and
3.   Chemical equilibrium (equality of chemical potential, i.e., no chemical reaction).

1.34.1   Thermal Equilibrium

If the temperature of the system does not change with respect to time and has the same value at all points in the system, such type of equilibrium is called thermal equilibrium.

1.34.2   Mechanical Equilibrium

A system is in mechanical equilibrium if there is no unbalanced force in the system and also between the system and its surroundings. In other words, the pressure in the system is the same at all points and does not change with respect to time; such type of equilibrium is called mechanical equilibrium.

1.34.3   Chemical Equilibrium

If there is no chemical reactions occurring in the system and the chemical composition is the same throughout the system and does not vary with respect to time, such type of equilibrium is called chemical equilibrium.

When a system is simultaneously in a state of thermal equilibrium, mechanical equilibrium, and chemical equilibrium, such type of system is said to be in a state of **thermodynamic equilibrium**. If a system which is in a state of thermal equilibrium and mechanical equilibrium but not in chemical equilibrium, it is called **metastable equilibrium**.

Fig. 1.44 Sign convention for heat

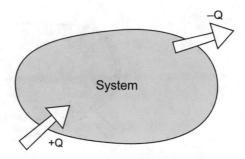

1.35 Heat

Heat is defined as the form of energy which is transferred due to the temperature difference between a system and its surroundings (or between two systems). It cannot be stored in the system or body. It is transferred spontaneously only from higher temperature to lower temperature. It is denoted by Q, and its SI unit is joule (J) or kilojoule (kJ).

Sign Convention

Heat transfer to a system is taken as $+ve$ while heat transfer from a system is taken as $-ve$ (Fig. 1.44).

1.36 Work

Work is defined as the form of energy which is equal to the product of the force and the distance moved in the direction of the force. As we know, the concept of heat is applicable only due to existence of the temperature difference, i.e., no heat transfer when no temperature difference. Like heat, the concept of work is applicable only due to the distance moved, i.e., no work when no distance moved. It is denoted by W.

Mathematically,

Work:

$$W = \text{Force} \times \text{Distance moved in the direction of force}$$
$$W = F.x$$

The SI unit of work is joule (J),

$$1\ J = 1\ Nm$$

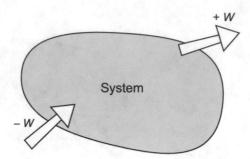

Fig. 1.45 Sign convention for work

Sign Convention

Work done on a system is taken as $-ve$ while work done by a system is taken as $+ve$ (Fig. 1.45).

Note: The sign convention adopted for work is reversed as the sign convention adopted for heat.

Similarities between heat and work:

1.   Both heat and work represent energy crossing the boundary of a system, i.e., both heat and work are boundary phenomena.
2.   Both heat and work are not stored in a system.
3.   Both are associated with a process, not a state.
4.   Both are path functions and have inexact differentials, i.e., they are depending upon the path followed by a process.

Dissimilarities between heat and work:

1.   Heat is a low-grade energy whereas work is a high-grade energy.
2.   Heat is transferred due to temperature difference only, whereas work is equal to the product of the force and the distance moved in the direction of the force.
3.   If a system is in a stable equilibrium state, then no work interaction between the system and its surroundings can take place, whereas there is no such restriction for heat interaction, for example, a gas contained in a rigid container at high pressure and temperature. The rigidness of the container provides an upper limit to the volume of the system. In this case, no work interaction will occur. But due to the temperature difference between the system and the surroundings, heat interaction would take place.

1.37   Point Function and Path Function

1.37.1   Point Function

The parameters depending only on the initial and final states (or points) of a system are called point functions. Point functions have exact differentials because they depend on the end states only and are independent of the path followed during a process. It is denoted by the symbol d or Δ. For example, volume, temperature, pressure, and energy are point functions.

Let two processes A and B be maintained between two states 1 and 2 (Fig. 1.46).
For process A:
Change in volume:

$$dV_A = V_2 - V_1$$

For process B:

$$\text{Change in volume} : dV_B = V_2 - V_1$$

Thus, volume change during process A and process B is the same and is equal to the volume at the state 2 minus the volume at state 1, regardless of the path followed. Similarly,

$$\text{Change in temperature} : dT = T_2 - T_1$$
$$\text{Change in pressure} : dp = p_2 - p_1$$
$$\text{and}\quad \text{Change in energy} : dE = E_2 - E_1$$

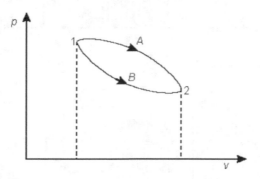

Fig. 1.46 Point function

1.37.2 Path Function

The parameters depending on both path followed during a process and end states of a system are called path functions. Path functions have inexact differentials because they depend on the path followed during a process. They are denoted by the symbol δ. For example, heat and work are path functions (Figs. 1.47 and 1.48).

Let two processes be maintained between states 1 and 2, via A and via B as shown in Fig. 1.47a.

For process 1–2 via A:

Work done:

$$\delta W_A = W_{1-2A}$$
$$= \text{Area under process } 1 - 2 \text{ } via \text{ } A.$$
$$\text{As shown in Fig. 1.47b}$$

For process 1–2 via B:

Work done:

$$\delta W_B = W_{1-2B}$$
$$= \text{Area under process } 1 - 2 \text{ } via \text{ } B.$$
$$\text{As shown in Fig. 1.44c}$$

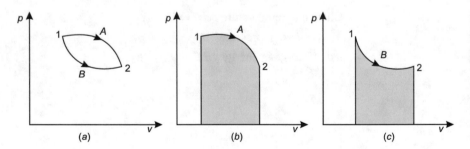

Fig. 1.47 Work—path function

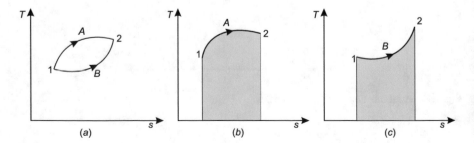

Fig. 1.48 Heat—path function

Thus, it is clear from Fig. 1.48b, c that work done for process 1–2 via A is not equal to work done for process 1–2 via B at the same end states. Hence, work is a path function that depends on the path followed during the process.

Similarly, for heat transfer:

For process 1–2 via A:

Heat transfer:

$$\delta Q_A = Q_{1-2A}$$
$$= \text{Area under process } 1 - 2 \ via \ A.$$
$$\text{As shown in Fig. 1.48b}$$

For process 1–2 via B:

Heat transfer:

$$\delta Q_B = Q_{1-2B}$$
$$= \text{Area under process } 1 - 2 \ via \ B.$$
$$\text{As shown in Fig. 1.48c}$$

Thus, it is clear from Fig. 1.48b, c, heat transfer for process 1–2 via A is not equal to heat transfer for process 1–2 via B at the same end states. Hence, heat is a path function that depends on the path followed during the process.

Remember

1.   Point functions are written as

$$dV \text{ or } \Delta V = V_2 - V_1$$
$$dp \text{ or } \Delta p = p_2 - p_1$$
$$dT \text{ or } \Delta T = T_2 - T_1$$

Point functions are not written as

$$\delta V = V_{1-2}$$
$$\delta p = p_{1-2}$$
$$\delta T = T_{1-2}$$

2.   Path functions are written as

$$\delta Q = Q_{1-2} \text{ or } Q_{12} \text{ or } {}_1 Q_2$$
$$\delta W = W_{1-2} \text{ or } W_{12} \text{ or } {}_1 W_2$$

Path functions are not written as

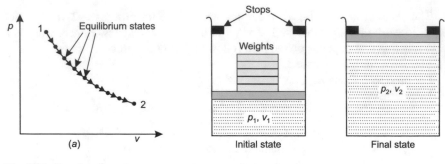

Fig. 1.49 Quasi-static process

$$dQ = Q_2 - Q_1$$
$$dW = W_2 - W_1$$

3. Area under process 1–2 (say) or curve in the *p-v* diagram shows work done.
4. Area under process 1–2 (say) or curve in the *T-s* diagram shows heat transfer.

1.38   Quasi-Static Process

If a very slow process passes through a number of states such that each state in the process is an equilibrium state at any instant, such a process is called a quasi-static process. It is also called a reversible process.

The quasi-static process is explained in a cylinder-piston arrangement as shown in Fig. 1.49. If the small weights on the piston are removed one by one very slowly, then at any instant of the upward movement of the piston, the change of the state from the thermodynamic equilibrium will be very small. Thus, every state passed through by the system will be in an equilibrium state. Such a process, which is a number of equilibrium points passing through the system, is called a quasi-static process.

1.39   Energy and Its Types

Energy is defined as the ability to do work. It can be classified into two categories:

1. Stored energy and
2. Transit energy.

1.39.1 *Stored Energy*

The energy that is contained within a system boundary is called stored energy. This energy is further classified into two categories:

(a) Macroscopic energy and
(b) Microscopic energy.

(a) **Macroscopic energy**: This form of energy is possessed by a system at a macroscopic level (i.e., considering a whole system). For example, kinetic energy and potential energy: for these energies, we will consider a total mass of fluid in a system. The energy of a system at the macroscopic level is discussed in this book.

(b) **Microscopic energy**: The microscopic form of energy is stored in the molecular structure of a system. It means mass of fluid at rest, where electrons move in an orbit of the atom. This energy is also called internal energy. The kinetic energy at a microscopic level is better explained in statistical thermodynamics, but it is not explained in this book.

Some of the stored energies are explained below according to the subject point of view:

(i) Kinetic energy,
(ii) Potential energy, and
(iii) Internal energy.

(i) **Kinetic energy**: Kinetic energy is the energy possessed by a system (or body) due to its motion (i.e., velocity). For example, a flowing fluid, a falling body, and the moving parts of machine, all have kinetic energy because of their motion. For a moving fluid, kinetic energy depends on mass and its velocity.

Mathematically,
Kinetic energy:

$$KE = \frac{MV^2}{2}$$

where

$M =$ mass of fluid, kg and
$V =$ velocity of fluid; m/s.

The SI unit is Joule or J,

$$1\,J = 1\,Nm$$

(ii) **Potential energy**: Potential energy is the energy possessed by a system (or body) due to its position above some assumed datum (i.e., reference level).

Consider a system of mass M at position z above some assumed datum level. Then the potential energy is mathematically expressed as

Potential energy:

$$PE = Mgz$$

where

$M =$ mass of fluid, kg,
$g =$ acceleration due to gravity $= 9.81$ m/s^2, and
$z =$ position of a system above datum level, m.

The SI unit is Joule or J.

(iii) **Internal energy**: As explained above, it is a microscopic form of energy stored in the molecular structure of a system. It means mass of fluid at rest, where electrons move in an orbit of the atom. It is denoted by U.

The total energy stored in the system is equal to the sum of kinetic, potential, and internal energies.

Mathematically,

Total energy:

$$E = \text{kinetic energy} + \text{potential energy} + \text{internal energy}$$

or

$$E = KE + PE + U$$
$$E = \frac{MV^2}{2} + Mgz + U \text{ kJ}$$

For unit mass, the above equations is written as

$$e = \frac{V^2}{2} + gz + u \text{ kJ/kg}$$

When the mass of fluid at rest in a system (i.e., kinetic energy) is zero and a system lies on the datum level (i.e., potential energy is zero), then the total energy is equal to the internal energy.

That is,

$$E = U$$

For unit mass,

$$e = u$$

1.39.2 Transit Energy

The energy that is capable of crossing the boundary of the system is called transit energy, for example, heat, work, and electrical energy.

1.40 Ideal and Real Gases

The ideal gas equation of state $pv = RT$ can be derived from the postulates of the kinetic theory of gases with two important assumptions:

(i)   there is little or no attraction of force between molecules.
(ii) volume occupied by molecules themselves is negligible.

When a gas obeys the ideal gas equation of state, then it is called an ideal or perfect gas. When a gas does not obey the ideal gas equation of state, then it is called a real gas.

When pressure is very low (or of very low density), the intermolecular attraction and volume of the molecules compared to the total volume of the gas are not much significant. Then, a real gas obeys very closely the ideal gas equation of state. But at high pressure (or high density), the force of attraction and repulsion among molecules increases, and also the volume of molecules approaches the total volume of gas. So, then real gases deviate from the ideal gas equation of state. For a real gas, the ideal gas equation of state ($pv = RT$) should be modified by using the compressibility factor for the gas at given pressure and temperature.

Compressibility factor is defined as the ratio of the actual volume of gas to the volume predicted by the ideal gas equation of state. It is denoted by the letter Z.

Mathematically,

$$\text{Compressibility factor: } Z = \frac{\text{Actual volume of gas}}{\text{Volume predicted by ideal gas equation of state}}$$

$$\text{OR}$$

$$= \frac{\text{Density predicted by ideal gas equation}}{\text{Actual density of gas}}$$

$$Z = \frac{v_a}{v} \text{ or } v = \frac{v_a}{Z}$$

$$\therefore$$

$$p\frac{v_a}{Z} = RT$$

or

$$pv_a = ZRT$$

For an ideal gas, $Z = 1$.

The deviation of Z from unity is a measure of the deviation of the actual relation from the ideal gas equation of state.

1.41 Law of Corresponding States

For a certain gas, the compressibility factor Z is a function of T and p, and the plot can be made of lines of the constant temperature of coordinates of p and Z as shown in Fig. 1.50.

For each substance, there is a compressibility chart factor. For different substances, there are different compressibility chart factors. In such a way, the overall process becomes wide and rough. It is more advantageous if one chart could be used for all substances. The general shape of the vapour dome of all the constant temperature lines on the p-v plane is similar for all substances, but its scale may be different. This similarity can be used to develop dimensionless properties such as reduced pressure (p_r), reduced temperature (T_r), and reduced volume (V_r). The dimensionless properties are also called reduced properties.

Reduced pressure (p_r) is defined as the ratio of the existing pressure to critical pressure.

Mathematically,

Reduced pressure:

$$p_r = \frac{p}{p_c}$$

Similarly, reduced temperature:

$$T_r = \frac{T}{T_c}$$

Fig. 1.50 Compressibility factor chart

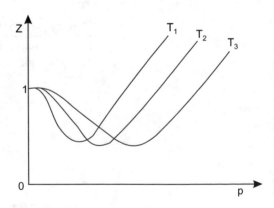

and reduced volume:

$$V_r = \frac{V}{V_c}$$

The chart drawn in reduced parameters is called a generalized compressibility chart. The above compressibility factor chart for each substance is redrawn into reduced parameters for all the substances (Fig. 1.51).

$$Z = f(p_r, T_r)$$

When T_r is plotted as a function of p_r and Z, it is known as the generalized compressibility chart. This chart is commonly used where detailed data on a particular gas are required but its critical properties are available.

The relation among the reduced properties p_r, T_r, and V_r is known as the law of corresponding states. It can be derived from the various equation of states, such as Van der Waals, Berthelet, and Dieterici.

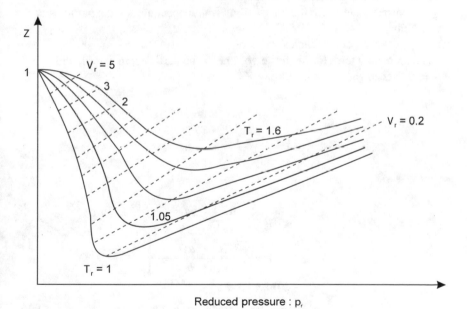

Fig. 1.51 Compressibility chart

1.42   Van Der Waals Equation of State

The equation of state of the real gas is derived from the ideal gas equation by considering the following facts:

1.   The particles of a real gas occupy a finite volume.
2.   The particles of a real gas are surrounded by force fields which cause them to interact with one another.

The most celebrated equation of state for real gases, which was derived from considerations 1 and 2 above, is the Van der Waals equation, which, for 1 mol of gas, is written as

$$\left(p + \frac{a}{v^2}\right)(v - b) = \overline{R}T \tag{1.14}$$

where

p is the measured pressure of the gas, N/m^2,
$\frac{a}{v^2}$ is a correction term for the interactions which occur among the particles of the gas, N/m^2,
$v = \frac{V}{n}$, volume per unit moles, m^3/kmol,
b is a correction term for the finite volume of the particles, m^3/kmol, and.
$\overline{R}$ = universal gas constant, J/kmol K

$$\left(p + \frac{a}{v^2}\right) = \frac{\overline{R}T}{v - b}$$

or

$$p = \frac{\overline{R}T}{v - b} - \frac{a}{v^2}$$

$$p = \frac{\overline{R}Tv^2 - a(v - b)}{(v - b)v^2}$$

$$p(v - b)v^2 = \overline{R}Tv^2 - a(v - b)$$

$$pv^3 - pbv^2 = \overline{R}Tv^2 - av + ab \tag{1.15}$$

or

$$pv^3 - (pb + \overline{R}T)v^2 + av - ab = 0$$

Therefore, v^3 for given values of p and T has three roots of which only one needs to be real. For low temperature, i.e., $T < T_c$, shown in Fig. 1.52, three real roots exist

Fig. 1.52 Isotherms lines in
p-v phase diagram

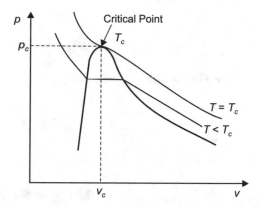

for a certain range of pressure (also certain range of volume). As the temperature increases, the three real roots approach each other (i.e., come near) and at the critical temperature they become equal (i.e., three roots coincide). Above this temperature, only one real root exists for all values of p. The critical isothermal T_c at the critical state in the p-v curve is where the three roots of the Van der Waals equation coincide.

At critical state,

$$T = T_c, p = p_c \text{ and } v = v_c$$

The Van der Waals Eq. (1.15) becomes

$$p_c = \frac{\overline{R}T_c}{v_c - b} - \frac{a}{v_c^2} \tag{1.16}$$

At critical state, the slope is not only zero but also changes, and the first and second derivatives of p_c w.r.t. v_c at T_c = constant is zero.

$$\left(\frac{\partial p_c}{\partial v_c}\right)_{T_r} = \frac{-\overline{R}T_c}{(v_c - b)^2} + \frac{2a}{v_c^3} = 0 \tag{1.17}$$

and

$$\left(\frac{\partial^2 p_c}{\partial v_c^2}\right) = \frac{-2\overline{R}T_c}{(v_c - b)^3} + \frac{6a}{v_c^4} = 0 \tag{1.18}$$

From Eq. (1.17)

$$\frac{\overline{R}T_c}{(v_c - b)^2} = \frac{2a}{v_c^3}$$

or

$$\overline{R}T_c v_c^3 = 2a(v_c - b)^2 \tag{1.19}$$

From Eq. (1.18)

$$\frac{2\overline{R}T_c}{(v_c - b)^2} = \frac{6a}{v_c^4}$$

or

$$2\overline{R}T_c v_c^4 = 6a(v_c - b)^3 \tag{1.20}$$

Dividing Eq. (1.20) by Eq. (1.19), we have

$$\frac{2\overline{R}T_c v_c^4}{\overline{R}T_c v_c^3} = \frac{6a(v_c - b)^3}{2a(v_c - b)^2}$$

$$\frac{2}{3}v_c = v_c - b$$

or

$$b = \frac{v_c}{3}$$

Substituting the value of b in Eq. (1.19), we have

$$\overline{R}T_c v_c^3 = 2a\left(v_c - \frac{v_c}{3}\right)^2$$

$$\overline{R}T_c v_c^3 = 2a \times \frac{4}{9}v_c^2$$

$$\overline{R}T_c v_c = \frac{8}{9}a$$

or

$$\overline{R} = \frac{8}{9T_c v_c}$$

Now substituting the values of $\overline{R} = \frac{8}{9}\frac{a}{T_c v_c}$ and $b = \frac{v_c}{3}$ in Eq. (1.16), we have

$$p_c = \frac{8}{9}\frac{a}{T_c v_c}\frac{T_c}{\left(v_c - \frac{v_c}{3}\right)} - \frac{a}{v_c^2}$$

$$p_c = \frac{8}{9}\frac{a}{v_c \times \frac{2}{3}v_c} - \frac{a}{v_c^2}$$

$$p_c = \frac{4a}{3v_c^2} - \frac{a}{v_c^2}$$

$$p_c = \frac{4a - 3a}{3v_c^2}$$

$$p_c = \frac{a}{3v_c^2}$$

or

$$a = 3p_c v_c^2$$

Then,

$$\overline{R} = \frac{8}{9} \times \frac{3p_c v_c^2}{T_c v_c}$$

$$\overline{R} = \frac{8}{3}\frac{p_c v_c}{T_c}$$

Substituting the values of a, b, and $\overline{R}$ in Eq. (1.14), we have

$$\left(p + \frac{3p_c v_c^2}{v^2}\right)\left(v - \frac{v_c}{3}\right) = \frac{8}{3} \times \frac{p_c v_c T}{T_c}$$

or

$$\left(\frac{p}{p_c} + 3\frac{v_c^2}{v^2}\right)\left(\frac{v}{v_c} - \frac{1}{3}\right) = \frac{8}{3} \times \frac{T}{T_c}$$

$$\left(p_r + \frac{3}{v_r^2}\right)(3v_r - 1) = 8T_r$$

This expression is called the law of the corresponding state because it reduces the properties of all gases to a single formula.

Summary

1. Thermodynamics is the science that deals with the relation among heat, work, and properties of the system in equilibrium.
2. **Continuum**. Materials such as solids, liquids, and gases consist of discrete molecules separated by empty space. However, certain physical phenomena, such as analysis of fluid flow problem, are made by assuming the concept that matter exists as a continuum, i.e., matter is viewed as a continuum, homogeneous matter with no holes in between the molecules.
3. **Standard Temperature and Pressure (STP)**. It refers to the condition of the standard atmospheric pressure of 760 mm of mercury (1.01325 bar) and a temperature of 15 °C (or 288 K).
4. **Normal Temperature and Pressure (NTP)**. It refers to the condition of the atmospheric pressure of 760 mm of mercury (1.01325 bar) and a temperature of 0 °C (or 273 K).

5. **System.** A system is defined as the quantity of matter (or region in space) chosen for a study.

6. **Open System.** In this type of system, mass of fluid flow can cross the boundary of the system. If mass of fluid flow can cross the boundary of the system, it means energy (in the form of heat or work) must transfer with the mass of fluid also. In a simple way, it is the open system defined as the system in which both mass and energy can cross the boundary of the system.

7. **Closed System.** A closed system is also known as a control mass; it consists of a fixed amount of mass, and no mass can cross its boundary. But energy, in the form of heat or work, can cross the boundary, and the volume of closed system either to be fixed or changed.

8. **Isolated System.** The isolated system is one in which both mass and energy cannot cross the boundary of the system. Even there is no chemical reaction that takes place within the isolated system.

9. **Intensive Property.** The properties that are independent of the mass (or size) of the system are called intensive properties. Temperature, pressure, density, viscosity, specific heat, and thermal conductivity are examples of intensive properties.

10. **Extensive Property.** The properties that depend on the mass (or size) of the system are called extensive properties. Energy, enthalpy, and entropy are examples of extensive properties.

11. **Phase.** A phase is a quantity of matter which is uniform throughout both in chemical composition and physical structure.

12. **Homogeneous System.** A system which consists of a single phase is called a homogeneous system.

13. **Heterogeneous System.** A system which consists of two or more phases is called a heterogeneous system.

14. **Pure Substance.** A pure substance is a single substance which has an unvarying molecular structure during the process of energy transfer. It may exist in more than one phase, but the chemical composition is the same in all phases.

15. **State.** The state of a system is the specific condition of a system at any instant of time to be described or measured by its properties such as temperature—T, pressure—p, and volume—V.

16. **Process.** A process occurs when the system undergoes a change or series of changes in the state of the system.

17. **Cycle.** When the initial and final states of a series of processes of the system are identical, it is called a cycle.

18. **Equation of State.** The equation in temperature, pressure, and volume (or specific volume or density) which is used to describe the condition or state of the system is called an equation of state.

For a perfect (or ideal) gas, the equation of state is

$$\boxed{pV = mRT}$$
$$p\frac{V}{m} = RT$$
$$\boxed{pv = RT}$$
$$\frac{p}{\rho} = RT$$
$$\boxed{p = \rho RT}$$
$$\boxed{pV = n\overline{R}T}$$

19. **Gas Constant**: *R.* It is defined as the ratio of the universal gas constant $(\overline{R})$ to the molecular weight of a perfect gas.

Mathematically,
Gas constant:

$$R = \frac{\text{Universal gas constant:} \overline{R}}{\text{Molecular weight:} M}$$
$$R = \frac{\overline{R}}{M}$$

where

$$\overline{R} = 8.314 \text{ kJ/kmol K}$$

20. **Thermodynamics Process:**

(i) *Isothermal process* $[T = C]$. The isothermal process takes place at constant temperature. The volume (or specific volume) of the gas varies inversely with pressure at constant temperature in this type of process.

Mathematically,

$$v \propto \frac{1}{p} \quad \text{at } T = C$$

or

$$pv = C$$

The isothermal process follows Boyle's law which states that the specific volume of a perfect gas is inversely proportional to the absolute pressure when the temperature is kept constant.

(ii) *Isobaric process* $[p = C]$. The isobaric process takes place at constant pressure. The volume (or specific volume) of the gas varies directly with the temperature at constant pressure in this type of process.

Mathematically,

$$v \propto T \quad \text{at } p = C$$

or

$$\frac{v}{T} = C$$

The isobaric process follows Charles' law which states that the specific volume of a perfect gas is directly proportional to the absolute temperature when the pressure is kept constant.

(iii)   *Isochoric process (or isometric process)* $[V = C]$. The isochoric process takes place at constant volume. The pressure of the gas varies directly with the temperature at constant volume in this type of process.

Mathematically,

$$p \propto T \quad \text{at } V = C$$

or

$$\frac{p}{T} = C$$

This process follows Gay-Lussac's law which states that the absolute pressure of a perfect gas is directly proportional to the absolute temperature when the volume is kept constant.

(iv)   *Adiabatic process* $[pv^{\gamma} = C]$. When there is no heat transfer between the system and surroundings during a process, it is known as an adiabatic process. The properties, pressure, temperature, and volume of the system will vary during this process. Normally, this process is considered an ideal process and also called reversible adiabatic (or adiabatic isentropic). In this process, the entropy of the system will remain constant without the transfer of heat to or from the surroundings.

This process is represented mathematically as

$$pv^{\gamma} = C$$

where

$p =$ pressure of gas,
$v =$ specific volume, and
$\gamma =$ c_p/c_v, adiabatic index.
        $= 1.4$ for air.

p-v-T relationship:

$$\frac{T_2}{T_1} = \left(\frac{p_2}{p_1}\right)^{\frac{\gamma-1}{\gamma}} = \left(\frac{v_1}{v_2}\right)^{\gamma-1}$$

(v)   *Polytropic process* $[pv^n = C]$. The polytropic process is a reversible process like the adiabatic isentropic process but the difference is that in the case of polytropic process, the entropy change occurs due to reversible heat transfer. This process can be represented by the equation

$$pv^n = C$$

where

$n =$ polytropic index

p-v-T relationship:

$$\frac{T_2}{T_1} = \left(\frac{p_2}{p_1}\right)^{\frac{n-1}{n}} = \left(\frac{v_1}{v_2}\right)^{n-1}$$

21.   **Dalton's Law of Partial Pressure.** It states that the pressure of a mixture of ideal gases is equal to the sum of the partial pressure of constituent gases at constant temperature and volume.

Mathematically,

$$p = p_A + p_B$$

where

$p =$ pressure of the mixture of gases,
$p_A =$ partial pressure of gas A in the gas mixture, and
$p_B =$ partial pressure of gas B in the gas mixture.

22.   **Amagat's Law of Partial Volume.** It states that the volume of a mixture of ideal gases is equal to the sum of the volumes of each gas that would occupy the mixture at constant pressure and temperature.

Mathematically,

$$V = V_A + V_B$$

where

$V =$ volume of the mixture of gases,

$V_A =$ partial volume of gas A in the gas mixture, and
$V_B =$ partial volume of gas B in the gas mixture.

23. **Density:**

$$\rho = \frac{\text{Mass of fluid:} m}{\text{Volume of fluid:} V}$$
$$\rho = \frac{m}{V}$$

where

$$\rho = 1000 \text{ kg/m}^3 \text{ for water at } 4\,^\circ\text{C}$$
$$= 13600 \text{ kg/m}^3 \text{ for mercury at NTP}$$

24. **Specific Volume:**

$$v = \frac{\text{Volume of fluid:} V}{\text{Mass of fluid:} m}$$
$$v = \frac{V}{m} = \frac{1}{m/V}$$
$$v = \frac{1}{\rho}$$

Specific volume is the reciprocal of the density.

25. **Specific Weight:**

$$w = \frac{\text{Weight of fluid} : W}{\text{Volume of fluid} : V} = \frac{mg}{V} = \rho g$$

26. **Specific Gravity:**

(i)   *Specific gravity for liquid:*

$$S_l = \frac{\text{Density of liquid} : \rho}{\text{Density of water} : \rho_{\text{water}}} = \frac{\rho}{1000}$$

(ii)   *Specific gravity for gas:*

$$S_g = \frac{\text{Density of gas} : \rho}{\text{Density of air} : \rho_{\text{air}}} = \frac{\rho}{1.29}$$

where

$$\rho_{\text{air}} = 1.29 \text{ kg/m}^3 \text{ at NTP}$$

27. **Temperature**. It is defined as a measure of the intensity of hotness or coldness of the body. It is measured by a thermometer in degree centigrade (°C) and degree Fahrenheit (°F) scales.

28. **Pressure**. It is defined as normal force exerted by a fluid per unit area. It is also known as the intensity of pressure.

Mathematically,
Pressure:

$$p = \frac{\text{Force} : F}{\text{Area} : A}$$

The SI unit of pressure is N/m^2 or Pa.

29. **Pressure Head:**

$$h = \frac{p}{\rho g} = \frac{p}{w}$$

where

$p =$ intensity of pressure.

30. **Pascal's Law.** It states that the intensity of pressure at any point in a fluid at rest is the same in all directions.

31. **Absolute Pressure.**

Absolute pressure = atmospheric pressure + gauge pressure

$$p_{\text{abs}} = p_{\text{atm}} - p_{\text{g}}$$

32. **Absolute Pressure.**

Absolute pressure = atmospheric pressure − vacuum pressure

$$p_{\text{abs}} = p_{\text{atm}} - p_{\text{vac}}$$

33. **Thermal Equilibrium.** If the temperature of the system does not change with respect to time and has the same value at all point in the system. Such type of equilibrium is called thermal equilibrium.

34. **Mechanical Equilibrium.** A system is in mechanical equilibrium if there is no unbalanced force in the system and also between the system and its surroundings.

35. **Chemical Equilibrium.** A system is in chemical equilibrium if there is no chemical reactions occurring in the system and chemical composition is the same throughout the system.

36. **Point Function.** The parameters depending only on the initial and final states of a system are called point functions.

For example, volume, temperature, pressure, and energy are point functions.

37. **Path Function.** The parameters depending on both path followed during a process and end states of a system are called path functions. For example, Heat and work are path functions.

38. **Quasi-Static Process**. If a very slow process passes through a number of states such that each state in the process is an equilibrium state at any instant, such a process is called a quasi-static process. It is also called a reversible process.

39. **Van der Waals Equation of State**. The Van der Walls equation for one mole of a real gas is

$$\left(p + \frac{a}{v^2}\right)(v - b) = \overline{R}T$$

where

p is the measured pressure of the gas, N/m^2,

$\frac{a}{v^2}$ is a correction term for the interactions which occur among the particles of the gas, N/m^2,

$v = \frac{V}{n}$, volume per unit moles, m^3/kmol, and.

b is a correction term for the finite volume of the particles, m^3/kmol.

$\overline{R} =$ universal gas constant, J/kmol K.

40. **Heat**. It is defined as the form of energy which is transferred due to the temperature difference between a system and its surroundings (or between two systems). It cannot be stored in the system or body. According to the second law of thermodynamics, it can be transferred spontaneously only from higher temperature to lower temperature.

41. **Work**. It is defined as the form of energy which is equal to the product of the force and the distance moved in the direction of the force.

Assignment-1

1. What is thermodynamics?
2. Explain the difference between 'Microscopic' and 'Macroscopic' points of view of thermodynamics.
3. What do you understand by NTP and STP? What are their values?
4. What do you understand by continuum?
5. Define a thermodynamic system. Explain its different types.
6. Define the following terms:

    (a) System,
    (b) Surroundings,
    (c) Boundary, and
    (d) Universe.

7. Explain the terms 'closed system' and 'open system'.
8. Explain the isolated system with an example.
9. What is the difference between intensive and extensive properties?
10. Explain, with examples, what microscopic and macroscopic points of view to study the subject of thermodynamics are.
11. What is a thermodynamic system? What is the difference between an open system and a closed system?

12. What are intensive and extensive properties? Explain with examples each.
13. Explain the terms:

    (a) State,
    (b) Path,
    (c) Cycle process, and
    (d) Equilibrium.

14. When is the energy crossing the boundaries of a closed system of heat and when that of work? State whether the following is heat transfer or work transfer:

    (a) A gas in a piston-cylinder device is compressed, and as a result its temperature rises.
    (b) A room is heated by an iron that is left plugged in (take the entire room, including the iron, as the system).

15. What is the equation of state? Write its three different forms.
16. Define the isothermal, isobaric, and isochoric processes.
17. What is an adiabatic process? What is an adiabatic system?
18. What is the difference between adiabatic and polytropic processes?
19. State the following laws:

    (a) Boyle's law,
    (b) Charles' law,
    (c) Avogadro's law,
    (d) Dalton's law of partial pressure, and
    (e) Amagat's law.

20. Define the following fluid properties:

    (a) Specific weight,
    (b) Density,
    (c) Specific gravity, and
    (d) Specific volume.

21. What is specific gravity? How is it related to density?
22. Is the state of the air in an isolated room completely specified by the temperature and the pressure? Explain.
23. What do you mean by the intensity of pressure? State its dimension and units.
24. Explain pressure head.
25. State and prove Pascal's law.
26. Explain atmospheric pressure. What is the value of atmospheric pressure head in terms of a mercury column and a water column?
27. What is the difference between gauge pressure and absolute pressure?
28. Explain why some people experience more bleeding and some others experience shortness of breath at high elevations?
29. Someone claims that the absolute pressure in a liquid of constant density doubles when the depth is doubled. Do you agree? Explain.

30.   Consider two identical fans, one at sea level and the other on top of a high mountain, running at the same speeds. How would you compare (*a*) the volume flow rates and (*b*) the mass flow rates of these two fans?
31.   Explain mechanical, chemical, and thermal equilibria.
32.   For a system to be in thermodynamic equilibrium, do the temperature and the pressure have to be the same everywhere?
33.   What is the metastable equilibrium?
34.   What are the similarities and dissimilarities between heat and work?
35.   A room is heated as a result of solar radiation coming in through the windows. Is this a heat or work interaction for the room?
36.   In what forms can energy cross the boundaries of a closed system?
37.   What are point and path functions? Give some examples.
38.   What is a quasi-static process? What is its importance in engineering?
39.   What is the difference between the macroscopic and microscopic forms of energy?
40.   What is mechanical energy? How does it differ from thermal energy? What are the forms of mechanical energy of fluid stream?
41.   Write comprehensive technical notes on the generalized compressibility chart.
42.   What is a compressibility factor? State the significance.

Assignment-2

1.   A quantity of a gas occupies 0.18 m^3 at a pressure of 6.2 bar. It is expanded to a pressure of 1.5 bar at constant temperature. Determine the volume of gas after expansion.
   [**Ans**. 0.744 m^3]
2.   A certain volume of a gas at STP is heated until its temperature becomes 400 °C and its volume is doubled. Determine the final pressure.
   [**Ans**. 114.41 kPa]
3.   12 L of air at a pressure of 8 bar and a temperature of 65 °C is expanded to a pressure of 2.5 bar. The temperature after the expansion is 32 °C. Determine the final volume of air.
   [**Ans**. 34.65 L]
4.   A gas of molecular weight of 44 occupies 180 L at a pressure of 1.5 bar and a temperature of 25 °C. Determine the mass of the gas.
   [**Ans**. 0.4795 kg]
5.   Determine the intensity of pressure in water at a depth of 6 m below the free surface of water.
   [**Ans**. 58.86 kPa]
6.   Convert intensity of pressure of 4 MPa into equivalent pressure head of oil of specific gravity 0.82.
   [**Ans**. 497.25 m]
7.   A hydraulic press has a ram of 500 mm diameter and a plunger of 70 mm diameter. Find the weight lifted by the hydraulic press when the force applied at the plunger is 1000 N.
   [**Ans**. 51.02 kN]

8.   An open tank contains water up to a depth of 3 m and above it an oil of specific gravity 0.9 for a depth of 1 m. Find the intensity of pressure

   (i)   at the interface of the low liquids and
   (ii) at the bottom of the tank.

   Ans. (i) 8.829 kPa and (ii) 38.259 kPa]

9.   An open tank contains mercury up to a depth of 0.5 m and above it water up to a depth of 1 m and above water an oil of specific gravity 0.89 for a depth of 0.5 m. Find the intensity of pressure

   (i)   at the interface of the oil and water,
   (ii) at the interface of the water and mercury, and
   (iii) at the bottom of the tank.

   [**Ans.** (i) 4.365 kPa, (ii) 14.175 kPa, and (iii) 80.88 kPa]

10. The diameters of the plunger and ram of a hydraulic press are, respectively, 60 mm and 140 mm. If a load of 6 kN is applied to the plunger to lift a load W on the ram, determine the value of W when

   (i)   the plunger and the ram are at the same level and
   (ii) the plunger is 400 mm above the ram assuming that water is used in the hydraulic press.

   [**Ans.** (i) 32.666 kN and (ii) 32.729 kN]

11. Determine the vacuum pressure in metres of water when the absolute pressure is 0.5434 bar. Assume atmospheric pressure as 10.30 m of water.
   [**Ans.** 4.76 m of water]

12. At a point 8 m below the free surface of water, determine the pressure at the point in

   (i)   kPa and
   (ii) mm of Hg.

   If atmospheric pressure is 760 mm of Hg, determine the absolute pressure in

   (iii) m of water,
   (iv) bar, and
   (v)   mm of Hg.

   [**Ans.** (i) 78.48 kPa, (ii) 588.23 mm of Hg, (iii) 18.336 m of water, (iv) 1.798 bar, and (v) 1348.23 mm of Hg.]

Chapter 2
Zeroth Law of Thermodynamics

Nomenclature

The following is a list of the nomenclature introduced in this chapter:

R	kJ/kgK	Gas constant
m	kg	Mass
T	°C or K	Temperature
R	–	Reservoir
p_{atm}	kPa	Atmospheric pressure
ρ_{Hg}	kg/m^3	Density of mercury
g	m/s^2	Acceleration due to gravity
h	m	Difference in mercury level
V	m^3	Volume

2.1 Introduction

The zeroth law of thermodynamics was formulated by R. H. Fowler in 1931. This law was formulated approximately half a century after the formulation of the first and second laws of thermodynamics. It was named the zeroth law since it should have preceded the first and the second laws of thermodynamics. This law states that if two bodies are in thermal equilibrium with a third body, they are also in thermal equilibrium with each other. This law gives the concept of thermal equilibrium (*i.e.*, the equality of temperature), and it is used as a basic law for temperature measurement.

© The Author(s) 2022
S. Kumar, *Thermal Engineering Volume 1*,
https://doi.org/10.1007/978-3-030-67274-4_2

2.2 Temperature

Temperature is defined as a measure of the intensity of hotness or coldness of a body. A body is said to be at a high temperature (or hot) if it shows a high level of thermal energy in it. Similarly, a body is said to be at a low temperature (or cold) if it shows a low level of thermal energy in it. Temperatures are measured by a thermometer in degree centigrade or Celsius and degree Fahrenheit scales.

1. **Degree Centigrade or Celsius Scale:** The freezing point of water on this scale is marked as zero, and the boiling point of water as 100. The interval between these two points has 100 equal parts, and each part represents 1 degree Celsius (or 1 °C). This scale is widely used by scieists and engineers.

2. **Degree Fahrenheit Scale:** In this scale, the freezing point of water is marked as 32 and the boiling point of water as 212. The interval between these two points has 180 equal parts, and each part represents 1 degree Fahrenheit (or 1°F). This scale is widely used by doctors.

Relation between Celsius and Fahrenheit scales is given by

$$\frac{°C - 0}{100 - 0} = \frac{°F - 32}{212 - 32}$$

or

$$\frac{°C}{100} = \frac{°F - 32}{180}$$

or

$$°C = \frac{100}{180°}(°F - 32)$$

or

$$°C = \frac{5}{9}(°F - 32)$$

2.2.1 Absolute Temperature Scale

When gases are cooled below the melting point of ice (i.e., below 0 °C), at -273.15 °C temperature, they will exist at zero pressure and perfect vacuum. If the temperature at this state is termed as absolute zero, then the temperature at any other point from absolute zero is called absolute temperature, and such a scale of temperature is known as the Kelvin scale of temperature, designated by K.

Relation between absolute temperature and degree Celsius is given by

$$T \text{ K} = T_1 \,^\circ\text{C} + 273.15$$

where

T = absolute temperature,
T_1 = degree Celsius temperature, and

$$T \text{ K} \approx T_1 \,^\circ\text{C} + 273$$

It is common practice to round the constant 273.15 to 273.

2.3 Zeroth Law of Thermodynamics

It states that **if two systems (or bodies) are in thermal equilibrium with a third system (or body), they are also in thermal equilibrium with each other**.

Suppose systems A and B are in thermal equilibrium as shown in Fig. 2.1(a) and are brought in contact with a third system C as shown in Fig. 2.1(b). Now after contact, all three systems come in thermal equilibrium among themselves (i.e., system A will be in thermal equilibrium with system B without direct thermal contact between the two, system B will be in thermal equilibrium with system C, and also system A will be in thermal equilibrium with system C).

This law was formulated by R. H. Fowler in 1931. This law was formulated approximately half a century after the formulation of the first and second laws

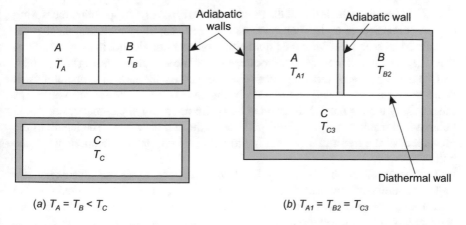

(a) $T_A = T_B < T_C$ (b) $T_{A1} = T_{B2} = T_{C3}$

Fig. 2.1 Demonstration of the Zeroth Law

of thermodynamics. **It was named the zeroth law since it should have preceded the first and second laws of thermodynamics.**

This law has its applications in temperature measurements. When we put a thermometer in our mouth to check the body temperature, there is a rise in the mercury level of the thermometer. After some time, we observe that no change in the mercury level of the thermometer occurs during this operation. Then, we say that our body is in thermal equilibrium with the given thermometer.

Let system A be the human body where the temperature is measured, system C be the glass wall of the mercury thermometer, and system B be the mercury in the thermometer, *i.e.*, the human body comes in thermal equilibrium with the mercury in the thermometer through the glass wall of the thermometer.

2.4   Temperature Measurement

The zeroth law of thermodynamics is used for the measurement of temperature. The instrument is used for the measurement of temperature, called the thermometer. Many types of thermometers are available in the market for the measurement of temperature. Some types of thermometers are listed below according to the subject point of view:

1.   Liquid thermometers,
2.   Gas thermometers,
3.   Electrical resistance thermometer, and
4.   Thermoelectric thermometer.

2.5   Liquid Thermometers

The liquid thermometer is one of the most common types of a temperature measuring device. It consists of a glass stem, indicating scale, capillary tube, and a spherical or cylindrical bulb filled with the liquid either mercury or alcohol having red colour (Fig. 2.2). The size of the capillary tube depends on the size of the sensing bulb, liquid, and the desired temperature range of the instrument. An increase in temperature will cause the liquid to expand and rise up in the capillary tube. Since the volume of the capillary tube is much less than the bulb, the relatively small changes of liquid volume will result in a significant liquid rise in the capillary tube. The length of the movement of the free surface of the liquid column indicates the temperature of the bulb.

The thermometer bulb is usually filled with mercury. It has the advantages of a uniform coefficient of expansion over a wide range of temperature and remains liquid over a large range of its freezing and boiling points are -38.9 °C and 356.7 °C, respectively, does not wet the wall of the tube and is good conductor of heat.

Fig. 2.2 Mercury in glass thermometer

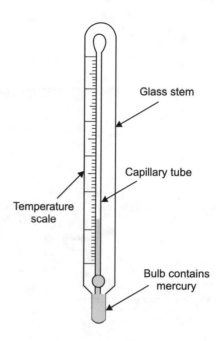

2.6   Gas Thermometers

If the working substance a gas, then the thermometers are called gas thermometers. Nitrogen or helium, chemically inert gases, is used in gas thermometers. They have good values for their coefficient of expansion and low specific heats as compared to liquids so even a small change in temperature can be calculated accurately. The gas thermometers are not suitable for routine work as compared to liquid thermometers, because they are large in size and need some calculations.

The following two types of gas thermometers are discussed:

(i)   Constant volume gas thermometers and
(ii)   Constant pressure gas thermometers.

2.6.1   Constant Volume Gas Thermometer

The gas thermometer is based on the equation of state for the ideal gas (Figs. 2.3 and 2.4),

$$pV = mRT$$

A fixed volume bulb containing a gas of known mass m and gas constant R is used as the temperature-sensing element. The pressure of the gas is read using a pressure gauge.

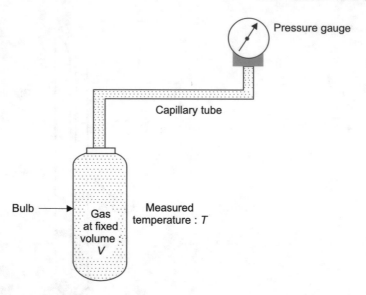

Fig. 2.3 Ideal-gas thermometer

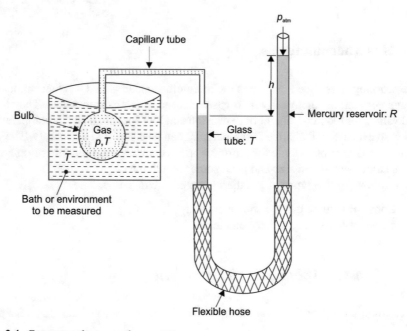

Fig. 2.4 Constant volume gas thermometer

Although it is possible to calculate T knowing p, m, and R, there could be uncertainty in the value of m and R. Instead, we recognize that under constant volume and mass, we must have

$$\frac{T_1}{T_2} = \frac{p_1}{p_2}$$

So making a calibration measurement using a known $T_2 = T_{ref}$ and the corresponding $p_2 = p_{ref}$, we have

$$\frac{T}{T_{ref}} = \frac{p}{T_{ref}}$$

or

$$T = \frac{T_{ref}}{p_{ref}} p$$

where

$$T_{ref} = \text{constant and}$$

$\therefore$

$$T = Cp$$
$$T \propto p$$

The constant volume gas thermometer can read the minimum temperature approaching 1 K.

Another arrangement of a constant volume gas thermometer is shown in Fig. 2.4. It consists of five main parts as follows:

1. Glass bulb (i.e., temperature-sensing element) which contains the inert gas at constant volume.
2. Capillary tube, where one end of the capillary tube is connected to the bulb and the other end to the vertical glass tube T.
3. Vertical glass tube whose top end is connected to the capillary tube and the bottom end to the flexible hose. The level of mercury is maintained constant irrespective of the change in pressure in the glass bulb by the movement of mercury reservoir R.
4. Flexible hose, one end of which is connected to the glass tube T and the other end to the mercury reservoir R.
5. Mercury reservoir, the lower end of which is connected to the flexible hose and the upper end is open to the atmosphere. The volume of the gas in the bulb is

kept constant by raising or lowering the mercury reservoir R to keep the mercury level in the glass tube T.

The thermometer is calibrated by using an ice water bath and a steam-water bath. The pressures of the mercury under each situation are recorded. The volume of the gas is kept constant by maintaining the constant level of the mercury in the glass tube T by adjusting the mercury reservoir R. The information is plotted as shown in Fig. 2.5.

To find the temperature of the system, the gas bulb is placed in thermal contact with the system. In a short time, the bulb comes in thermal equilibrium with the system. The gas in the bulb expands, on being heated, pushing the mercury downward in glass tube T. The mercury reservoir R is then adjusted so that the mercury level in the glass tube T comes back to the original position. Then, the absolute pressure of the gas,

$$p = p_{atm} + \rho_{Hg}gh \tag{2.1}$$

where

p_{atm} is atmospheric pressure,
ρ_{Hg} is the density of mercury, and
h is the difference in the mercury level.

Since the absolute pressure (p) calculated from Eq. (2.1) of the system, the temperature is read from the graph on the basis of p for the specific gas.

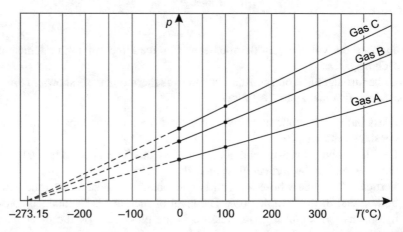

Fig. 2.5 p versus T plots of experimental data obtained from a constant volume gas thermometer using three different gases

2.6.2 Constant Pressure Gas Thermometer

This thermometer is based on Charles' law which states that the volume of a given mass of gas is directly proportional to its absolute temperature.
   Mathematically,

$$V \propto T \quad \text{at } p = C$$

$$\frac{V}{T} = \text{constant}$$

$$\frac{V_1}{T_1} = \frac{V_2}{T_2} \quad \text{or} \quad \frac{T_2}{T_2} = \frac{V_2}{V_2}$$

So making a calibration measurement using a known $T_2 = T_{ref}$ and the corresponding $V_2 = V_{ref}$, we have

$$\frac{T}{T_{ref}} = \frac{V}{V_{ref}}$$

$$T = \frac{T}{V_{ref}} V$$

where $\quad \dfrac{T_{ref}}{V_{ref}}$ constant

$$\therefore \quad T = CV$$

The constant pressure gas thermometer is large in size and not easy to operate as compared with a constant volume gas thermometer.

2.7 Electric Resistance Thermometer

Electric thermometer, also called resistance temperature detectors or resistive thermal devices, work on the principle of change in resistance of some materials with changing temperature. As they are almost invariably made of platinum, they are often called platinum resistance thermometers. Thus, resistance is the thermometric property used in these thermometers. It consists of a platinum wire wound in a double spiral on a mica plate.
   The principle of the Wheatstone bridge is employed in these thermometers as shown in Fig. 2.6.
   The platinum wire has a set of compensating leads having exactly similar resistance. The platinum wire along with the compensating leads is enclosed in a sealed glazed porcelain tube.

Fig. 2.6 Electric resistance
thermometer

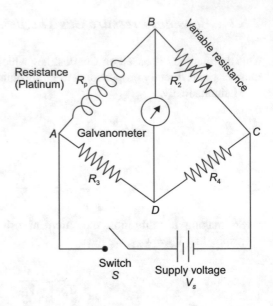

In a Wheatstone bridge shown in Fig. 2.6, R_p is the resistance corresponding to
the platinum wire, R_3 and R_4 are fixed resistances, and R_2 is a variable resistance.

The network gets energized when voltage V_s is applied to the bridge by closing
the switch S. Current flowing from the battery gets divided at points A. Part of this
current I_1 flows through R_p and R_2 to point C; the other part I_2 flows through R_3 and
R_4 to point C.

In equilibrium condition when no current is indicated on the galvanometer
connected between the bridge points B and D, the potential at B equals the potential
at D,

$$\text{i.e.,} \quad V_{ab} = V_{ad} \Rightarrow I_1 R_p = I_2 R_3$$

$$\text{and} \quad V_{bc} = V_{dc} \Rightarrow I_1 R_2 = I_2 R_4.$$

Eliminating the current from these relations, we get

$$R_p = R_2 \left(\frac{R_2}{R_4} \right)$$

Generally, there is a change in resistance of the platinum wire due to temperature which is to be determined. This change will unbalance the bridge, and the
galvanometer would give deflection. The resistance R_2 is then adjusted to regain the
balance.

In the measurement of temperature, the relationship between the temperature and
electric resistance is usually non-linear and is described by a higher order polynomial

$$R_p(T) = R_0(1 + A.T. + B.T^2 + C.T^2 + \ldots.)$$

where R_0 is the nominal resistance at a specified temperature (at 0 °C). The number of higher order terms considered is a function of the required accuracy of measurement.

The coefficients A, B, and C depend on the conductor material and basically define the temperature-resistance relationship.

For temperature ranging from 0 °C to 850 °C, the second-order polynomial is used, i.e.,

$$R_p(T) = R_0(1 + AT + BT^2) \tag{2.2}$$

The coefficients are as follows:

$$A = 3.9083 \times 10^{-3}\,°C^{-1}$$
$$B = 5.775 \times 10^{-7}\,°C^{-1}$$

and Eq. (2.2) is utilized for temperature measurement.

2.8 Thermoelectric Thermometer

A thermoelectric thermometer works on the principle of the Seebeck effect. According to the Seebeck effect, when two conductors of dissimilar metals M_1 and M_2 are joined together to form a circuit (thermocouple) and two unequal temperatures T_1 and T_2 are imposed at two junctions, an electric current flows through the circuit or e.m.f. is produced. The current produced in this way is called thermoelectric current while the e.m.f. produced is called thermo e.m.f. The measurement of temperature is done in this thermometer by knowing the e.m.f. produced.

A thermoelectric thermometer is shown in Fig. 2.7 where two dissimilar conductors Cu and Fe are joined by soldering or welding to form a circuit. One junction of this thermocouple is kept at the ice point while another junction is kept in the oil bath.

A sensitive galvanometer is connected with the thermocouple as shown in Fig. 2.7.

When the oil bath is heated, e.m.f. is produced due to the Seebeck effect. Further oil in the bath is heated to some another temperature which is measured by some calibrated thermometer, and it is seen that for different temperatures of oil, different e.m.f. is produced and the graph is plotted between the temperature of the oil bath and e.m.f. produced.

In order to use this thermocouple for temperature measurement, a cold junction shall still be kept at the ice point while a hot junction is kept in contact with the bath whose temperature is to be measured. To measure the temperature depending upon the e.m.f. produced, a graph (plotted between e.m.f. produced and temperature of oil bath) is used and the corresponding temperature is noted from this graph.

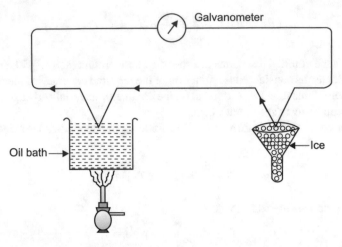

Fig. 2.7 Thermoelectric thermometer

The advantage of this thermocouple is that it comes to thermal equilibrium with the system whose temperature is to be measured, quite rapidly because its mass is very small.

Summary

1. **Temperature:** It is defined as a measure of the intensity of hotness or coldness of a body. Temperatures are measured by a thermometer in degree centigrade or Celsius and degree Fahrenheit scales.
2. **Relation between degree Celsius (°C) and Fahrenheit (°F):**

$$^\circ C = \frac{5}{9}(^\circ F - 32)$$

3. **Relation between absolute temperature (K) and degree Celsius (°C):**

$$T\,K = T_1\,^\circ C + 273$$

   where

   $T =$ absolute temperature and
   $T_1 =$ degree Celsius temperature

4. **Zeroth Law of Thermodynamics:** It states that if two systems are in thermal equilibrium with a third system, they are also in thermal equilibrium with each other.
5. **Application of Zeroth Law of Thermodynamics:** This law is used for temperature measurement.

6. **Temperature Measurement:** The zeroth law of thermodynamics is used for the measurement of temperature. The instrument is used for the measurement of temperature, called the thermometer. Many types of thermometers are available in the market for the measurement of temperature. Some types of thermometers are listed below according to the subject point of view:

(a) Liquid thermometers
(b) Gas thermometers,
(c) Electrical resistance thermometer, and
(d) Thermoelectric thermometer.

Assignment-1

1. What is temperature?
2. What is the relation between Celsius and Fahrenheit scales?
3. Normal temperature of a human body is 37 °C. What is it in kelvin?
4. State the zeroth law of thermodynamics.
5. Is it possible to measure a temperature without invoking the zeroth law of thermodynamics? If not why?
6. What is a constant volume gas thermometer? Why is it preferred to a constant pressure gas thermometer?
7. Describe with a neat sketch of an electrical resistance thermometer.
8. Describe with a neat sketch a thermoelectric thermometer.

Chapter 3
First Law of Thermodynamics

Nomenclature

The following variables are introduced in this chapter:

Q	kJ	Heat transfer
q	kJ/kg	Specific heat transfer
W	kJ	Work
w	kJ/kg	Specific work
U	kJ	Internal energy
u	kJ/kg	Specific internal energy
$\oint$	–	Cycle integral
η	–	Efficiency
p	kPa	Pressure
V	m^3	Volume
v	m^3/kg	Specific volume
PMM I	–	Perpetual motion machine of the first kind
H	kJ	Enthalpy
h	kJ/kg	Specific enthalpy
ρ	kg/m^3	Density
R	kJ/kgK	Gas constant
T	K	Temperature
c_v	kJ/kgK	Specific heat at constant volume
c_p	kJ/kgK	Specific heat at constant pressure
γ	–	Adiabatic index
C	kJ/K	Heat capacity
m	kg	Mass
n	–	Polytropic index

© The Author(s) 2022
S. Kumar, *Thermal Engineering Volume 1*,
https://doi.org/10.1007/978-3-030-67274-4_3

3.1   Introduction

The first law of thermodynamics is also called the **law of conservation of energy**. It means that energy can neither be created nor be destroyed, but it can be converted from one form to another. This statement of the first law of thermodynamics is essentially true and universally accepted. Actually, thermodynamics begins with this law. For a cyclic process, the algebraic sum of heat transfer is equal to the algebraic sum of work done. It is so because heat and work are completely interchangeable forms of energy. This statement of the first law of thermodynamics for cyclic processes was highly criticized by the second law of thermodynamics. We will study all aspects of the first law of thermodynamics in this chapter.

3.2   First Law of Thermodynamics

The first law of thermodynamics is also called the **law of conservation of energy**. According to this law, **energy can neither be created nor destroyed but can be converted from one form of energy to another**. Once a student asked his teacher, sir, I am confused with the statement of the first law of thermodynamics which says, "Energy can neither be created nor destroyed" but sir, we create heat energy, when we ignite the coal. The teacher replied, "energy is never created, this is absolutely right". Actually, we do not create energy from coal when igniting it. We are just converting chemical energy already present in coal in the form of heat energy. We cannot create heat energy from ash, why? Because there is no form of energy present in ash (Fig. 3.1).

3.3   Statement of the First Law of Thermodynamics

(i)   **The first law for a process**

It states that **the heat interaction is equal to the sum of the change in internal energy and work interaction**.
  Mathematically,

Heat interaction = change in internal energy + work interaction

$$\delta Q = dU + \delta W$$

Fig. 3.1 First law of thermodynamic

For process 1–2,

$$q_{1-2} = (U_2 - U_1) + W_{1-2}$$

where

Q_{1-2} is heat interaction in kJ. If heat is transferred to the system, a $+ve$ sign is assigned to Q_{1-2}. If heat is transferred from the system, a $-ve$ sign is assigned to Q_{1-2};

$(U_2 - U_1)$ is the change in internal energy in kJ. The $+ve$ sign shows that the change in internal energy increases and the $-ve$ sign indicates the change in internal energy decreases;

W_{1-2} is work interaction in kJ. If work is done by the system, a $+ve$ sign is assigned to it, whereas if work is done on the system, it is assigned a $-ve$ sign.

For unit mass,

$$q_{1-2} = (u_2 - u_1) + w_{1-2}$$

where

q_{1-2} is specific heat transfer in kJ/kg, which is defined as the heat transfer per unit mass,

i.e.,

$$q_{1-2} = \frac{Q_{1-2}}{m}$$

$(u_2 - u_1)$ is the change in specific internal energy in kJ/kg, which is defined as the change in internal energy per unit mass,

i.e.,

$$(u_2 - u_1) = \frac{U_2 - U_1}{m}$$

w_{1-2} is the specific work in kJ/kg, which is defined as the work done per unit mass,

i.e.,

$$w_{1-2} = \frac{W_{1-2}}{m}$$

(ii)   **The first law for a cyclic process**

It states that **when a system undergoes a cyclic process, the algebraic sum of the heat transfer is equal to the algebraic sum of the work transfer**. It is so because work and heat are mutually convertible from one to another.

Mathematically,

$$\Sigma \delta Q = \Sigma \delta W$$
$$\text{OR}$$
$$\oint \delta Q = \oint \delta W$$

where the symbol $\oint$ is called 'cycle integral' or 'integral around the cycle'.

3.4 $\oint dU = 0$ for Cyclic Process

Let the cyclic process be 1–2–1, which contains two processes, i.e., 1–2 and 2–1, as shown in Fig. 3.2.

$$\text{Process } 1-2, \quad Q_{1-2} = (U_2 - U_1) + W_{1-2} \tag{3.1}$$

Fig. 3.2 Cyclic process

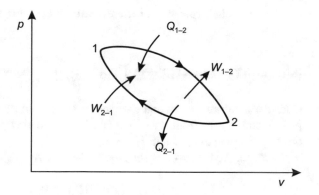

Process 2–1, $-Q_{2-1} = (U_1 - U_2) - W_{2-1}$ (3.2)

Adding Eqs. (3.1) and (3.2), we get

$$Q_{1-2} - Q_{2-1} = (U_2 - U_1) + (U_1 - U_2) + W_{1-2} - W_{2-1}$$

$$\oint \delta Q = \oint \delta U + \oint \delta W$$

$$\oint \delta Q - \oint \delta W = \oint dU$$

$$0 = \oint dU \quad \because \oint \delta Q = \oint \delta W$$

or

$$\oint dU = 0$$

OR

$$\Sigma dU = 0$$

also

$$\oint dU = (U_2 - U_1) + (U_1 - U_2)$$

$$\oint dU = U_2 - U_1 + U_1 - U_2$$

$$\oint dU = 0$$

Thus, **the algebraic sum of the internal energy for a cyclic process is zero**.

3.5   Internal Energy—A Property of the System

Consider a closed system that changes its state from state 1 to state 2 by following path A, and returns from state 2 to state 1 by following path B which is represented on the p–v plane as shown in Fig. 3.3.

According to the first law for process 1–2 via A

$$Q_A = \Delta U_A + W_A$$

or

$$Q_A - W_A = \Delta U_A \qquad (3.3)$$

and for process 2–1 via B

$$Q_B = \Delta U_B + W_B$$

or

$$Q_B - W_B = \Delta U_B \qquad (3.4)$$

According to the first law for cyclic process 1–2–1 via A to B,

$$\oint \delta Q = \oint \delta W$$

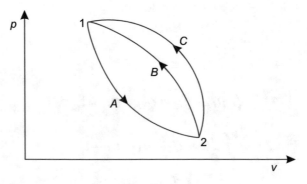

Fig. 3.3 Internal energy—a property of the system

$$Q_A + Q_B = W_A + W_B$$

or

$$Q_A - W_A = W_B - Q_B$$

$$Q_A - W_A = -(Q_B - W_B) \tag{3.5}$$

From Eqs. (3.3), (3.4), and (3.5), we get

$$\Delta U_A = -\Delta U_B \tag{3.6}$$

Similarly, the system returns from state 2 to state 1 by following path C instead of path B.

The first law for process 2–1 via C

$$Q_C = \Delta U_C + W_C$$

or

$$Q_C - W_C = \Delta U_C \tag{3.7}$$

The first law for cyclic process 1–2–1 via A to via C,

$$\oint \delta Q = \oint \delta W$$

$$Q_A + Q_C = W_A + W_C$$

$$Q_A - W_A = W_C - Q_C$$

or

$$Q_A - W_A = -(Q_C - W_C) \tag{3.8}$$

From Eqs. (3.3), (3.7), and (3.8), we get

$$\Delta U_A = -\Delta U_C \tag{3.9}$$

From Eqs. (3.6) and (3.9), we get

$$\Delta U_B = \Delta U_C \tag{3.10}$$

Equation (3.10) shows that the change in internal energy between two states of the system is the same, whether the system follows path *B* or path *C*. That is, the internal energy is independent of the path followed by the system. So, the internal energy is a point function and a property of the system.

3.6 Perpetual Motion Machine of the First Kind—PMM I

A perpetual motion machine of the first kind is an imaginary machine that produces work continuously without any input (no energy is supplied to it), i.e., PMM I is an energy-creating machine, as shown in Fig. 3.4a. On the other hand, PMM I consumes work continuously without producing any form of energy at the output. So, PMM I is an energy-destroying machine, as shown in Fig. 3.4b.

The first law of thermodynamics is the law of conservation of energy, i.e., energy is neither created nor destroyed, but can be converted from one form to another, but PMM I is either creating energy or destroying energy. It means that PMM I violates the first law of thermodynamics. Hence, it is an impossible machine.

Remember

- PMM I is an imaginary machine.
- PMM I violates the first law of thermodynamics.
- PMM I is either an energy-creating or energy-destroying machine.
- PMM I is an impossible machine.

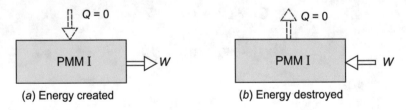

(*a*) Energy created   (*b*) Energy destroyed

Fig. 3.4 Perpetual motion machine of the first kind

3.7 Enthalpy

Enthalpy is a property of the system (i.e., point function). In the analysis of systems that involve fluid flow, it is defined as the sum of the internal energy and product of pressure and volume.

Mathematically, Enthalpy:

$$H = U + pV \tag{3.11}$$

Specific enthalpy:

$$h = \frac{H}{m}$$

∴

$$h = \frac{U + pV}{m} = \frac{U}{m} + \frac{pV}{m}$$

$$\boldsymbol{h = u + pv} \tag{3.12}$$

where

$$u = \frac{U}{m}, \text{ specific internal energy; kJ/kg,}$$

$$v = \frac{V}{m}, \text{ specific volume; } \frac{m^3}{kg}$$

$$= \frac{1}{\rho}$$

∴

$$\boldsymbol{h = u + \frac{p}{\rho}}$$

where $pv\left(\text{or } \frac{p}{\rho}\right)$ = flow energy, also called the flow work, which is the pressure energy per unit mass needed to move the fluid and maintain the flow.

Hence, for an open system, it becomes convenient to consider enthalpy instead of the internal energy of the fluid at the inlet and output so that the flow work is included.

Hence, for an open system, it becomes convenient to consider enthalpy instead of the internal energy of the fluid at the inlet and outlet so that the flow work is included.

For a non-flow process, $pv = RT$ (equation of state), substituting in Eq. (3.12), we get

$$h = u + RT \qquad \begin{array}{l} \because \quad R = c_p - c_v \\ \text{or } c_p = c_v + R \end{array}$$

$$h = c_y T + RT = (c_v + R)T$$

$$h = c_p T \ \text{kJ/kg}$$

where

c_p = specific heat at constant pressure, kJ/kgK,
T = Temperature, K.

Enthalpy:

$$H = mh$$

$$H = mc_p T$$

For a non-flow process, the change in enthalpy:

$$dH = mc_p dT$$

where

$$dH = H_2 - H_1$$

and

$$dT = T_2 - T_1$$

For a perfect gas, the change in enthalpy is defined as the heat addition at constant pressure.

3.8 Internal Energy

According to the first law of thermodynamics, if the heat energy (Q) is supplied to the system, some amount of supplied energy gets stored in the molecular structure of the system and the remaining part is converted into useful work. The energy stored in the molecular structure of the system is called internal energy or microscopic energy (or hidden energy). For ideal gases, the internal energy is the only function of absolute temperature, and its value is zero at the absolute zero temperature.

The internal energy is a property of the system (i.e., point function), and its value at a temperature T is given by.

Internal energy:

$$U = mc_v T \text{ kJ}$$

where

m = mass of the ideal gas in kg,
c_v = specific heat at constant volume in kJ/kgK, and
T = temperature of the gas in K.

Specific internal energy:

$$u = \frac{U}{m} \text{ kJ/kg}$$

Change in internal energy:

$$dU = mc_v dT$$

Change in specific internal energy:

$$du = \frac{dU}{m} = c_v dT$$

The change in internal energy is defined as the heat addition at constant volume. According to the first law of thermodynamics,

$$Q = dU + pdV$$

$$Q_V = dU \quad |\because dV = 0$$

or

$$dU = Q_V = mc_v dT$$

Change in specific internal energy:

$$du = q_V = c_v dT$$

Specific internal energy is defined as the internal energy per unit mass.

3.9 Specific Heat

Specific heat is the property of a substance. It is defined as the amount of heat that is required to raise the unit temperature of the unit mass of the substance.

Mathematically,

Heat transfer:

$$Q = mcdT$$

where

m	= mass of the substance in kg,
c	= specific heat of the substance in kJ/kgK, and
dT	= change in temperature in °C or K.

Liquids and solids have one specific heat value, but gases have two different types of specific heat as.

(i)   Specific heat at constant pressure, c_p;
(ii) Specific heat at constant volume, c_v.

(i) Specific heat at constant pressure: c_p

It is defined as the amount of heat that is required to raise the unit temperature of the unit mass of the gas at constant pressure. It is denoted by c_p.

Mathematically,

Heat supply at constant pressure:

$$Q_p = mc_p dT$$

By definition of specific heat,

$$m = 1\text{kg}, \quad dT = 1\text{K}$$

Then,

$$c_p = Q_p$$

Specific heat transfer:

$$q_p = \frac{Q_p}{m} = \frac{mc_p dT}{m}$$

$$q_p = c_p dT$$

or

$$c_p = \frac{q_p}{dT} \tag{3.13}$$

According to the first law of thermodynamics,
For process,

$$Q = dU + pdV$$

In specific form,

$$q = du + pdv \tag{3.14}$$

By definition of enthalpy,
Enthalpy:

$$H = U + pV$$

where

H = enthalpy of the gas in kJ,
U = internal energy in kJ,
p = pressure of the gas in kPa, and
V = volume of gas flow in m^3.

Specific enthalpy:

$$h = u + pv$$

By differentiating the above equation at constant pressure,

$$dh = du + pdv$$

From Eq. (3.14), $du + pdv = q$ and by substituting this in the above equation, we get

$$dh = q$$

or

$$dh = q_p$$

or

$$q_p = dh$$

Substituting $q_p = dh$ in Eq. (3.13), we get

$$c_p = \left(\frac{dh}{dT}\right)_P$$

The specific heat at constant volume can be also defined as the partial derivative of specific enthalpy with respect to temperature at constant pressure.

(ii)   **Specific heat at constant volume: c_v**

It is defined as the amount of heat that is required to raise the unit temperature of the unit mass of the gas at constant volume. It is denoted by c_v.

Mathematically,

Heat supply at constant volume:

$$Q_V = mc_v dT$$

By definition of specific heat,

$$m = 1 \text{ kg}, \quad dT = 1 \text{ K}$$

Then

$$c_v = Q_V$$

Specific heat transfer:

$$q_V = \frac{Q_V}{m} = \frac{mc_v dT}{m}$$

$$q_V = c_v dT \qquad\qquad (3.15)$$

According to the first law of thermodynamics,

For process,

$$Q = dU + pdV$$

$$Q_V = dU + 0 \quad |\because V = C$$

or

$$Q_V = dU$$

In specific form,

$$q_V = du \qquad\qquad (3.16)$$

Equating Eqs. (3.15) and (3.16), we get

$$c_v dT = du$$

or

$$c_v = \left(\frac{du}{dT}\right)_V$$

The specific heat at constant volume can be also defined as the partial derivative of specific internal energy with respect to temperature at constant volume.

3.10 Heat or Thermal Capacity: $C = mc$

Heat capacity is defined as the product of mass and specific heat. It is denoted by the capital letter C. It is also called thermal capacity.
Mathematically,
Heat capacity:

$$C = \text{mass} \times \text{specific heat} = mc$$

The SI unit of heat capacity is kJ/K or kJ/°C.
We know that the heat transfer:

$$Q = mc dT$$

where

$$mc = C, \text{heat capacity}$$

$\therefore$

$$Q = CdT$$

or Heat capacity:

$$C = \frac{Q}{dT}$$

The heat capacity of a system is also defined as the ratio of the heat added to or withdrawn from the system to the change in temperature of the system.

The concept of heat capacity is only used when the addition of heat to or withdrawal of heat from the system produces a temperature change; the concept is not used when a phase change is involved. For example, if the system is a mixture of ice and water at 1 atm pressure and 0 °C, then the addition of heat simply melts some of the ice, and no change in temperature occurs. In such a case, the heat capacity, as defined, would be infinite.

3.11   Relations Among c_p, c_v, R, and γ

(i)   Relation among c_p, c_v, and R

We know that the enthalpy

$$H = U + pV$$

$$H = U + mRT \quad |\because pV = mRT$$

By differentiating, we get

$$dH = dU + mRdT$$

$$mc_pdT = mc_vdT + mRdT \quad \left|\begin{array}{l} \because dH = mc_pdT \\ \text{and } dU = mc_vdT \end{array}\right.$$

or

$$c_p = c_v + R$$

or

$$c_p - c_v = R \tag{3.17}$$

where R is the gas constant which is defined as the difference between the specific heat at constant pressure and specific heat at constant volume.

(ii) Relation among c_p, R, and γ

From Eq. (3.17),

$$c_p - c_v = R$$

$$c_p\left(1 - \frac{c_v}{c_p}\right) = R$$

$$c_p\left(1 - \frac{1}{c_p/c_v}\right) = R$$

$$c_p\left(1 - \frac{1}{\gamma}\right) = R$$

where $\gamma = \frac{c_p}{c_v}$, adiabatic index which is defined as the ratio of specific heat at constant pressure and the specific heat at constant volume.

$\therefore$

$$c_p\left(\frac{\gamma - 1}{\gamma}\right) = R$$

or

$$c_p = \frac{\gamma R}{\gamma - 1}$$

(iii) Relation among c_v, R, and γ

Again from Eq. (3.17),

$$c_p - c_v = R$$

$$c_v\left(\frac{c_p}{c_v} - 1\right) = R$$

$$c_v(\gamma - 1) = R \quad \because \gamma = \frac{c_p}{c_v}$$

or

$$c_v = \frac{R}{\gamma - 1}$$

Remember

$$\bullet \; c_p - c_v = R \qquad \bullet \; c_p = \frac{\gamma R}{\gamma - 1} \qquad \bullet \; c_v = \frac{R}{\gamma - 1}$$

3.12 For Reversible Adiabatic Process, Prove That

(i) $TV^{\gamma-1} = C$
(ii) $pV^{\gamma} = C, and$
(iii) $Tp^{(1-\gamma)/\gamma}$

According to the first law of thermodynamic for a process,

$$\delta Q = dU + \delta W \tag{3.18}$$

where

$\delta W \quad = pdV$ for reversible process,
$\delta Q \quad = 0$ for adiabatic process, and
$\delta U \quad = mc_v dT.$

Equation (3.18) becomes

$$0 = mc_v dT + pdV$$

or

$$mc_v dT + pdV = 0$$

$$m\frac{R}{\gamma - 1}dT + pdV = 0 \;\; \left| \because c_v = \frac{R}{\gamma - 1} \right.$$

$$\frac{pV}{T} \times \frac{1}{\gamma - 1}dT + pdV = 0 \;\; \left| \therefore pV = mRT, \; or \; mR = \frac{pV}{T} \right.$$

or

$$\frac{1}{\gamma - 1}\frac{dT}{T} + \frac{dV}{V} = 0$$

or

$$\frac{dT}{T} + (\gamma - 1)\frac{dV}{V} = 0$$

On integrating, we get

$$\log_e T + (\gamma - 1)\log_e V = C$$
$$\log_e T + \log_e V^{\gamma-1} = C$$
$$\log_e TV^{\gamma-1} = C \qquad\qquad (3.19)$$
$$TV^{\gamma-1} = e^C$$
$$\boldsymbol{TV^{\gamma-1} = C}$$

From the equation of state,

$$pV = mRT$$

or

$$T = \frac{pV}{mR}$$

Substituting the above value of T in Eq. (3.19), we get

$$\frac{pV}{mR}V^{\gamma-1} = C$$

or

$$PVV^{\gamma-1} = mR \times C = \text{another constant}$$

$$\boldsymbol{PV^{\gamma} = C}$$

Again from the equation of state,

$$pV = mRT$$

$$V = \frac{mRT}{p}$$

Substituting the above value of V in Eq. (3.19), we get

$$T\left(\frac{mRT}{p}\right)^{\gamma-1} = C$$

$$T\frac{T^{\gamma-1}}{p^{\gamma-1}} = \frac{C}{(mR)^{\gamma-1}} = \text{another constant}$$

$$\frac{T^{\gamma}}{p^{\gamma-1}} = C$$

or

$$T^{\gamma}p^{-\gamma+1} = C$$

or

$$T^{\gamma}p^{1-\gamma} = C$$

or

$$Tp^{(1-\gamma)\gamma} = C^{1/\gamma} = C$$

$$\boldsymbol{Tp^{(1-\gamma)/\gamma} = C.}$$

3.13 Non-flow and Flow Processes

Non-Flow Process

A process occurring in a closed system which cannot permit the transfer of mass to cross its boundary is called a **non-flow process**. In a non-flow process, the energy crosses the boundary of the system in the form of heat and work, but there is no mass flow to the system or from the system. For example, the non-flow process is compression or expansion of gas inside the piston-cylinder arrangement (in IC engine).

Flow Process

A process occurring in an open system or through a control volume, which permits the transfer of mass to and from the system, is called a **flow process**. In this process, both mass and energy enter the system and leave the system simultaneously. For example, a flow process is flow through nozzle, diffuser, compressor, turbine, and heat exchanger.

3.14 Work Done During a Non-flow Process

Consider a piston-cylinder arrangement in which a fixed mass of gas expands from state 1 to state 2, as shown in Fig. 3.5.

Let A = cross-sectional area of the piston,

dx = small distance moved by the piston, and

p = constant pressure exerted by gas on the piston which moves through a small distance dx.

Work done in an infinitesimal non-flow process is given by

$$\delta W = \text{force} \times \text{distance} \quad \left|\begin{array}{l} \because p = \frac{F}{A} \\ \text{or } F = pA \end{array}\right.$$

$$\delta W = F\,dx = pA\,dx$$

$$\delta W = p\,dV \quad |\because dV = A\,dx \qquad (3.20)$$

That is, the work done in the differential form is equal to the product of the absolute pressure p and change in the volume dV of the system,

where p is the absolute pressure, which is always positive and dV is change in volume, which is positive during an expansion process (volume increasing) and negative during a compression process (volume decreasing). Thus, the work done

Fig. 3.5 Expansion work in a closed system (non-flow process)

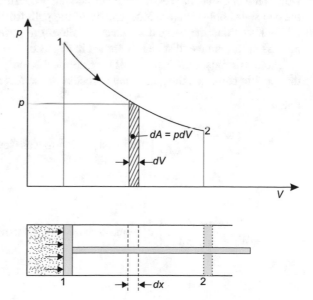

Fig. 3.6 The boundary work
done during a process
depends on the path followed
as well as the end states

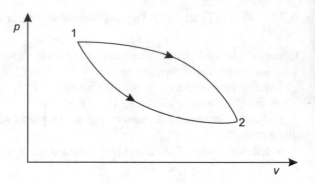

is positive during an expansion process and negative during a compression process.
This type of work is also called **boundary work or displacement work**.

Work done between the two states 1 and 2 is

$$W_{1-2} = \int_1^2 \delta W = \int_1^2 p \, dV \tag{3.21}$$

Note: Work done W_{1-2} is a path function (i.e., it depends on the path followed as
well as the end states). Pressure and volume are point functions (i.e., they depend on
the end states only) and are independent of the path followed by the process.

We know that the work done changes under the different path, followed by the
process at the same end states, as shown in Fig. 3.6.

Therefore, Eqs. (3.20) and (3.21) are applicable only when the system followed
the specific path, i.e., the quasi-static process or the reversible process,

that is,

$$\left.\begin{array}{c} \delta W = p \, dV \\ W_{1-2} = \int p \, dV \end{array}\right\} \text{ for quasi-static process (reversible process)}$$

and

$$\left.\begin{array}{c} \delta W \neq p \, dV \\ W_{1-2} \neq \int p \, dV \end{array}\right\} \text{ non-quasi-static process (irreversible process).}$$

3.15   Application of the First Law of Thermodynamics for Non-flow Processes (Closed System)

The first law of thermodynamics is applied for the different reversible non-flow processes and irreversible non-flow process, which are detailed as follows:

1.   Reversible non-flow processes. These types of process are as follows:

   (i)   Constant volume process (isometric or isochoric process),
   (ii)   Constant pressure process (isobaric process),
   (iii)   Constant temperature process (isothermal process),
   (iv)   Adiabatic isentropic process (frictionless and adiabatic process), and
   (v)   Polytropic process.

   All these processes are discussed in Chap. 1, article 1.15.
2.   Irreversible non-flow process. This type of process is as follows:

   (i)   Free expansion process

3.16   Constant Volume Process (Isometric or Isochoric Process)

This process takes place at constant volume and is governed by Gay-Lussac's law. According to this law, **the absolute pressure of perfect gas is directly proportional to the absolute temperature, when the volume is kept constant** (Fig. 3.7).
   Mathematically,

$$p \propto T \quad \text{at} \quad v = C$$

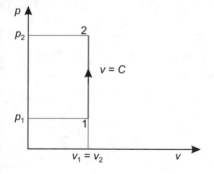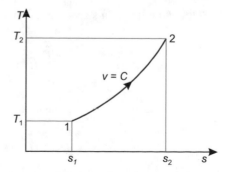

Fig. 3.7   Constant volume process in p–v and T–s diagrams

or

$$\frac{p}{T} = C$$

(a) **p–v–T relationship**:

For process 1–2,

$$\frac{p_1}{T_1} = \frac{p_2}{T_2} \text{ or } \frac{T_2}{T_1} = \frac{p_2}{p_1}$$

(b) **Work done**:

We know that the work done:

$$\delta W = p dV$$

Integrating between states 1 and 2 gives

$$\int_1^2 \delta W = \int p dV$$

$$W_{1-2} = 0 \quad \because V_1 = V_2 = C$$

(c) **Change in internal energy**:

We know that the change in internal energy:

$$dU = mc_v dT$$

Integrating between states 1 and 2 gives

$$\int_1^2 dU = mc_v \int_1^2 dT$$

or

$$U_2 - U_1 = mc_v(T_2 - T_1)$$

where

m	= mass of the gas in kg,
c_v	= specific heat at constant volume in kJ/kgK,

$T_2 - T1$ = change in temperature in K or °C, and
$U_2 - U1$ = change in internal energy in kJ.

Specific change in internal energy:

$$u_2 - u_1 = \frac{U_2 - U_1}{m}, \text{change in internal energy per unit mass}$$
$$= \frac{mc_v dT}{m} = c_v dT \text{ kJ/kg}$$

(d)   **Heat transfer**:

We know that

$$\delta Q = dU + \delta W$$

Integrating between states 1 and 2 gives

$$\int_1^2 \delta Q = \int_1^2 dU + \int_1^2 \delta W$$

$$Q_{1-2} = (U_2 - U_1) + W_{1-2}$$

or

$$Q_{1-2} = U_2 - U_1 \quad \because W_{1-2} = 0$$
$$= mc_v(T_2 - T_1) \text{ kJ}$$

Specific heat transfer:

$$q_{1-2} = \frac{Q_{1-2}}{m} = \frac{mc_v(T_2 - T_1)}{m}$$

$$q_{1-2} = c_v(T_2 - T_1)\text{kJ/kg}$$

(e)   **Change in enthalpy**:

We know that

$$H = U + pV$$

On differentiating, we get

$$dH = dU + V dp \quad \because V = C$$

From the equation of state,

$$pV = mRT$$

On differentiating at constant V, m, and R, we get

$$V dp = mRdT$$

$\therefore$

$$dH = dU + mRdT$$

$$dH = mc_v dT + mRdT = m(c_v + R)dT$$

$$dH = mc_p dT \quad \left| \begin{array}{l} \because R = c_p - c_v \\ \text{or } c_p = c_v + R \end{array} \right.$$

Integrating between states 1 and 2 gives

$$\int_1^2 dH = mc_p \int_1^2 dT$$

$$\boldsymbol{H_2 - H_1 = mc_p(T_2 - T_1)\,\text{kJ}}$$

Change in specific enthalpy,

$$h_2 - h_1 = \frac{H_2 - H_1}{m}, \text{ change in enthalpy per unit mass}$$

$$h_2 - h_1 = \frac{mc_p(T_2 - T_1)}{m}$$

$$\boldsymbol{h_2 - h_1 = c_p(T_2 - T_1)\,\text{kJ/kg}.}$$

3.17   Constant Pressure Process (Isobaric Process)

This process takes place at constant pressure and is governed by Charles' law. According to this law, **the specific volume of a perfect gas is directly proportional to the absolute temperature, when the pressure is kept constant** (Fig. 3.8).

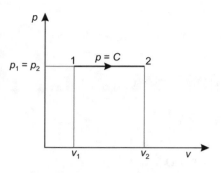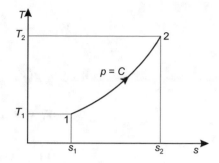

Fig. 3.8 Constant pressure process in *p–v* and *T–s* diagrams

Mathematically,

$$v \propto T \quad \text{at} \quad p = C$$

or

$$\frac{v}{T} = C$$

(a)   ***p–v–T* relationship**:

For process 1–2,

$$\frac{v_1}{T_1} = \frac{v_2}{T_2}$$

or

$$\frac{T_2}{T_1} = \frac{v_2}{v_1}$$

(b)   **Work done**:

We know that the work done:

$$\delta W = p \, dV.$$

Integrating between states 1 and 2 gives

$$\int_1^2 \delta W = \int_1^2 p \, dV$$

$$W_{1-2} = p \int_{1}^{2} dV \quad \because p = C$$

$$W_{1-2} = p(V_2 - V_1) = pV_2 - pV_1$$

$$W_{1-2} = mRT_2 - mRT_1$$

$$W_{1-2} = mR(T_2 - T_1)$$

(c)   **Change in internal energy**:

We know that the change in internal energy,

$$dU = mc_v dT$$

Integrating between states 1 and 2 gives

$$\int_{1}^{2} dU = mc_v \int_{1}^{2} dT$$

$$U_2 - U_1 = mc_v(T_2 - T_1)$$

(d)   **Heat transfer**:

We know that

$$\delta Q = dU + \delta W$$

Integrating between states 1 and 2 gives

$$\int_{1}^{2} \delta Q = \int_{1}^{2} dU + \int_{1}^{2} \delta W$$

$$Q_{1-2} = (U_2 - U_1) + W_{1-2}$$

$$Q_{1-2} = mc_v(T_2 - T_1) + mR(T_2 - T_1)$$

$$Q_{1-2} = m(c_v + R)(T_2 - T_1) \quad \left| \begin{array}{l} \because c_p - c_v = R \\ \text{or } c_p = c_v + R \end{array} \right.$$

$$Q_{1-2} = mc_p(T_2 - T_1)$$

(e) **Change in enthalpy**:

We know that

$$H = U + pV$$

On differentiating, we get

$$dH = dU + pdV \quad |\because p = C$$

$$dH = mc_v dT + mRdT$$
$$= m(c_v + R)dT$$

$$dH = mc_p dT$$

Integrating between states 1 and 2 gives

$$\int_1^2 dH = mc_p \int_1^2 dT$$

$$\boldsymbol{H_2 - H_1 = mc_p(T_2 - T_1)}.$$

3.18 Constant Temperature Process (Isothermal Process)

This process takes place at constant temperature and is governed by Boyle's law. According to this law, **the specific volume of a perfect gas is inversely proportional to the absolute pressure when the temperature is kept constant** (Fig. 3.9).
 Mathematically,

$$v \propto c\frac{1}{p} \quad \text{at} \quad T = C$$

or

$$\boldsymbol{pv = C}$$

(a) *p–v–T* **relationship**:

For process 1–2,

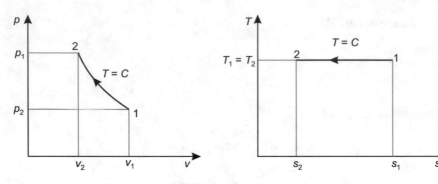

Fig. 3.9 Constant temperature process in *p–v* and *T–s* diagrams

$$p_1 v_1 = p_2 v_2$$

or

$$\frac{v_2}{v_1} = \frac{p_1}{p_2}$$

(b)   **Work done**:

Work done:

$$\delta W = p dV$$

Integrating between states 1 and 2 gives

$$\int_1^2 \delta W = \int_1^2 p dV$$

$$W_{1-2} = \int_1^2 p dV$$

We know that

$$pV = C$$

or

$$p = \frac{C}{V}$$

$\therefore$

$$W_{1-2} = \int_{1}^{2} C \frac{dV}{V} = C \int_{1}^{2} \frac{dV}{V} = C \left[\log_e V \right]_1^2$$

$$= C \left[\log_e V_2 - \log_e V_1 \right] = C \log_e \frac{V_2}{V_1}$$

$$= p_1 V_1 \log_e \frac{V_2}{V_1} = p_2 V_2 \log_e \frac{V_2}{V_1} \quad |\because pV = C$$

$$= p_1 V_1 \log_e \frac{p_1}{p_2} = p_2 V_2 \log_e \frac{p_1}{p_2}$$

$$W_{1-2} = mRT_1 \log_e \frac{V_2}{V_1} = mRT_1 \log_e \frac{p_1}{p_2}$$

(c)   **Change in internal energy**:

We know that the change in internal energy,

$$dU = mc_v dT$$

Integrating between states 1 and 2 gives

$$\int_{1}^{2} dU = mc_v \int_{1}^{2} dT$$

$$U_2 - U_1 = 0 \quad |\because T = C$$

Thus, for ideal gas, internal energy is a function of temperature only. That is,

$$U = f(T) \textbf{for ideal gas}$$

(d)   **Heat transfer**:

We know that

$$\delta Q = dU + \delta W$$
$$\delta Q = 0 + \delta W$$
$$\delta Q = \delta W$$

Integrating between states 1 and 2 gives

$$\int_1^2 \delta Q = \int_1^2 \delta W$$

$$Q_{1-2} = W_{1-2} = p_1 V_1 \log_e \frac{V_2}{V_1}$$

$$= mRT_1 \log_e \frac{V_2}{V_1} = mRT_1 \log_e \frac{p_1}{p_2}$$

(e)   **Change in enthalpy**:

We know that

$$H = U + pV$$

$$H = U + C \quad \because pV = C$$

On differentiating, we get

$$dH = dU + 0$$

$$dH = dU$$

Integrating between states 1 and 2 gives

$$\int_1^2 dH = \int_1^2 dU$$

$$H_2 - H_1 = U_2 - U_1$$

$$H_2 - H_1 = 0 \quad \because U_1 = U_2$$

or

$$H_1 = H_2$$

Thus, for ideal gas, enthalpy is also a function of temperature only as internal energy.

That is,

$$H = f(T) \quad \textbf{for ideal gas.}$$

3.19   Adiabatic Isentropic Process (Frictionless and Adiabatic Process)

In this process, the following conditions are maintained:

(i)   no heat transfer between the system and surroundings and
(ii)   the entropy of the system remains constant.

Adiabatic isentropic process follows the law of $pv^\gamma = C$, where γ is an adiabatic index (Fig. 3.10).

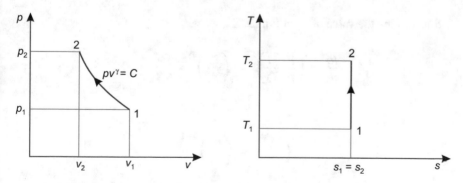

Fig. 3.10   Adiabatic isentropic process in p–v and T–s diagrams

(a) *p–v–T* **relationship**:

For process 1–2,

$$p_1 v_1^{\gamma} = p_2 v_2^{\gamma}$$

or

$$\frac{p_2}{p_1} = \left(\frac{v_1}{v_2}\right)^{\gamma} \tag{3.22}$$

From the equation of state,

$$pv = RT$$

$$\frac{pv}{T} = R = C$$

For process 1–2,

$$\frac{p_1 v_1}{T_1} = \frac{p_2 v_2}{T_2}$$

or

$$\frac{v_1}{v_2} = \frac{p_2}{p_1} \frac{T_1}{T_2}$$

Substituting the value of $\frac{v_1}{v_2}$ in Eq. (3.22), we get

$$\frac{p_2}{p_1} = \left(\frac{p_2}{p_1} \frac{T_1}{T_2}\right)^{\gamma} = \left(\frac{p_2}{p_1}\right)^{\gamma} \times \left(\frac{T_1}{T_2}\right)^{\gamma}$$

or

$$\left(\frac{T_2}{T_1}\right)^{\gamma} = \left(\frac{p_2}{p_1}\right)^{\gamma-1}$$

or

$$\frac{T_2}{T_1} = \left(\frac{p_2}{p_1}\right)^{\frac{\gamma-1}{\gamma}} \tag{3.23}$$

From Eqs. (3.22) and (3.23), we get $\frac{T_2}{T_1} = \left[\left(\frac{v_1}{v_2}\right)^{\gamma}\right]^{\frac{\gamma-1}{\gamma}}$

$$\frac{T_2}{T_1} = \left(\frac{v_1}{v_2}\right)^{\gamma-1} \tag{3.24}$$

From Eqs. (3.23) and (3.24), we get

$$\frac{T_2}{T_1} = \left(\frac{p_2}{p_1}\right)^{\frac{\gamma-1}{\gamma}} = \left(\frac{v_1}{v_2}\right)^{\gamma-1}$$

(b) **Work done**:

We know that work done:

$$\delta W = p\,dV$$

Integrating between states 1 and 2 gives

$$\int_1^2 \delta W = \int_1^2 p\,dV$$

$$W_{1-2} = \int_1^2 p\,dV$$

We know that

$$pV^\gamma = C$$

or

$$p = \frac{C}{V^\gamma}$$

$\therefore$

$$W_{1-2} = \int_1^2 \frac{C\,dV}{V^\gamma} = C\int_1^2 V^{-\gamma}dV$$

$$= C\left[\frac{V^{-\gamma+1}}{-\gamma+1}\right]_1^2 = C\frac{\left[V_2^{1-\gamma} - V_1^{1-\gamma}\right]}{1-\gamma}$$

$$= \frac{CV_2^{1-\gamma} - CV_1^{1-\gamma}}{1-\gamma} \qquad [\because p_1V_1^\gamma = p_2V_2^\gamma = C]$$

$$= \frac{p_2 V_2^{\gamma} V_2^{1-\gamma} - p_1 V_1^{\gamma} V_1^{1-\gamma}}{1 - \gamma}$$

$$= \frac{p_2 V_2 - p_1 V_1}{1 - \gamma} = \frac{p_1 V_1 - p_2 V_2}{\gamma - 1}$$

$$= \frac{m R T_1 - m R T_2}{\gamma - 1} = \frac{m R (T_1 - T_2)}{\gamma - 1}$$

(c)   **Change in internal energy**:

We know that the change in internal energy,

$$dU = mc_v dT$$

Integrating between states 1 and 2 gives

$$\int_1^2 dU = mc_v \int_1^2 dT$$

$$U_2 - U_1 = mc_v(T_2 - T_1)$$

$$= \frac{m R (T_2 - T_1)}{\gamma - 1} \quad \left| \because c_v = \frac{R}{\gamma - 1} \right.$$

$$= \frac{-m R (T_1 - T_2)}{\gamma - 1}$$

$$U_2 - U_1 = -W_{1-2}$$

(d)   **Heat transfer**:

We know that

$$\delta Q = dU + \delta W$$

Integrating between states 1 and 2 gives

$$\int_1^2 dH = \int_1^2 dU + \int_1^2 \delta W$$

$$Q_{1-2} = (U_2 - U_1) + W_{1-2}$$

$$Q_{1-2} = -W_{1-2} + W_{1-2}$$

$$Q_{1-2} = 0$$

(e)   **Change in enthalpy**:

We know that

$$H = U + pV$$

On differentiating, we get

$$dH = dU + p\,dV + V\,dp$$

Integrating between states 1 and 2 gives

$$\int_1^2 dH = \int_1^2 dU + \int_1^2 p\,dV + \int_1^2 V\,dp$$

$$H_2 - H_1 = (U_2 - U_1) + W_{1-2} + \int_1^2 V\,dp$$

$$H_2 - H_1 = -W_{1-2} + W_{1-2} + \int_1^2 \frac{C^{1/\gamma}\,dp}{p^{1/\gamma}} \quad \left| \begin{array}{l} \because pV^{\gamma} = C \\ V = \frac{C^{1/\gamma}}{p^{1/\gamma}} \end{array} \right.$$

$$= C^{1/\gamma} \int_1^2 p^{-1/\gamma}\,dp = C^{1/\gamma} \left[\frac{p^{-\frac{1}{\gamma}+1}}{-\frac{1}{\gamma}+1} \right]_1^2$$

$$= C^{1/\gamma} \frac{\left[p_2^{-\frac{1}{\gamma}+1} - p_1^{-\frac{1}{\gamma}+1} \right]}{\frac{-1+\gamma}{\gamma}}$$

$$= C^{1/\gamma} \frac{\gamma \left[p_2^{-\frac{1}{\gamma}+1} - p_1^{-\frac{1}{\gamma}+1} \right]}{\gamma - 1}$$

$$= \frac{\gamma}{\gamma - 1} \left[C^{1/\gamma} p_2^{-\frac{1}{\gamma}+1} - C^{1/\gamma} p_1^{-\frac{1}{\gamma}+1} \right]$$

$$= \frac{\gamma}{\gamma - 1} \left[V_2 p_2^{1/\gamma} p_2^{-\frac{1}{\gamma}+1} - V_1 p_1^{1/\gamma} p_2^{-\frac{1}{\gamma}+1} \right]$$

$$= \frac{\gamma}{\gamma - 1} [p_2 V_2 - p_1 V_1]$$

$$= \frac{\gamma}{\gamma - 1} [mRT_2 - mRT_1]$$

$$= \frac{m\gamma R}{\gamma - 1} [T_2 - T_1]$$

$$H_2 - H_1 = mc_p (T_2 - T_1) \quad \left| \because c_p = \frac{\gamma R}{\gamma - 1} \right.$$

3.20   Polytropic Process

In the polytropic process, the following changes take place:

(i)   Entropy of the process changes.
(ii)   Both heat and work transfer take place.

The polytropic process follows the law of $pv^n = C$, where n is a polytropic index (Fig. 3.11).

(a)   *p–v–T* relationship:

The relations for the polytropic process are derived in a similar way as discussed for the adiabatic process (i.e., replaced n instead of γ)

$$\frac{T_2}{T_1} = \left(\frac{p_2}{p_1}\right)^{\frac{n-1}{n}} = \left(\frac{V_1}{V_2}\right)^{n-1}$$

(b)   **Work done**:

We know that the work done: $\delta W = pdV$
Integrating between states 1 and 2 gives

$$\int_1^2 \delta W = \int_1^2 pdV$$

$$W_{1-2} = \int_1^2 pdV$$

We know that

$$pV^n = C$$

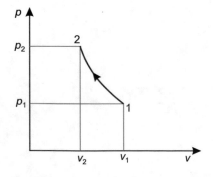 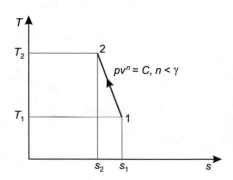

Fig. 3.11 Polytropic process in *p–v* and *T–s* diagrams

or

$$p = \frac{C}{V^n}$$

∴

$$W_{1-2} = \int_1^2 \frac{C\,dV}{V^n} = C \int_1^2 V^{-n}\,dV$$

$$= C \left[\frac{V^{-n+1}}{-n+1} \right]_1^2 = C \left[\frac{V_2^{-n+1} - V_1^{-n+1}}{1-n} \right]$$

$$= \frac{CV_2^{-n+1} - CV_1^{-n+1}}{1-n} \quad [\because p_1 V_1^n = p_2 V_2^n = C]$$

$$= \frac{p_2 V_2^n V_2^{-n+1} - p_1 V_1^n V_1^{-n+1}}{1-n}$$

$$= \frac{p_2 V_2 - p_1 V_1}{1-n} = \frac{p_1 V_1 - p_2 V_2}{n-1}$$

$$= \frac{mR(T_1 - T_2)}{n-1}$$

$$= \frac{mR(T_1 - T_2)}{n-1} == \frac{mR(T_2 - T_1)}{1-n}$$

(c)   **Change in internal energy**:

We know that the change in internal energy,

$$dU = mc_v\,dT$$

Integrating between states 1 and 2 gives

$$\int_1^2 dU = mc_v \int_1^2 dT$$

$$U_2 - U_1 = mc_v(T_2 - T_1)$$

$$U_2 - U_1 = \frac{mR}{\gamma - 1}(T_2 - T_1) \quad \left(\because c_v = \frac{R}{\gamma - 1} \right)$$

$$= \frac{mR(T_2 - T_1)}{\gamma - 1}$$

Multiplying and dividing by $(1 - n)$, we have

$$U_2 - U_1 = \frac{mR(T_2 - T_1)(1 - n)}{(\gamma - 1)(1 - n)}$$

$$= \left(\frac{1 - n}{\gamma - 1}\right) \frac{mR(T_2 - T_1)}{1 - n}$$

$$U_2 - U_1 = \left(\frac{1 - n}{\gamma - 1}\right) \times W_{1-2}$$

$$U_2 - U_1 = \left(\frac{1 - n}{\gamma - 1}\right) \times \textbf{work done}$$

(d)   **Heat transfer**:

We know that

$$\delta Q = dU + \delta W$$

Integrating between states 1 and 2 gives

$$\int_1^2 \delta Q = \int_1^2 dU + \int_1^2 \delta W$$

$$Q_{1-2} = (U_2 - U_1) + W_{1-2}$$

$$= \left(\frac{1 - n}{\gamma - 1}\right) \times W_{1-2} + W_{1-2}$$

$$= \left(\frac{1 - n}{\gamma - 1} + 1\right) \times W_{1-2}$$

$$= \left(\frac{1 - n + \gamma + 1}{\gamma - 1}\right) \times W_{1-2}$$

$$= \left(\frac{\gamma - n}{\gamma - 1}\right) \times W_{1-2}$$

$$Q_{1-2} = \left(\frac{\gamma - n}{\gamma - 1}\right) \times \text{work done}$$

$$= \left(\frac{\gamma - n}{\gamma - 1}\right) \times \frac{mR(T_2 - T_1)}{1 - n}$$

$$= \left(\frac{\gamma - n}{1 - n}\right) \times \frac{mR(T_2 - T_1)}{\gamma - 1}$$

$$= \left(\frac{\gamma - n}{1 - n}\right) \times mc_v(T_2 - T_1) \quad \left(\because c_v = \frac{R}{\gamma - 1}\right)$$

$$= \left(\frac{\gamma - n}{1 - n}\right) \times (U_2 - U_1)$$

$$Q_{1-2} = \left(\frac{\gamma - n}{1 - n}\right) \times \text{change in internal energy}$$

OR

$$Q_{1-2} = (U_2 - U_1) + W_{1-2}$$

$$= mc_v(T_2 - T_1) + \frac{mR(T_2 - T_1)}{1 - n}$$

$$= \frac{mR}{\gamma - 1}(T_2 - T_1) + \frac{mR(T_2 - T_1)}{1 - n} \quad \left(\because c_v = \frac{R}{\gamma - 1}\right)$$

$$= \left[\frac{1 - n}{\gamma - 1} + 1\right]\frac{mR(T_2 - T_1)}{1 - n}$$

$$= \left[\frac{1 - n + \gamma - 1}{\gamma - 1}\right]\frac{mR(T_2 - T_1)}{1 - n}$$

$$= \left(\frac{\gamma - n}{\gamma - 1}\right)\frac{mR(T_2 - T_1)}{1 - n}$$

$$= \left(\frac{\gamma - n}{\gamma - 1}\right) \times W_{1-2} \tag{3.25}$$

where

$$W_{1-2} = \frac{mR(T_2 - T_1)}{1 - n}, \text{ polytropic work.}$$

From Eq. (3.25), we have

$$Q_{1-2} = \left(\frac{\gamma - n}{1 - n}\right)\frac{mR}{\gamma - 1}(T_2 - T_1)$$

$$= mc_n(T_2 - T_1)$$

where

$$c_n = \left(\frac{\gamma - n}{1 - n}\right)\frac{R}{\gamma - 1}, \text{ polytropic specific heat}$$

$$c_n = \left(\frac{\gamma - n}{1 - n}\right)c_v \quad \left[\because \frac{R}{\gamma - 1} = c_v\right]$$

$$c_n = -\left(\frac{\gamma - n}{n - 1}\right)c_v$$

(e) **Change in enthalpy**:

We know that

$$H = U + pV$$

On differentiating, we get

$$dH = dU + pdV + Vdp$$

Integrating between states 1 and 2 gives

$$\int_1^2 dH = \int_1^2 dU + \int_1^2 pdV + \int_1^2 Vdp$$

$$H_2 - H_1 = (U_2 - U_1) + W_{1-2} + \int_1^2 Vdp$$

$$= Q_{1-2} + \int_1^2 Vdp \quad \left[\because Q_{1-2} = (U_2 - U_1) + W_{1-2} \right.$$

$$= Q_{1-2} + \int_1^2 C^{1/n} \frac{dp}{p^{1/n}} \quad \left| \begin{array}{l} \because pV^n = C \\ \text{or} \quad V = \frac{C^{1/n}}{p^{1/n}} \end{array} \right.$$

$$= Q_{1-2} + C^{1/n} \int_1^2 p^{-1/n} dp$$

$$= Q_{1-2} + C^{1/n} \left[\frac{p^{-1/n+1}}{-\frac{1}{n}+1} \right]_1^2$$

$$= Q_{1-2} + C^{1/n} \left[\frac{p_2^{-\frac{1}{n}+1} - p_1^{-\frac{1}{n}+1}}{\frac{-1+n}{n}} \right]$$

$$= Q_{1-2} + nC^{1/n} \left[\frac{p_2^{-\frac{1}{n}+1} - p_1^{-\frac{1}{n}+1}}{n-1} \right]$$

$$= Q_{1-2} + n \left[\frac{C^{1/n} p_2^{-\frac{1}{n}+1} - C^{1n} p_1^{\frac{1}{n}+1}}{n-1} \right]$$

$$= Q_{1-2} + n \left[\frac{V_2 p_2^{1/n} p_2^{-\frac{1}{n}+1} - V_1 p_1^{1/n} p_1^{-\frac{1}{n}+1}}{n-1} \right]$$

$$= Q_{1-2} + n \left[\frac{p_2 V_2 - p_1 V_1}{n-1} \right]$$

$$= \left(\frac{\gamma - n}{\gamma - 1}\right) \times W_{1-2} + n\left[\frac{mRT_2 - mRT_1}{n - 1}\right]$$

$$= \left(\frac{\gamma - n}{\gamma - 1}\right) \times \frac{mR(T_2 - T_1)}{1 - n} - n\frac{mR(T_2 - T_1)}{1 - n}a$$

$$= \left[\left(\frac{\gamma - n}{\gamma - 1}\right) - n\right]\frac{mR(T_2 - T_1)}{1 - n}$$

$$= \left[\frac{\gamma - n - n\gamma + n}{\gamma - 1}\right]\frac{mR(T_2 - T_1)}{1 - n}$$

$$= \left(\frac{\gamma - n\gamma}{\gamma - 1}\right)\frac{mR(T_2 - T_1)}{1 - n}$$

$$= \frac{\gamma(1 - n)}{(\gamma - 1)}\frac{mR(T_2 - T_1)}{(1 - n)}$$

$$= \frac{m\gamma R}{\gamma - 1}(T_2 - T_1)$$

$$\boldsymbol{H_2 - H_1 = mc_p(T_2 - T_1)} \quad \left| \because c_p = \frac{\gamma R}{\gamma - 1}\right.$$

3.21 Free Expansion Process

Free expansion (unrestrained expansion) is a non-flow irreversible process. This process is demonstrated in Fig. 3.12, where an insulated system having two equal compartments is separated by a membrane. Let one compartment is filled with ideal gas and the other is having vacuum. When the membrane is ruptured and gas is allowed to occupy the whole volume of the system, then the gas expands to fill the complete volume of the system. The following are some characteristics of the free expansion process observed:

(a)   Heat transfer is zero, i.e., $Q_{1-2} = 0$, because the system is insulated.
(b)   Work done is zero, i.e., $W_{1-2} = 0$, because of free expansion (unrestrained expansion).

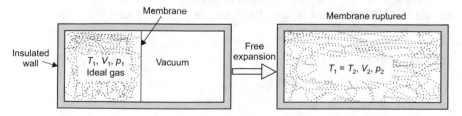

Fig. 3.12 Free expansion

(c)   According to the first law of thermodynamic,

$$Q_{1-2} = dU + W_{1-2}$$

$$0 = dU + 0$$

or

$$dU = 0$$

$$U_2 - U_1 = 0$$

$$U_2 = U_1 \qquad (3.26)$$

Thus, **the internal energy of the free expansion process remains constant**. For an ideal gas, the internal energy is a function of the temperature only. That is, $U = f(T)$.

We know that

$$U = mc_v T$$

$\therefore$ From Eq. (3.26), we have

$$mc_v T_1 = mc_v T_2$$

or

$$T_1 = T_2$$

Thus, the temperature of the free expansion process remains constant.

Problem 3.1: In a cyclic process, heat transfers are +15.3 kJ, −24.7 kJ, −4.2 kJ, and +32.5 kJ. What is the net work for this cycle process? (Table 3.1)

Solution: Given data:
    Heat transfers:

$$Q_1 = 15.3 \text{kJ},$$

$$Q_2 = -24.7 \text{kJ},$$

Table 3.1 Summary of non-flow processes for perfect gas

Sr. No	Process	p–V–T relation	Work done	Change in internal energy	Heat transfer	Change in enthalpy
1.	Constant volume process	$\dfrac{T_2}{T_1} = \dfrac{p_2}{p_1}$	0	$mc_v(T_2 - T_1)$	$mc_v(T_2 - T_1)$	$mc_p(T_2 - T_1)$
2.	Constant pressure process	$\dfrac{T_2}{T_1} = \dfrac{V_2}{V_1}$	$p(V_2 - V_1)$ OR $mR(T_2 - T_1)$	$mc_v(T_2 - T_1)$	$mc_p(T_2 - T_1)$	$mc_p(T_2 - T_1)$
3.	Constant temperature process	$\dfrac{V_2}{V_1} = \dfrac{p_1}{p_2}$	$p_1 V_1 \log_e \dfrac{V_2}{V_1}$ OR $mRT \log_e \dfrac{V_2}{V_1}$	0	$p_1 V_1 \log_e \dfrac{V_2}{V_1}$ OR $mRT \log_e \dfrac{V_2}{V_1}$	0
4.	Adiabatic isentropic process	$\dfrac{T_2}{T_1} = \left(\dfrac{p_2}{p_1}\right)^{\frac{\gamma-1}{\gamma}} = \left(\dfrac{V_1}{V_2}\right)^{\gamma-1}$	$\dfrac{p_1 V_1 - p_2 V_2}{\gamma - 1}$ $\dfrac{mR(T_1 - T_2)}{\gamma - 1}$	$mc_v(T_2 - T_1)$	0	$mc_p(T_2 - T_1)$
5.	Polytropic process	$\dfrac{T_2}{T_1} = \left(\dfrac{p_2}{p_1}\right)^{\frac{n-1}{n}} = \left(\dfrac{V_1}{V_2}\right)^{n-1}$	$\dfrac{p_1 V_1 - p_2 V_2}{n - 1}$ $\dfrac{mR(T_1 - T_2)}{n - 1}$	$mc_v(T_2 - T_1)$	$\left(\dfrac{\gamma - n}{\gamma - 1}\right) \times$ work done	$mc_p(T_2 - T_1)$

$$Q_3 = -4.2\text{kJ},$$

and

$$Q_4 = 32.5\text{kJ}.$$

For cyclic process:

$$\text{Net work} = \text{Net heat transfer}$$

$$W_{\text{net}}\text{or } \Sigma W = \Sigma Q$$

$$W_{\text{net}} = Q_1 + Q_2 + Q_3 + Q_4$$
$$= 15.3 - 24.7 - 4.2 + 32.5 = \mathbf{18.9\,kJ}$$

Problem 3.2: A domestic refrigerator is loaded with food and the door is closed. During a certain period, the machine consumes 1 kWh of energy and the internal energy of the system decreases by 5000 kJ. Determine the net heat transfer for the system.

Solution: Given data:
   Work:

$$W = -1\text{kWh} = -1\text{k}\frac{\text{J}}{\text{s}}\text{h}$$
$$= -1 \times 3600\frac{\text{kJs}}{\text{s}} \quad |\because 1\,\text{h} = 3600\,\text{s}$$
$$= -3600\,\text{kJ}$$

Internal energy:

$$\Delta U = -5000 \text{ kJ}$$

According to the first law of thermodynamics for a process,

$$Q = \Delta U + W = -5000 - 3600 = \mathbf{-8600\ kJ}$$

Problem 3.3: A slow chemical reaction takes place in a fluid at the constant pressure of 100 kPa. The fluid is surrounded by a perfect heat insulator during the reaction which begins at state 1 and ends at state 2. The insulation is then removed and 105 kJ of heat flows to the surroundings as the fluid goes to state 3. The following data are given for the fluid at states 1, 2, and 3.

State	Volume: V (m^3)	Temperature: T (°C)
1	0.003	20
2	0.3	370
3	0.06	20

Find U_2 and U_3, if $U_1 = 0$.

Solution: Given data:

$$p = 100\text{kPa}$$

Process 1–2,

$$V_1 = 0.003\,\text{m}^3,$$

$$V_2 = 0.3\,\text{m}^3,$$

$$Q_{1-2} = 0, \quad (\because \text{Adiabatic process})$$

$\therefore$

$$W_{1-2} = pdV$$

$$= p(V_2 - V_1) = 100 \times (0.3 - 0.003) = \mathbf{29.7\,kJ}$$

According to the first law of thermodynamic for process,

$$Q_{1-2} = (U_2 - U_1) + W_{1-2}$$

$$0 = (U_2 - U_1) + 29.7$$

or

$$U_2 - U_1 = -29.7\,\text{kJ}$$

$$U_2 - 0 = -29.7 \quad |\because U_1 = 0, \text{ given}$$

$$U_2 = \mathbf{-29.7\,kJ}$$

Process 2–3,

$$Q_{2-3} = -105 \text{ kJ and}$$

$$V_3 = 0.06 \text{ m}^3$$

Work:

$$W_{2-3} = pdV$$
$$= p(V_3 - V_2) = 100 \times (0.06 - 0.3) = -24 \text{ kJ}$$

The first law for process 2–3;

$$Q_{2-3} = (U_3 - U_2) + W_{2-3}$$

$$-105 = (U_3 - U_2) - 24$$

or

$$U_3 - U_2 = -81 \text{ kJ}$$

$$U_3 - (-29.7) = -81$$

$$U_3 + 29.7 = -81$$

or

$$U_3 = -110.7 \text{ kJ}$$

Problem 3.4: A banquet hall of space 300 m^3 is to be used for a dinner for 25 people. Each person occupies 0.072 m^3 of space and has an average heat transfer rate of 418 kJ/h. If the air-conditioning system of the hall fails, calculate the increase in internal energy, and temperature of air in the hall during the first 10 min of failure. Assume that the hall is well-insulated and receives no heat from outside.

Solution: Given data:
   Volume of hall:

$$V_H = 300 \text{ m}^3$$

Number of people:

$$n = 25$$

Volume of one person:

$$V_p = 0.072 \, \text{m}^3$$

Volume occupied by 25 people

$$= nV_p = 25 \times 0.072 = 1.8 \, \text{m}^3$$

Volume of air in the hall:

$$V = \text{Volume of hall} - \text{volume occupied by 25 person}$$
$$= 300 - 1.8 = 298.2 \, \text{m}^3$$

Let the human comfort condition in the room be maintained by the air-conditioning system at 25 °C and 101.325 kPa.

Mass of air in the hall:

$$m = \frac{pV}{RT}$$

where

$$p = 101.325 \text{kPa},$$

$$V = 298.2 \, \text{m}^3,$$

$$R = 0.287 \, \text{kJ/kgK for air,}$$

$$T = 25\,°\text{C} = (25 + 273)\text{K} = 298 \, \text{K, and}$$

$\therefore$

$$m = \frac{101.325 \times 298.2}{0.287 \times 298} = 353.28 \, \text{kg.}$$

Heat transfer rate of each person

$$= 418 \, \text{kJ/h}$$

Heat transfer rate of 25 people:

$$Q = 418 \times 25 = 10450 \, \text{kJ/h}$$

We know that the first law of thermodynamics for process,

$$Q = dU + W$$

or

$$Q = dU \quad \because W = 0$$

or

$$dU = Q$$
$$= 10450\,\text{kJ/h} = \frac{10450}{60}\,\text{kJ/h} = 174.16\,\text{kJ/min}$$

Increase in internal energy during 10 min,

$$dU = 174.16 \times 10 = 1741.6\,\text{kJ}$$

also

$$dU = mc_v dT$$

$\therefore$

$$1741.6 = 353.28 \times 0.718 \times dT$$

or

$$dT = 6.86\,°\text{C} \approx \mathbf{7\,°C}$$

Hence, an increase in temperature of air during the first 10 min is **7 °C**.

Problem 3.5: A piston and cylinder machine containing a fluid system has a stirring device in the cylinder. The piston is frictionless, and it is held down against the fluid due to the atmospheric pressure of 101.325 kPa. The stirring device is turned 10,000 revolutions with an average torque against the fluid of 1.275 Nm. Meanwhile, the piston of 0.6 m diameter moves out 0.8 m. Find the net work transfer for the system (Fig. 3.13).

Solution: Given data:
Atmospheric pressure:

$$p_{atm} = 101.325 \text{kPa}$$

Fig. 3.13 Schematic for
Problem 3.5

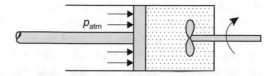

Number of revolution of the stirring device: N

$$N = 10000$$

Average torque:

$$T = 1.275 \, \text{Nm}$$

∴ Work done by the stirring device on the system:

$$W_{\text{Stirring}} = 2\pi NT = 2 \times 3.14 \times 10000 \times 1.275 \text{Nm}$$

$$= 80070 \text{Nm or J} = 80.7 \, \text{kJ}$$

Diameter of the piston:

$$d = 0.6 \, \text{m}$$

∴ Area of the piston:

$$A = \frac{\pi}{4}d^2 = \frac{3.14}{4} \times (0.6)^2 = 0.2826 \, \text{m}^2$$

Distance moved by the piston:

$$L = 0.8 \, \text{m}$$

Work done by the system on the surrounding:

$$W_{1-2} = p_{\text{atm}}dV = p_{\text{atm}}AL$$

$$= 101.325 \times 0.2826 \times 0.8 = 22.907 \, \text{kJ}$$

Net work transfer for the system:

$$W = W_{\text{Stirring}} + W_{1-2} = -80.7 + 22.907 = \mathbf{-57.793 \, kJ}$$

Problem 3.6: A closed system consists of a fluid inside a cylinder fitted with a frictionless piston. The fluid was stirred by a paddle wheel. 120 kJ of mechanical work was supplied along with 40 kJ of energy in the form of heat. At the same time, the piston moved in such a way that pressure remained constant at 200 kPa and the volume changed from 2 to 4 m³. Do calculations for the change in internal energy and enthalpy of the fluid system (Fig. 3.14).

Solution: Given data:

Work done by a paddle on the system:

$$W_p = -120 \, \text{kJ}$$

Heat supplied to the system:

$$Q = 40 \, \text{kJ}$$

Pressure of fluid:

$$p = 200 \, \text{kPa}$$

Change in volume:

$$dV = V_2 - V_1 = 4 - 2 = 2 \, \text{m}^3$$

∴ Displacement work by the system:

$$W_{1-2} = pdV = p(V_2 - V_1) = 200 \times (4 - 2) = 400 \, \text{kJ}$$

Net work:

$$W = W_{1-2} + W_p = 400 - 120 = 280 \, \text{kJ}$$

According to first law of thermodynamics for a process,

$$Q = dU + W$$

$$40 = dU + 280$$

Fig. 3.14 Schematic for
Problem 3.6

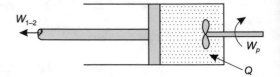

or

$$dU = -240\,\text{kJ}$$

By definition of enthalpy:

$$H = U + pV$$

On differentiating, we get

$$dH = dU + pdV + Vdp$$

$$dH = dU + pdV + 0 \quad \because p = C$$
$$= -240 + 400 = \mathbf{160\,kJ}$$

Problem 3.7: The gas in the system receives heat which causes expansion against a constant pressure of 2 bar. An agitator in the system is driven by an electric motor using 100 W. For 4 kJ of heat supplied the volume increase of the system in 30 s is 0.06 m³. Determine the net change in energy of the system (Fig. 3.15).

Solution: Given data:
Pressure:

$$p = 2\,\text{bar} = 200\text{kPa}$$

Work done by an electric motor:

$$W_M = -100\,\text{W} = -0.1\,\text{kW}$$

Heat supplied:

$$Q = 4\,\text{kJ}$$

Process duration:

$$t = 30\,\text{s}$$

Increase in volume:

Fig. 3.15 Schematic for Problem 3.7

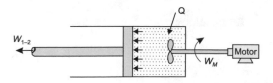

$$dV = 0.06\,\text{m}^3$$

Work done by an electric motor on the system in 30 s,

$$W_M = -0.01 \times 30\,\text{kJ} = -0.30\,\text{kJ}$$

Displacement work by the system:

$$W_{1-2} = pdV = 200 \times 0.06 = 12\,\text{kJ}$$

Net work:

$$W = W_{1-2} + W_M = 12 - 0.30 = 11.70\,\text{kJ}$$

According to the first law of thermodynamic for a process,

$$Q = dE + W$$

$$4 = dE + 11.70$$

or

$$dE = -\,7.7\,\text{kJ}$$

Problem 3.8: A spherical balloon of 1 m diameter contains a gas at 200 kPa and 300 K. The gas inside the balloon is heated until the pressure reaches 450 kPa. During the process of heating, the pressure of the gas inside the balloon is proportional to the diameter of the balloon. Make calculations for the work done by the gas inside the balloon.

Solution: Given data:
   At initial condition:
   Diameter of balloon:

$$D_1 = 1\,\text{m}$$

Pressure:

$$p_1 = 200\text{kPa}$$

Temperature:

$$T_1 = 300\,\text{K}$$

At final condition:

$$p_2 = 450\text{kPa}$$

$$p \propto D \quad \text{given condition}$$

$$p = CD$$

At initial state 1,

$$p_1 = CD_1$$

$$200 = C \times 1$$

or

$$C = 200\text{kPa/m}$$

At state 2,

$$p_2 = CD_2$$

$$450 = 200 \times D_2$$

or

$$D_2 = 2.25\,\text{m}$$

Volume of a spherical balloon:

$$V = \frac{\pi}{6}D^3$$

On differentiating, we get

$$dV = \frac{\pi}{6}3D^2 dD$$

$$dV = \frac{\pi}{2}D^2 dD$$

We know that the work done:

$$\delta W = p d V$$

$$\delta W = C D \times \frac{\pi}{2} D^2 d D$$

$$\delta W = \frac{\pi C}{2} D^3 d D$$

Integrating between states 1 and 2 gives

$$\int_1^2 \delta W = \frac{\pi C}{2} \int_1^2 D^3 d D$$

$$W_{1-2} = \frac{\pi}{2} C \left[\frac{D^4}{4} \right]_1^2 = \frac{\pi}{2} C \left[\frac{D_2^4 - D_1^4}{4} \right]$$

$$= \frac{3.14}{2} \times 200 \left[\frac{(2.25)^4 - (1)^4}{4} \right] = \mathbf{1933.37\,kJ}$$

Problem 3.9: When a system is taken from state a to state b along path acb as shown in Fig. 3.16, 80 kJ of heat flows into the system, and the system does 38 kJ of work.

How much heat flows into the system along path adb, if the work done is 14 kJ?

When the system is returned from b to a along the curved path, the work done on the system is 20 kJ. Does the system absorb or liberate heat, and how much of the heat is absorbed or liberated?

If $U_a = 0$ and $U_d = 24$ kJ, find the heat absorbed in the path ad and db.

Solution: Given data:

For path acb:

$$Q_{acb} = 80\,kJ$$

Fig. 3.16 Schematic for Problem 3.9

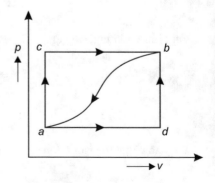

$$W_{acb} = 38\,\text{kJ}$$

We know

$$Q_{acb} = (U_b - U_a) + W_{acb}$$

$$80 = U_b - U_a + 38$$

or

$$U_b - U_a = 80 - 38 = 42\,\text{kJ}$$

For path *adb*:

$$W_{adb} = 14\,\text{kJ}$$

$$Q_{adb} = (U_b - U_a) + W_{adb} = 42 + 14 = \mathbf{56\,kJ}$$

also

$$W_{adb} = W_{ad} + W_{db}$$

$$W_{adb} = W_{ad} + 0 \quad |\because W_{db} = 0 \text{ at } V = C$$

$$W_{adb} = W_{ad} = 14\,\text{kJ}$$

For path *ba*:

$$W_{ba} = -20\,\text{kJ}$$

$$Q_{ba} = (U_a - U_b) + W_{ab} = -(U_b - U_a) + W_{ab}$$
$$= -42 - 20 = \mathbf{-62\,kJ}$$

The $-ve$ sign shows that the system liberates 62 kJ of heat.
Now

$$U_a = 0$$

$$U_d = 24\,\text{kJ}$$

We know

$$U_b - U_a = 42$$

$$U_b - 0 = 42$$

$$U_b = 42\,\text{kJ}$$

For path *ad*:

$$
\begin{aligned}
Q_{ad} &= (U_d - U_a) + W_{ad} \\
&= (24 - 0) + 14 \quad |\cdot\; W_{ad} = W_{adb} \\
&= \mathbf{38\,kJ}
\end{aligned}
$$

For path *db*:

$$Q_{db} = (U_b - U_d) + W_{db} = (42 - 24) + 0 = \mathbf{18\,kJ}$$

Problem 3.10: Consider a system taken from state 1 to state 2 along the path 1–a–2, 100 kJ of heat flows into the system and the system does 40 kJ of work (Fig. 3.17).

(a)   Determine the heat that flows into the system along the path 1–b–2 if it is accompanied by 20 kJ of work transfer by the system.
(b)   Calculate the work done and heat exchanged if the system returns to the initial state 1 along the straight path 2–1.
(c)   Calculate for the heat absorbed in the process 1–b and b–2 if internal average values at state points 1 and b are given as 0 and 50 kJ, respectively.

Solution: Given data:
   For path 1–a–2:

$$Q_{1a2} = 100\,\text{kJ}$$

Fig. 3.17 Schematic for Problem 3.10

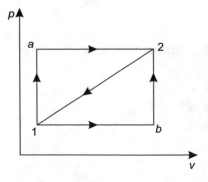

$$W_{1a2} = 40\,\text{kJ}$$

We know that

$$Q_{1a2} = (U_2 - U_1) + W_{1a2}$$

$$100 = (U_2 - U_1) + 40$$

or

$$U_2 - U_1 = 100 + 40 = 60\,\text{kJ}$$

(a)   For path 1–b–2:

$$W_{1b2} = 20\,\text{kJ}$$

We known that

$$Q_{1b2} = (U_2 - U_1) + W_{1b2}$$
$$= 60 + 20 = \mathbf{80\,kJ}$$

(b)   Work done under cycle 1–b–2–a–1,

$$W_{1b2a1} = W_{1a2} - W_{1b2}$$
$$= 40 - 20 = 20\,\text{kJ}$$

Work done under cycle 1–b–2–1,

$$W_{1b21} = \frac{W_{1b21}}{2} = \frac{20}{2} = 10\,\text{kJ}$$

Work done for process 2–1,

$$W_{2-1} = -W_{1b2} - W_{1b21}$$
$$= -20 - 10 = \mathbf{-30\,kJ}$$

$$U_2 - U_1 = 60\,\text{kJ}$$

or

$$U_1 - U_2 = -60 \text{ kJ}$$

For process 2–1,

$$Q_{2-1} = (U_1 - U_2) + W_{2-1}$$
$$= -60 - 30 = -\mathbf{90\,kJ}$$

(c) $U_1 = 0$

$$U_b = 50 \text{ kJ}$$

$$U_2 - U_1 = 60$$

$$U_2 - 0 = 60$$

or

$$U_2 = 60 \text{ kJ}$$

$$W_{1b} = W_{1b2} - W_{b2}$$
$$= 20 - 0 = 20 \quad |\because W_{b2} = 0, V = C$$

For process 1–b,

$$Q_{1b} = (U_b - U_1) + W_{1b}$$
$$= (50 - 0) + 20 = 50 + 20 = \mathbf{70\,kJ}$$

For process b–2,

$$Q_{b2} = (U_2 - U_b) + W_{b2}$$
$$= (60 - 50) + 0 = \mathbf{10\,kJ}$$

Problem 3.11: The values of c_p and c_v for a gas are 1.00 kJ/kgK and 0.71 kJ/kgK, respectively. Find the density of this gas at 15 °C and 1.05 bar absolute pressure.

Solution: Given data:

$$c_p = 1.00 \text{ kJ/kgK},$$
$$c_v = 0.71 \text{ kJ/kgK},$$
$$T = 15\,°C = (15 + 273)\text{K} = 288 \text{ K, and}$$
$$p = 1.05\text{bar} = 105\text{kPa}.$$

Gas constant:

$$R = c_p - c_v = 1.00 - 0.71 = 0.29 \, \text{kJ/kgK}$$

We know that the equation of state,

$$pv = RT$$
$$\frac{p}{\rho} = RT$$

or

$$p = \rho RT$$

$$105 = \rho \times 0.29 \times 288$$

or

$$\rho = 1.257 \, \text{kg/m}^3$$

Problem 3.12: Air enters a compressor at 1 bar and 25 °C having a volume of 1.8 m³/kg and is compressed isothermally to 5 bar. Determine (i) work done, (ii) change in internal energy, and (iii) heat transferred (Fig. 3.18).

Solution: Given data:
Pressure:

$$p_1 = 1 \, \text{bar} = 100 \text{kPa}$$

Temperature:

$$T_1 = 25 \, ^\circ\text{C} = (25 + 273)\text{K} = 298 \, \text{K}$$

Specific volume:

Fig. 3.18 Isothermal
compression

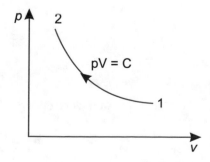

$$V_1 = 1.8 \, \text{m}^3/\text{kg}.$$

Pressure at state 2:

$$p_2 = 5\text{bar} = 500\text{kPa}$$

(i)   Work done:

$$w_{1-2} = p_1 v_1 \log_e \frac{v_2}{v_1}$$

$$= p_1 v_1 \log_e \frac{p_1}{p_2} \quad \because \text{Boyle's law} : \frac{v_2}{v_1} = \frac{p_1}{p_2}$$

$$= 100 \times 1.8 \log_e \frac{100}{500} = \mathbf{-289.69 \, kJ/kg}$$

(ii)   Change in internal energy:

$$u_2 - u_1 = 0$$

(iii)   Heat transfer:

$$q_{1-2} = w_{1-2} = \mathbf{-289.69 \, kJ/kg}$$

Problem 3.13: One kg of a gas at 240 °C expands adiabatically so that its volume is doubled and the temperature falls to 115 °C. The work done during expansion is 89.86 kJ. Determine the two specific heats.

Solution: Given data:

$$m = 1 \, \text{kg}$$
$$T_1 = 240 \, ^\circ\text{C} = (240 + 273)\text{K} = 513 \, \text{K}$$
$$T_2 = 115 \, ^\circ\text{C} = (115 + 273)\text{K} = 388 \, \text{K}$$
$$V_2 = 2V_1$$

or

$$\frac{V_2}{V_1} = 2, \text{ and}$$

$$W = 89.86 \, \text{kJ}$$

We know that work done during adiabatic process

$$W = \frac{p_1 V_1 - p_2 V_1}{\gamma - 1}$$

$$= \frac{mRT_1 - mRT_2}{\gamma - 1}$$

$$W = mc_v(T_1 - T_2)$$

$\therefore$

$$89.86 = 1 \times c_v(513 - 388) \quad \because c_v = \frac{R}{\gamma - 1}$$

$$89.86 = 1 \times c_v \times 125$$

or

$$c_v = \mathbf{0.7188\,kJ/kgK}$$

For adiabatic process 1–2,

$$T_1 V_1^{\gamma - 1} = T_2 V_2^{\gamma - 1}$$

or

$$\frac{T_1}{T_2} = \left(\frac{V_2}{V_1}\right)^{\gamma - 1}$$

$$\frac{513}{388} = (2)^{\gamma - 1}$$

Taking $\log_e$ on both sides, we get

$$\log_e 1.322 = (\gamma - 1)\log_e 2$$

or

$$\gamma - 1 = \frac{\log_e 1.322}{\log_e 2} = \frac{0.2791}{0.6931}$$

$$\gamma - 1 = 0.402$$

or

$$\gamma = 1.402$$

also

$$\gamma = \frac{c_p}{c_v}$$

$\therefore$

$$1.402 = \frac{c_p}{0.7188}$$

or

$$c_p = 1.007 \text{kJ/kgK}$$

Problem 3.14: A piston and cylinder machine contains a fluid system which passes through a complete cycle of four processes. During a cycle, the sum of all heat transfers is -170 kJ. The system completes 100 cycles per minute. Complete the following table showing the method of each item and compute the net rate of work output in kW.

Process	Q (kJ/min)	W (kJ/min)	ΔE (kJ/min)
a–b	0	2170	–
b–c	21,000	0	–
c–d	−2,100	–	−36,600
d–a	–	–	–

Solution: Given data:
    For process a–b:

$$Q_{ab} = 0$$

$$W_{ab} = 2170 \,\text{kJ/min}$$

The first law of thermodynamics for process a–b:

$$Q_{ab} = (\Delta E)_{ab} + W_{ab}$$

$$0 = (\Delta E)_{ab} + 2170$$

or

$$(\Delta E)_{ab} = -2170 \,\text{kJ/min}$$

For process b–c:

$$Q_{bc} = 21000\,\text{kJ}$$

$$W_{bc} = 0$$

The first law of thermodynamics for process b–c:

$$Q_{bc} = (\Delta E)_{bc} + W_{ab}$$

$$21000 = (\Delta E)_{bc} + 0$$

or

$$(\Delta E)_{bc} = 21000\,\text{kJ/min}$$

For process c d:

$$Q_{cd} = 2100\,\text{kJ/min}$$

$$(\Delta E)_{cd} = -36600\,\text{kJ/min}$$

The first law of thermodynamics for process c–d:

$$Q_{bc} = (\Delta E)_{bc} + W_{ab}$$
$$-2100 = -36600 + W_{cd}$$

or

$$W_{cd} = 34500\,\text{kJ/min}$$

For process d–a:

$$\sum_{Cycle} Q = -170\,\text{kJ}$$

The system completes 100 cycle/minute.

$\therefore$

$$\sum_{Cycle} Q = -170 \times 100\,\text{kJ/min} = -17000\ \text{kJ/min}$$

also

$$\sum_{Cycle} Q = Q_{ab} + Q_{bc} + Q_{cd} + Q_{da}$$

∴

$$-17000 = 0 + 21000 - 2100 + Q_{da}$$

or

$$Q_{da} = -35900 \, \text{kJ/min}$$

Since

$$\sum_{Cycle} \Delta E = 0$$

$$(\Delta E)_{ab} + (\Delta E)_{bc} + (\Delta E)_{cd} + (\Delta E)_{da} = 0$$

$$-2170 + 21000 - 36600 + (\Delta E)_{da} = 0$$

or

$$(\Delta E)_{da} = 17770 \, \text{kJ/min}$$

The first law of thermodynamics for process da:

$$Q_{da} = (\Delta E)_{da} + W_{da}$$
$$-35900 = 17770 + W_{da}$$
$$W_{da} = -53670 \, \text{kJ/min}$$

Now, the table is completed.

Process	Q (kJ/min)	W (kJ/min)	ΔE (kJ/min)
a–b	0	2170	−2170
b–c	21,000	0	21,000
c–d	−2,100	34,500	−36,600
d–a	−35,900	−53,670	17,770

Net rate of work output:

$$\sum_{Cycle} W = W_{ab} + W_{bc} + W_{cd} + W_{da}$$

$$= 2170 + 0 + 34500 - 53670 = -17000 \, \text{kJ/min}$$

$$= -\frac{17000}{60} \, \text{kJ/s} = -\mathbf{283.33 \, kW}$$

Problem 3.15:   A gas having a value of γ equal to 1.66 is expanded reversibly from the same initial state (i) isothermally and (ii) adiabatically, such that the pressure ratio is 5 in each case. Calculate the ratio of work done for conditions (i) and (ii) (Fig. 3.19).

Solution: Given data:
   Adiabatic index:

$$\gamma = 1.66$$

   Pressure ratio:

$$\frac{p_1}{p_2} = 5$$

(i)   Process 1–2, isothermal

Work done:

$$W_{1-2} = p_1 V_1 \log_e \frac{V_2}{V_1} = m R T_1 \log_e \frac{p_1}{p_2}$$
$$= m R T_1 \log_e 5 = 1.609 m R T_1$$

(ii)   Process $1 - 2'$, adiabatic

Work done:

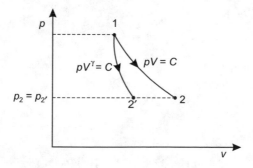

Fig. 3.19 *p-V* diagram for Problem 3.15

$$W_{1-2'} = \frac{p_1 V_1 - p_{2'} V_{2'}}{\gamma - 1} = \frac{mR(T_1 - T_{2'})}{\gamma - 1}$$

$$= \frac{mRT_1}{\gamma - 1}\left(1 - \frac{T_{2'}}{T_1}\right)$$

where

$$\frac{T_{2'}}{T_1} = \left(\frac{p_{2'}}{p_1}\right)^{\frac{\gamma-1}{\gamma}}$$

$$W_{1-2'} = \frac{mRT_1}{\gamma - 1}\left[1 - \left(\frac{p_{2'}}{p_1}\right)^{\frac{\gamma-1}{\gamma}}\right] = \frac{mRT_1}{1.66 - 1}\left[1 - \left(\frac{1}{5}\right)^{\frac{1.66-1}{1.66}}\right]$$

$$= \frac{mRT_1}{1.66}\left[1 - (0.2)^{0.397}\right]$$

$$= \frac{mRT_1}{0.66}[1 - 0.527] = 0.716mRT_1.$$

Ratio of work done for conditions (i) and (ii),

$$\frac{W_{1-2}}{W_{1-2'}} = \frac{1.609mRT_1}{0.716mRT_1} = \mathbf{2.246}$$

Problem 3.16: 4 kg of air is contained within a piston-cylinder arrangement. It undergoes a process for which the pressure–volume relationship is $pV^{1.5} = C$. The initial pressure is 3 bar, the initial volume is 0.1 m³, and the final volume is 0.2 m³. The change in the specific internal energy of air in the process is $u_2 - u_1 = -4.6$ kJ/kg. Determine the net heat transfer for the process.

Solution: Given data:
    Mass of air:

$$m = 4\,\text{kg}$$

Process following the law:

$$pV^{1.5} = C$$

Initial pressure:

$$p_1 = 3\,\text{bar} = 3 \times 10^2\,\text{kPa}$$

Initial volume:

$$V_1 = 0.1\,\text{m}^3$$

Final volume:

$$V_2 = 0.2\,\text{m}^3$$

$$u_2 - u_1 = -4.6\,\text{kJ/kg}$$

For process 1–2,

$$p_1 V_1^{1.5} = p_2 V_2^{1.5}$$

$$3 \times (0.1)^{1.5} = p_2 \times (0.2)^{1.5}$$

or

$$p_2 = 3 \times \left(\frac{0.1}{0.2}\right)^{1.5} = 1.06\,\text{bar} = 1.06 \times 10^2\,\text{kPa}$$

Work done by the system:

$$
\begin{aligned}
W_{1-2} &= \frac{p_2 V_2 - p_1 V_1}{1 - n} \\
&= \frac{1.06 \times 10^2 \times 0.2 - 3 \times 10^2 \times 0.1}{1 - 1.5} = \frac{-8.8}{-0.5} = 17.6\,\text{kJ}
\end{aligned}
$$

Net heat transfer for the process:

$$
\begin{aligned}
Q_{1-2} &= (U_2 - U_1) + W_{1-2} \\
&= m(u_2 - u_1) + W_{1-2} \\
&= 4 \times (-4.6) + 17.6 = -18.4 + 17.6 = \mathbf{-0.8\,kJ}
\end{aligned}
$$

The $-ve$ sign shows the heat transfer from the system.

Problem 3.17: The pressure–volume relation for a non-flow reversible process is $p = (8 - 4\,V)$ bar, where V is in m^3. If 130 kJ of work is supplied to the system, calculate the final pressure and volume of the system. Take the initial volume as 0.5 m^3.

Solution: Given data:
  Pressure:

$$p = (8 - 4V)\,\text{bar}$$

$$= (8 - 4V) \times 10^2\,\text{kPa}$$

Work supplied to the system:

$$W_{1-2} = -130 \, \text{kJ}$$

Initial volume:

$$V_1 = 0.5 \, \text{m}^3$$

we have

$$W_{1-2} = \int_1^2 p \, dV$$

$$W_{1-2} = \int_1^2 (8 - 4V)0^2 dV$$

$$W_{1-2} = 10^2 \left[8V - \frac{4V^2}{2} \right]_1^2$$

$$W_{1-2} = 10^2 \left[8V_2 - 2V_2^2 - 8V_1 + 2V_1^2 \right]$$

$$-130 = 10^2 \left[8V_2 - 2V_2^2 - 8 \times 0.5 + 2 \times (0.5)^2 \right]$$

$$-1.30 = 8V_2 - 2V_2^2 - 4 + 0.5$$

$$-1.30 = 8V_2 - 2V_2^2 - 3.5$$

or

$$2V_2^2 - 8V_2 + 2.2 = 0$$

$$V_2 = \frac{-b \pm \sqrt{b^2 - 4ac}}{2a}$$

$$= \frac{8 \pm \sqrt{64 - 4 \times 2 \times 2.2}}{2 \times 2} = \frac{8 \pm 6.81}{4}$$

Possible value,

$$V_2 = \frac{8 - 6.81}{4} = 0.2975 \, \text{m}^3$$

Final volume of the system is 0.2975 m³.
Final pressure of the system:

$$p_2 = (8 - 4V^2) \text{ bar} = 8 - 4 \times 0.2975 = \mathbf{6.81\,bar}$$

Problem 3.18: Air during a reversible process is compressed from initial pressure 12 kN/m² to 6 times the initial pressure. Due to this compression, volume of air decreases from initial volume 4 m³ to 1.8 m³. Determine

(i)   Law of the process and
(ii) Work done in compressing the air.

Solution: Given data:

$$p_1 = 12\,\text{kN/m}^2,$$
$$p_2 = 6p_1 = 6 \times 12 = 72\,\text{kN/m}^2,$$
$$V_1 = 4\,\text{m}^3, \text{ and}$$
$$V_2 = 1.8\,\text{m}^2.$$

We know that compression follows:

(i)   Isothermal process,
(ii)   Adiabatic process, and
(iii) Polytropic process.
(i)   Isothermal process:

$$pV = C$$
$$p_1 V_1 = p_2 V_2$$
$$12 \times 4 = 72 \times 1.8$$
$$4.8 \neq 129.6$$

Thus, compression process does not follow the law

$$PV = C$$

(i)   Adiabatic process:

$$pV^\gamma = C$$

$$p_1 V_1^\gamma = p_2 V_2^\gamma$$

$$12 \times (4)^{1.4} = 7.2 \times (1.8)^{1.2}$$

$$83.57 \neq 163.94$$

Thus, also the compression process does not follow the law of $pv^\gamma = C$.

Let $V_2' = $ volume at the end of compression when compression follows the law $pv^\gamma = C$

$$p_1 V_1^\gamma = p_2 \left(V_2'\right)^\gamma$$

$$12 \times (4)^{1.4} = 72 \times \left(V_2'\right)^{1.4}$$

or

$$V_2' = 1.112 \, \text{m}^3$$

It means $V_2 > V_2'$, thus the process is reversible polytropic and follows the law $pV^n = C$,

where

$$n > \gamma.$$

(iii)   Polytropic process:

$$pV^n = C$$

$$p_1 V_1^n = p_2 V_2^n$$

$$\left(\frac{V_1}{V_2}\right)^n = \frac{p_2}{p_1}$$

Taking $\log_e$ on both sides, we have (Fig. 3.20)

$$n \log_e \frac{V_1}{V_2} = \log_e \frac{p_2}{p_1}$$

$$n \times \log_e \frac{4}{1.8} = \log_e \frac{72}{12}$$

$$n = 2.24$$

Fig. 3.20 *p-V* diagram for
Problem 3.18

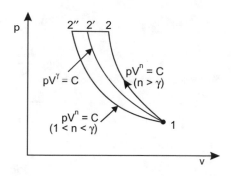

(i)   The compression follows the law $pV^n = C$, where $n > \gamma$.
(ii) Work done:

$$W_{1-2} = \frac{p_2 V_2 - p_1 V_1}{1 - n}$$
$$= \frac{72 \times 1.8 - 12 \times 4}{1 - 2.24} = \mathbf{-56.80\,kJ}$$

The $-ve$ sign shows the work done on the system.

Problem 3.19: 2.5 m³ of air at a pressure of 9 bar expands to a volume of 12.5 m³. The final pressure is 1.5 bar and the expansion is polytropic. Determine the heat supplied and the change in internal energy.

Solution: Given data:

$$V_1 = 2.5\,\mathrm{m^3},$$
$$p_1 = 9\,\mathrm{bar} = 900\,\mathrm{kPa},$$
$$V_2 = 12.5\,\mathrm{m^3}, \text{ and}$$
$$p_2 = 1.5\,\mathrm{bar} = 150\,\mathrm{kPa}.$$

We know that the polytropic law,

$$pV^n = C$$

For process 1–2,

$$p_1 V_1^n = p_2 V_2^n$$

$$\frac{p_1}{p_2} = \left(\frac{V_2}{V_1}\right)^n$$

Taking $\log_e$ on both sides, we get

$$\log_e \frac{p_1}{p_2} = n \log_e \frac{V_2}{V_1}$$

$$\log_e \frac{900}{150} = n \log_e \frac{12.5}{2.5}$$

$$\log_e 6 = n \log_e 5$$

or

$$n = \frac{\log_e 6}{\log_e 5} = 1.113$$

Work done:

$$W_{1-2} = \frac{p_1 V_1 - p_2 V_2}{n - 1}$$

$$= \frac{900 \times 2.5 - 150 \times 12.5}{1.113 - 1} = 3318.58 \, \text{kJ}$$

Heat supplied:

$$Q_{1-2} = \left(\frac{\gamma - n}{\gamma - 1} \right) \times \text{Work done}$$

$$= \left(\frac{1.4 - 1.113}{1.4 - 1} \right) \times 3318.58 = \mathbf{2881.08 \, kJ}$$

According to the first law for process,

$$Q_{1-2} = dU + W_{1-2}$$

$$2381.08 = dU + 3318.58$$

or

$$dU = -\mathbf{937.5 \, kJ}$$

Problem 3.20: The characteristic equation of a certain gas is $pv = 300 \, T$, where p in Pa, v in m^3/kg, T in kelvin, and c_v of the gas is 0.75 kJ/kgK. One kg of this gas at a pressure of 4.2 bar is absolutely expanded from a volume of 0.14 m^3 to a volume of 0.84 m^3, and the equation of the curve of expansion is $pV^{1.31} = $ constant. Find how many thermal units are interchanged between the gas and the cylinder walls during expansion and the direction in which the flow takes place.

Solution: Given data:

Equation of state,

$$pv = 300\,T$$

where

$$R = 300\,\text{J/kgK} = 0.30\,\text{kJ/kgK}$$

$$c_v = 0.75\,\text{kJ/kgK,}$$

$$m = 1\,\text{kg,}$$

$$p_1 = 4.2\,\text{bar} = 420\text{kPa,}$$

$$V_1 = 0.14\,\text{m}^3\text{, and}$$

$$V_2 = 0.84\,\text{m}^3.$$

Polytropic law of expansion,

$$pV^{1.31} = \text{constant}$$

where

$$n = 1.31$$

For process 1–2,

$$p_1 V_1^{1.31} = p_2 V_2^{1.31}$$

$$420 \times (0.14)^{1.31} = p_2 \times (0.84)^{1.31}$$

or

$$p_2 = 40.16\text{kPa}$$

$$c_p - c_v = R$$

$$c_p = R + c_v$$

$$= 0.30 + 0.75 = 1.50 \, \text{kJ/kgK}$$

Adiabatic index:

$$\gamma = \frac{c_p}{c_v} = \frac{1.05}{0.75} = 1.4$$

Work done:

$$W = \frac{p_1 V_1 - p_2 V_2}{n - 1}$$
$$= \frac{420 \times 0.14 - 40.16 \times 0.89}{1.31 - 1} = 80.85 \, \text{kJ}$$

Heat transfer:

$$Q = \left(\frac{\gamma - n}{\gamma - 1}\right) \times W$$
$$= \left(\frac{1.4 - 1.31}{1.4 - 1}\right) \times 80.85 = \mathbf{18.19 \, kJ}$$

The $+ve$ sign indicates the heat supplied to the gas.

Problem 3.21: A closed system having air of 50 kg has an initial velocity of 10 m/s. The velocity of air increases to 30 m/s, and its elevation also by 40 m. During this process, the system receives 30,000 J of heat and 4500 J of work. If the system delivers 0.002 kWh of electrical energy, determine the change in internal energy of the system.

Solution: Given data:

$$m = 50 \, \text{kg}$$
$$V_1 = 10 \, \text{m/s}$$
$$V_2 = 30 \, \text{m/s}$$
$$z_2 - z_1 = 40 \, \text{m}$$
$$Q_{1-2} = 30000 \, \text{J} = 30 \, \text{kJ}$$
$$W_{1-2} = -4500 \, \text{J} = -4.5 \, \text{kJ}$$

Delivers electrical energy

$$= 0.002 \text{kWh} = 0.002 \text{kJh/s}$$
$$= 0.002 \times 3600 \, \text{kJ} = 7.2 \, \text{kJ}$$

Net work done:

$$W_{net} = W_{1-2} + \text{Delivers electrical energy}$$
$$= -4.5 + 7.2 = 2.7\,\text{kJ}$$

We know that the first law for process

$$Q_{1-2} = dE + W_{net}$$

where

$$dE = dU + d(KE) + d(PE)$$

$\therefore$

$$Q_{1-2} = dU + d(KE) + d(PE) + W_{net}$$

$$Q_{1-2} = dU + \frac{1}{2}m\left(V_2^2 - V_1^2\right) + mg(z_2 - z_1) + W_{net}$$

If Q_{1-2}, dU and W_{net} are in kJ, then $\frac{1}{2}m\left(V_2^2 - V_1^2\right)$ and $mg\,(z_1 - z_2)$ are made in kJ by dividing 1000.

$\therefore$

$$Q_{1-2} = dU + \frac{m}{2000}\left(V_2^2 - V_1^2\right) + \frac{mg(z_2 - z_1)}{1000} + W_{net}$$

$$30 = dU + \frac{50}{2000}\left(30^2 - 10^2\right) + \frac{50 \times 9.81 \times 40}{1000} + 2.7$$

$$30 = dU + 20 + 19.62 + 2.7$$

or

$$dU = -12.32\,\text{kJ}$$

Problem 3.22: One mole of an ideal gas at 0.5 MPa and 300 K is heated at constant pressure till the volume is doubled and then allowed to expand at constant temperature till the volume is doubled further. Determine the net work done. Sketch the processes on p–V and T–s diagrams (Fig. 3.21).

Solution: Given data:
   Number of moles:

$$n = 1$$

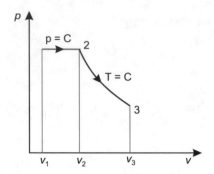 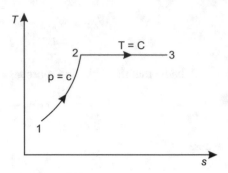

Fig. 3.21 *p-V* and *T-s* diagrams for Problem 3.22

$$p_1 = 0.5\text{MPa} = 500\text{kPa} = p_2$$

$$T_1 = 300\,\text{K}$$

Let

$$V_1 = V$$

and

$$V_2 = 2V_1 = 2V,$$

$$V_3 = 2V_2 = 4V,\, and$$

$$T_2 = T_3.$$

Work done during process 1–2:

$$W_{1-2} = \int_1^2 p\,dV = p\int_1^2 dV = p[V]_1^2$$
$$= p(V_2 - V_1)$$

$$\begin{aligned}
W_{1-2} &= p(2V - V)\\
&= pV\\
&= p_1 V_1 \quad \because p_1 = p_2 = p,\, V_1 = V\\
&= n\bar{R}T_1 \quad \because p_1 V_1 = n\bar{R}T_1
\end{aligned}$$

where $n = 1$, number of moles.

$\bar{R} = 8.314$ J/mol K, universal gas constant.

∴

$$W_{1-2} = 1 \times 8.314 \times 300\,\text{J} = 2494.2\,\text{J} = 2.494\,\text{kJ}.$$

Work during isothermal process 2–3,

$$W_{2-3} = n\bar{R}T_2 \log_e \frac{V_3}{V_2}$$

Applying the Charles law for process 1–2,

$$\frac{V_2}{T_2} = \frac{V_1}{T_1}$$
$$\frac{2V}{T_1} = \frac{V}{300}$$

or

$$T_2 = 600\,\text{K}$$

∴

$$W_{2-3} = 1 \times 8.314 \times 600 \log_e \frac{4V}{2V}$$
$$= 4988.4 \log_e 2 = 3457.69\,\text{J} = 3.457\,\text{kJ}$$

Net work done:

$$W_{\text{net}} = W_{1-2} + W_{2-3}$$
$$= 2.494 + 3.457 = \mathbf{5.951\,kJ}$$

Problem 3.23: 0.06 m^3 of air at 300 K and 1 bar is compressed adiabatically to 10 bar. It is then cooled at constant volume and further expanded isothermally so as to reach the condition from where it started. Evaluate (a) Pressure at the end of constant cooling, (b) Change in internal energy during constant volume process, and (c) Net work done and heat transferred during the cycle. Assume air to be a perfect gas (Fig. 3.22).

Solution: Given data:
1–2, adiabatic process,
2–3, isochoric process,
3–1, isothermal process,

$$V_1 = 0.06\,\text{m}^3,$$

Fig. 3.22 *p-V* diagram for
Problem 3.23

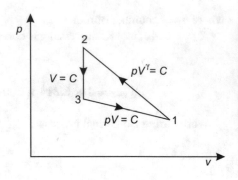

$$T_1 = 300 \, \text{K},$$

$$p_1 = 1 \, \text{bar},$$

$$p_2 = 10 \, \text{bar},$$

$$p_1 V_1 = m R T_1$$

$$100 \times 0.06 = m \times 0.287 \times 300$$

or

$$m = 0.0696 \, \text{kg}$$

Process 1–2,

$$\frac{T_2}{T_1} = \left(\frac{p_2}{p_1} \right)^{\frac{\gamma-1}{\gamma}}$$

$$T_2 = T_1 \left(\frac{p_2}{p_1} \right)^{\frac{\gamma-1}{\gamma}}$$

$$= 300 \left(\frac{10}{1} \right)^{\frac{1.4-1}{1.4}} = 578.25 \, \text{K} \quad \gamma = 1.4 \text{ for air}$$

and

$$\frac{T_2}{T_1} = \left(\frac{V_1}{V_2} \right)^{\gamma-1}$$

$$\frac{578.25}{300} = \left(\frac{0.06}{V_2}\right)^{1.4-1}$$

or

$$\frac{0.06}{V_2} = 5.158$$

or

$$V_2 = 0.01163 \, \text{m}^3$$

Process 2–3,

$$\frac{p_2}{T_2} = \frac{p_3}{T_3}$$

$$\frac{p_2}{T_2} = \frac{p_3}{T_1} \quad \because T_1 - T_3$$

$$\frac{10}{578.25} = \frac{p_3}{300}$$

or

$$p_3 = \textbf{5.188 bar}$$

(a) Pressure at the end of constant cooling:
$p_3 = \textbf{5.188 bar}$.

(b) Change in internal energy during constant volume process:

$$U_3 - U_2 = mc_v(T_3 - T_2)$$
$$= 0.0696 \times 0.718(300 - 578.25) = \textbf{-13.90 kJ}$$

(c) For process 1–2,

$$W_{1-2} = \frac{mR(T_1 - T_2)}{\gamma - 1} = mc_v(T_1 - T_2)$$
$$= 0.0696 \times 0.718(300 - 578.25) = -13.90 \, \text{kJ}$$

and

$$Q_{1-2} = 0$$

For process 2–3,

$$W_{2-3} = 0$$

and

$$Q_{2-3} = mc_v(T_3 - T_2) = mc_v(T_1 - T_2) \quad |\because T_3 = T_1$$
$$= W_{1-2} = -13.90 \, \text{kJ}$$

For process 3–1,

$$W_{3-1} = mRT_3 \log_e \frac{V_1}{V_3}$$

$$= 0.0696 \times 0.287 \times 300 \log_e \left(\frac{0.06}{0.01163} \right) = 9.83 \, \text{kJ}$$

and

$$Q_{3-1} = W_{3-1} = 9.83 \, \text{kJ/kg}$$

Net work done during the cycle:

$$W_{net} = W_{1-2} + W_{2-3} + W_{3-1} = -13.90 + 0 + 9.83 = -4.07 \, \text{kJ}$$

Net heat transfer during the cycle:

$$Q_{net} = Q_{1-2} + Q_{2-3} + Q_{3-1} = 0 - 13.90 + 9.83 = -4.07 \, \text{kJ}$$

also for cycle,

$$W_{net} = Q_{net} = -4.07 \, \text{kJ}$$

Problem 3.24: A certain mass of air is initially at 267 °C and 7 bar occupies 0.21 m³. The air is expanded at constant pressure such that volume becomes three times the initial volume. A polytropic process with $n = 1.3$ is then carried out, followed by an isothermal process which completes the cycle. Determine

(a)   heat rejected and received during each process and
(b)   net work done during the cycle.

   Sketch the cycle on p–v and T–s diagrams (Fig. 3.23).

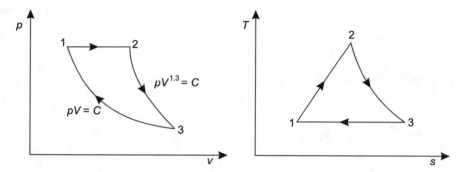

Fig. 3.23 p-V and T-s diagrams for Problem 3.24

Solution: Given data:
   1–2, constant pressure process,
   2–3, polytropic process,
   3–1, isothermal process,

$$T_1 = 267\,°\text{C} = (267 + 273)\text{K} = 540\,\text{K},$$

$$p_1 = 7\,\text{bar} = 700\text{kPa},$$

$$V_1 = 0.21\,\text{m}^3,$$

$$V_2 = 3V_1 = 3 \times 0.21 = 0.63\,\text{m}^3,\ \text{and}$$

$$T_3 = T_1 = 540\,\text{K}.$$

At state 1,

$$p_1 V_1 = mRT_1$$

$$700 \times 0.21 = m \times 0.287 \times 540$$

or

$$m = 0.9485\,\text{kg}$$

Process 1–2,

$$\frac{T_2}{T_1} = \frac{V_2}{V_1}$$

$$\frac{T_2}{540} = 3$$

or

$$T_2 = 3 \times 540 = 1620\,\text{K}$$

Process 2–3,

$$\frac{T_2}{T_3} = \left(\frac{p_2}{p_3}\right)^{\frac{n-1}{n}}$$

or

$$\frac{p_2}{p_3} = \left(\frac{T_2}{T_3}\right)^{\frac{n}{n-1}} = \left(\frac{1620}{540}\right)^{\frac{1.3}{1.3-1}} = (3)^{4.33} = 116.39$$

$$\frac{p_2}{p_3} = 116.39$$

$$\frac{p_1}{p_3} = \frac{p_2}{p_3} = 116.39 \quad |\because p_1 = p_2$$

(a)   Heat rejected and received during each process.

For process 1–2,

$$Q_{1-2} = mc_p(T_2 - T_1)$$
$$= 0.9485 \times 1.005(1620 - 540) = \mathbf{1029.5\,kJ}$$

The $+ve$ sign shows heat received during the process.
For process 2–3,

$$Q_{2-3} = mc_v\left(\frac{\gamma - n}{1 - n}\right)(T_3 - T_2)$$

$$= 0.9485 \times 0.718\left(\frac{1.4 - 1.3}{1 - 1.3}\right)(540 - 1620)$$

$$= \mathbf{245.17\,kJ}$$

The $+ve$ sign shows heat received during the process.
For process 3–1,

$$Q_{3-1} = mRT_3 \log_e\left(\frac{p_3}{p_1}\right)$$

$$= 0.9485 \times 0.287 \times 540 \log_e\left(\frac{1}{116.39}\right) = \mathbf{-699.26\,kJ}$$

The $-ve$ sign shows the heat rejected during the process.

(b)   Net work done during the cycle:

The first law of thermodynamic for cycle,

$$\oint w = \oint Q = Q_{1\,2} + Q_{2-3} + Q_{3-1}$$

$$= 1029.5 + 245.17 - 699.26 = \mathbf{575.41kJ}$$

Problem 3.25: 0.2 m³ of an ideal gas at a pressure of 2 MPa and 600 K is expanded isothermally to 5 times the initial volume. It is then cooled to 300 K at constant volume and then compressed back polytropically to its initial state. Determine the net work done and heat transferred during the cycle (Fig. 3.24).

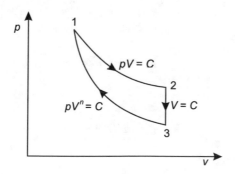

Fig. 3.24 *p-V* diagram for Problem 3.25

Solution: Given data:

$$V_1 = 0.2\,\text{m}^3,$$

$$p_1 = 2\text{MPa} = 2 \times 10^3 \text{kPa},$$

$$T_1 = 600\,\text{K},$$

$$V_2 = 5V_1$$
$$= 5 \times 0.2 = 1\,\text{m}^3, \text{and}$$

$$T_3 = 300\,\text{K}.$$

At state 1,

$$p_1 V_1 = mRT_1$$

$$2 \times 10^3 \times 0.2 = m \times 0.287 \times 600$$

or

$$m = 2.32\,\text{kg}$$

Process 1–2 isothermal:

$$p_1 V_1 = p_2 V_2$$

$$2 \times 10^3 \times 0.2 = p_2 \times 1$$

or

$$p_2 = 400\,\text{kPa}$$

Work done:

$$W_{1-2} = mRT_1 \log_e \frac{V_2}{V_1}$$

Assume an ideal gas is air

$$\text{For air}: c_p = 1.005\,\text{kJ/kgK}$$
$$c_v = 0.718\,\text{kJ/kgK}$$
$$R = 0.287\,\text{kJ/kgK}$$
$$\gamma = 1.4$$

$$W_{1-2} = 2.32 \times 0.287 \times 600 \log_e \frac{1}{0.2} = 642.96\,\text{kJ}$$

Heat transfer:

$$Q_{1-2} = W_{1-2} = 642.96\,\text{kJ}$$

Process 2–3 isochoric:

$$\frac{p_2}{T_2} = \frac{p_3}{T_3}$$

$$\frac{400}{600} = \frac{p_3}{300} \quad \because T_2 = T_1 = 600\,\text{K}$$

or

$$p_3 = 200\,\text{kPa}$$

Work done:

$$W_{2-3} = 0$$

Heat transfer:

$$Q_{2-3} = mc_v(T_3 - T_2) = 2.32 \times 0.718(300 - 600) = -499.72\,\text{kJ}$$

Process 3–1 polytropic:
Work done:

$$W_{3-1} = \frac{mR(T_1 - T_3)}{1 - n}$$

$$p_3 V_3^n = p_1 V_1^n$$

$$\left(\frac{V_3}{V_1}\right)^n = \frac{p_1}{p_3}$$

Taking $\log_e$ on both sides, we have

$$n \log_e \left(\frac{V_3}{V_1} \right) = \log_e \left(\frac{p_1}{p_3} \right)$$

$$n \log_e \left(\frac{1}{0.2} \right) = \log_e \left(\frac{2 \times 10^3}{200} \right) \quad \because V_3 = V_2$$

$$n \times 1.609 = 2.302$$

or

$$n = 1.43$$

$\therefore$ Work done:

$$W_{3-1} = \frac{2.32 \times 0.287 \times (600 - 300)}{1 - 1.43} = -464.53 \, \text{kJ}$$

Heat transfer:

$$\begin{aligned}
Q_{3-1} &= dU + W_{3-1} \\
&= mc_v(T_3 - T_1) + W_{3-1} \\
&= 2.32 \times 0.718(600 - 300) - 464.53 \\
&= 499.72 - 464.53 = 35.19 \, \text{kJ}
\end{aligned}$$

(i)   Net work done:

$$\begin{aligned}
W_{\text{net}} &= W_{1-2} + W_{2-3} + W_{3-1} \\
&= 642.96 + 0 - 464.53 = 178.43 \, \text{kJ}
\end{aligned}$$

(ii)   Net heat transfer:

$$\begin{aligned}
Q_{\text{net}} &= Q_{1-2} + Q_{2-3} + Q_{3-1} \\
&= 642.96 - 499.72 + 35.19 = \textbf{178.43 kJ}
\end{aligned}$$

Problem 3.26: Air undergoes a cyclic process in a cylinder and piston arrangement. First, air at 1 bar and 27 °C is compressed adiabatically to 10 bar then expanded isothermally up to initial pressure, then brought to the initial condition under constant

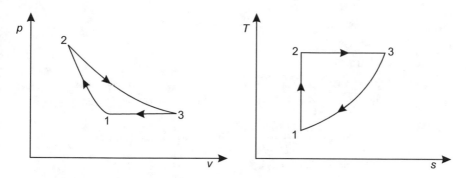

Fig. 3.25 *p-V* and *T-s* diagram for Problem 3.26

pressure. Determine the heat transfer, work transfer, and change in internal energy for each process and for the complete cycle. Draw the *p–v* and *T–s* diagrams (Fig. 3.25).

Solution: Given process:

Process 1–2, adiabatic compression, follow $pv^\gamma = C$,
Process 2–3 isothermal expansion, follow $pv = C$, and.
Process 3–1 isobaric, $\frac{v}{T} = C$.

For process: 1–2

$$p_1 = 1\,\text{bar} = 1 \times 10^5 \text{N/m}^2$$

$$T_1 = 27\,^\circ\text{C} = (27 + 273)\text{K} = 300\,\text{K}$$

$$p_2 = 10\,\text{bar} = 10 \times 10^5 \text{N/m}^2$$

We know that the relation between temperature and pressure

$$\frac{T_2}{T_1} = \left(\frac{p_2}{p_1}\right)^{\frac{\gamma-1}{\gamma}} \text{ for adiabatic}$$

$$T_2 = T_1\left(\frac{p_2}{p_1}\right)^{\frac{\gamma-1}{\gamma}}$$

$$T_2 = 300 \times \left(\frac{10 \times 10^5}{1 \times 10^5}\right)^{\frac{1.4-1}{1.4}} \quad |\because \gamma = 1.4 \text{ for air}$$

$$= 300(10)^{0.2857} = 579.19\,\text{K}.$$

(i) Heat transfer: $q_{1-2} = 0$, because no heat transfer in the adiabatic process.
(ii) Work transfer:

$$W_{1-2} = \frac{mR(T_2 - T_1)}{1 - \gamma}$$

Specific work transfer:

$$
\begin{aligned}
w_{1-2} &= \frac{R(T_2 - T_1)}{1 - \gamma} \quad |\because m = 1 \text{ kg} \\
&= \frac{0.287(579.19 - 300)}{1 - 1.4} \quad |\because R = 0.287 \text{ kJ/kgK} \\
&= -200.32 \text{ kJ/kg}
\end{aligned}
$$

The $-ve$ sign shows the work done on the system.

(iii) Change in specific internal energy: $u_2 - u_1$

$$q_{1-2} = (u_2 - u_1) + w_{1-2}$$

$$0 = (u_2 - u_1) - 200.32$$

or

$$u_2 - u_1 = 200.32 \text{ kJ/kg}$$

For process: 2–3

$$p_3 = p_1 = 1 \times 10^5 \text{N/m}^2$$

(i) Heat transfer:

$$
\begin{aligned}
q_{2-3} &= RT_2 \log_e \frac{p_2}{p_3} \\
&= 0.287 \times 579.19 \log_e \frac{10 \times 10^5}{1 \times 10^5} = 382.75 \text{ kJ/kg}
\end{aligned}
$$

(ii) Specific work:

$$w_{2-3} = q_{2-3} = 382.75 \text{ kJ/kg}$$

(iii) Change in specific internal energy:

$$u_2 - u_1 = 0$$

For process: 3–1

(i) Heat transfer:

$$q_{3-1} = c_p(T_1 - T_3) = 1.005 \times (300 - 579.19)$$
$$= \mathbf{-280.58 \ kJ/kg}$$

(ii) Specific work:

$$w_{3-1} = p(v_1 - v_3) = R(T_1 - T_3) = 0.287(300 - 579.19)$$
$$= \mathbf{-80.12 \ kJ/kg}$$

(iii) Change in specific internal energy: $u_1 - u_3$

$$q_{1-3} = (u_1 - u_3) + w_{1-3}$$

$$-280.58 = (u_1 - u_3) - 80.12$$

or

$$u_1 - u_3 = \mathbf{-200.46 \ kJ/kg}$$

For complete cycle: 1–2–3–1

(i) Heat transfer:

$$q_{1-2-3-1} = q_{1-2} + q_{2-3} + q_{3-1} = 0 + 382.75 - 280.58$$
$$= \mathbf{102.17 \ kJ/kg}$$

(ii) Specific work:

$$w_{1-2-3-4} = w_{1-2} + w_{2-3} + w_{3-1}$$

$$= -200.32 + 382.75 - 80.12$$
$$= \mathbf{102.31\ kJ/kg}$$
$$= q_{1-2-3-1}$$

(iii)   Change in specific internal energy is zero.

Problem 3.27: The Lenoir cycle, which approximates the operation of a pulse jet, is shown in Fig. 3.26 on the pressure-volume plane. The processes are as follows:

(a)   1–2, isochoric, heat addition of 230 kJ/kg;
(b)   2–3, adiabatic expansion, change in internal energy $= -75$ kJ/kg;
(c)   3–1, isobaric compression, heat rejection of 190 kJ/kg.

   Determine.

(i)   the net work output of the cycle and
(ii)   the change in internal energy during process 3–1.

Solution: Given data:
   Heat addition:

$$q_{1-2} = 230\ kJ/kg$$

Change in internal energy for process 2–3: $u_3 - u_2 = -75$ kJ/kg
Heat rejection:

$$q_{3-1} = 190\,kJ/kg$$

Heat transfer during adiabatic process 2–3 is zero,

i.e.,

$$q_{2-3} = 0$$

Fig. 3.26 Lenoir cycle

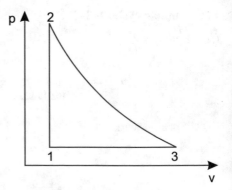

(i) Net work output of the cycle,

$$w_{net} = \text{Heat addition} - \text{Heat rejection}$$
$$= q_{1-2} - q_{3-1} = 230 - 190 = \mathbf{40\,kJ/kg}$$

(ii) Change in internal energy during process 3–1,

$$u_1 - u_3 =?$$

Work done for process 2–3:

$$w_{2-3} = -(u_3 - u_2) = -(-75) = 75\ \text{kJ/kg}$$

$$w_{net} = w_{2-3} + w_{3-1} + w_{1-2}$$

$$40 = 75 + w_{3-1} + 0$$

or

$$w_{3-1} = 40 - 75 = -35\ \text{kJ/kg}$$

For process 3–1,

$$q_{3-1} = (u_1 - u_3) + w_{3-1}$$

$$-190 = u_1 - u_3 - 35$$

or

$$u_1 - u_3 = -190 + 35 = \mathbf{-155\,kJ/kg}$$

Problem 3.28: A stationary system consisting of 2 kg of an ideal gas expands in an adiabatic process according to $pv^{1.2}$ constant. The initial conditions are 1 MPa and 200 °C, and the final pressure is 0.1 MPa. The properties of an ideal gas are given as follows:

$$u = 196 + 0.718\ T$$

$$pv = 0.287(T + 273.15)$$

where

u is the specific internal energy in kJ/kg,
T is the temperature in °C, and.
p is the pressure in kPa.

Determine W and ΔU for the process. Why is the work transfer not equal to ?

Solution: Given data:
    Mass:

$$m = 2 \text{ kg}$$

Adiabatic law follows:

$$pv^{1.2} = C$$

At initial condition:

$$p_1 = 1\text{MPa} = 10^3\text{kPa}$$

$$T_1 = 200\,°\text{C} = (200 + 273)\,\text{K} = 473\,\text{K}$$

At final condition:

$$p_2 = 0.1\text{MPa} = 100\text{kPa}$$

$$u = 196 + 0.718\,T$$

and

$$pv = 0.287(T + 273.15)$$

For adiabatic process,

$$\frac{T_2}{T_1} = \left(\frac{p_2}{p_1}\right)^{\frac{n-1}{n}}$$

$$\frac{T_2}{473} = \left(\frac{100}{1000}\right)^{\frac{1.2-1}{1.2}}$$

$$\frac{T_2}{473} = 0.6813$$

or

$$T_2 = 322.25 \, \text{K} = 49.25 \, ^\circ\text{C}$$

For a quasi-static process,

$$pv^{1.2} = C$$

Work:

$$W_{1-2} = \int_1^2 p\,dV$$

$$= m\left[\frac{p_1 v_1 - p_2 v_2}{n-1}\right]$$

where

$$p_1 v_1 = 0.287(T_1 + 273.15)$$
$$= 0.287(200 + 273.15) = 135.794 \, \text{kJ/kg},$$

$$p_2 v_2 = 0.287(T_2 + 273.15)$$
$$= 0.287(49.25 + 273.15) = 92.528 \, \text{kJ/kg, and}$$

$\therefore$

$$W_{1-2} = 2\left[\frac{135.794 - 92.528}{1.2 - 1}\right] = \mathbf{432.66 \, kJ}.$$

For adiabatic process:

$$pv^{1.2} = C \quad \text{(given)}$$

$$p_1 v_1^{1.2} = p_2 v_2^{1.2}$$

At state 1:

$$p_1 = 10^3 \text{kPa}$$

$$T_1 = 200\,°C$$

$$p_1 v_1 = 0.287(T_1 + 273.15)$$

$$10^3 \times v_1 = 0.287(200 + 273.15)$$

or

$$v_1 = 0.1357\,m^3/kg$$

$\therefore$

$$p_1 v_1^{1.2} = p_2 v_2^{1.2}$$

$$10^3 \times (0.1357)^{1.2} = 100 \times v_2^{1.2}$$

or

$$v_2^{1.2} = 0.9101$$

$$v_2 = (0.9101)^{1/1.2} = 0.9245\,m^3/kg$$

At state 2:

$$p_2 v_2 = 0.287(T_2 + 273.15)$$

$$100 \times 0.9245 = 0.287(T_2 + 273.15)$$

or

$$T_2 = 48.97\,°C$$

Change in internal energy: ΔU

$$
\begin{aligned}
\Delta U = U_2 - U_1 &= mu_2 - mu_1 = m(u_2 - u_1) \\
&= m(196 + 0.718 T_2 - 196 - 0.718 T_1) \\
&= m \times 0.718(T_2 - T_1) \\
&= 2 \times 0.718(48.97 - 200) = -216.87\,kJ
\end{aligned}
$$

The first law of thermodynamics for the process:

$$Q = \Delta U + W$$

But

$$Q = 0 \quad \text{(adiabatic process)}$$

$\therefore$

$$W = -\Delta U = -(-216.87) = 216.87 \text{ kJ}$$

This work is not equal to the $\int p\,dV$ work. Thus, this process is not quasi-static.

Problem 3.29: The properties of an ideal gas are given as follows:

$$u = 196 + 0.718T$$

$$pv = 0.287(T + 273.15)$$

where

u is the specific internal energy in kJ/kg,
T is temperature in °C,
p is pressure in kPa, and.
v is specific volume in m³/kg.

Determine the values of c_v and c_p.

Solution: Given data:

$$u = 196 + 0.718\,T$$

and

$$pv = 0.287(T + 273.15)$$

By definition of specific heat at constant volume: c_v

$$c_v = \left(\frac{du}{dT}\right)_V = \frac{d}{dT}(196 + 0.718\,T)$$
$$= 0 + 0.718 = 0.718 \text{ kJ/kgK}$$

By definition of specific enthalpy:

$$h = u + pv$$

$$h = 196 + 0.718\,T + 0.287(T + 273.15)$$

$$h = 196 + 0.718\,T + 0.287\,T + 78.39$$

$$h = 274.39 + 1.005\,T$$

By definition of specific heat at constant pressure: c_p

$$c_p = \left(\frac{dh}{dT}\right)_p = \frac{d}{dT}(274.39 + 1.005\,T)$$
$$= 0 + 1.005 = \mathbf{1.005\ kJ/kgK}$$

Problem 3.30: 8 kg gas expands in a cylinder-piston device from 1000 kPa, 1 m³ to 5 kPa according to $pv^{1.2} = $ constant. If the specific internal energy of the gas decreases by 40 kJ/kg, determine the heat transfer in magnitude and direction.

Solution: Given data:
 Mass of gas:

$$m = 8\ \text{kg}$$

At initial state:

$$p_1 = 1000\ \text{kPa}$$

$$V_1 = 1\ \text{m}^3$$

At final state:
Pressure:

$$p_2 = 5\ \text{kPa}$$

Change in specific internal energy:

$$\Delta u = -40\ \text{kJ/kg}$$

Change in internal energy:

$$\Delta U = m\,\Delta u = 8 \times (-40) = -320\ \text{kJ}$$

Work:

$$W_{1-2} = \frac{p_1 V_1}{n-1}\left[1 - \left(\frac{p_2}{p_1}\right)^{\frac{n-1}{n}}\right]$$

$$= \frac{100 \times 1}{1.2-1}\left[1 - \left(\frac{5}{1000}\right)^{\frac{1.2-1}{1.2}}\right]$$

$$= 5000[1 - 0.4136] = 2932 \text{ kJ}$$

The first law for the process:

$$Q_{1-2} = \Delta U + W_{1-2} = -320 + 2932 = \mathbf{2612 \text{ kJ}}$$

The +ve sign indicates the heat transfer to the system.

Problem 3.31: A piston-cylinder arrangement contains 0.05 m³ of a gas initially at 200 kPa. At this state, a linear spring that has a spring constant of 150 kN/m is touching the piston but exerting no force on it. Now the heat is transferred to the gas, causing the piston to rise and to compress the spring until the volume inside the cylinder doubles. If the cross-sectional area of the piston is 0.25 m², determine

(a) the final pressure inside the cylinder,
(b) the total work done by the gas, and
(c) the work done against the spring to compress it.

Solution: Given data:
At initial condition:

$$V_1 = 0.05 \text{ m}^3$$

$$p_1 = 200 \text{ kPa}$$

Spring constant:

$$K = 150 \text{ kN/m}^2$$

At final condition:

$$V_2 = 2V_1 = 2 \times 0.05 = 0.1 \text{ m}^3$$

Cross-sectional area of the piston:

$$A = 0.25 \text{ m}^2$$

Let dx be the small displacement of the piston, dp be the small change in pressure, and dV be the small change in volume between states 1 and 2 when heat is supplied to the gas (Fig. 3.27).

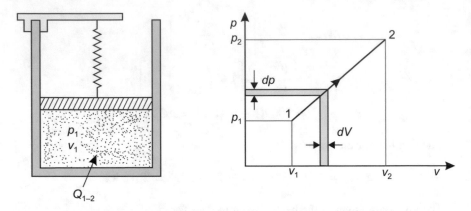

Fig. 3.27 Schematic and p-V diagram for Problem 3.31

At equilibrium condition.

Force applied by the gas on the piston $=$ force applied by the spring on the piston.

Change in pressure $\times$ area of the piston $=$ spring constant $\times$ displacement

$$dp\,A = K\,dx$$

$$dp = K\frac{dx}{A}$$

$$dp = K\frac{dx \cdot A}{A^2} \quad |\because dV = A\,dx$$

$$dp = \frac{K\,dV}{A^2} \tag{3.27}$$

Integrating between states 1 and 2 gives

$$\int_1^2 dp = \frac{K}{A^2}\int_1^2 dV$$

$$p_2 - p_1 = \frac{K}{A^2}(V_2 - V_1) \tag{3.28}$$

(a)   Final pressure inside the cylinder: p_2

$$p_2 - p_1 = \frac{K}{A^2}(V_2 - V_1)$$

$$p_2 - 200 = \frac{150}{(0.25)^2}(0.1 - 0.05)$$

$$p_2 - 200 = 120$$

or

$$p_2 = 200 + 120 = \textbf{320 kPa}$$

(b)   Total work done by the gas: W_{1-2}

We know that the work done:

$$\delta W = pdV$$

From Eq. (3.27),

$$dV = \frac{A^2}{K}dp.$$

∴

$$\delta W = \frac{A^2}{K}pdp$$

Integrating between states 1 and 2 gives

$$\int_1^2 \delta W = \frac{A^2}{K}\int_1^2 pdp$$

$$W_{1-2} = \frac{A^2}{K}\left[\frac{p^2}{2}\right]_1^2$$

$$W_{1-2} = \frac{A^2}{K}\left[\frac{p_2^2 - p_1^2}{2}\right]$$

$$= \frac{A^2}{2K}(p_2 - p_1)(p_2 + p_1)$$

From Eq. (3.28), we have $p_2 - p_1 = \frac{K}{A^2}(V_2 - V_1)$.

or

$$(p_2 - p_1)\frac{A^2}{K} = V_2 - V_1$$

Fig. 3.28 p-V diagram for
Problem 3.31

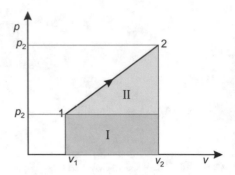

$\therefore$

$$W_{1-2} = \frac{(p_2 + p_1)}{2}(V_2 - V_1)$$

$$= \text{mean pressure during the process} \times \text{change in volume}$$

$$= \frac{(320 + 200)}{2}(0.1 - 0.05) = \mathbf{13 \ kJ}$$

(c)   Work done against the spring: W_S

Region I (rectangular area) represents the work done against the piston and atmosphere, and region II (triangular area) represents the work done against the spring (Fig. 3.28).

Thus,

$$W_S = \text{area of region II}$$

$$= \frac{1}{2}(V_2 - V_1)(p_2 - p_1)$$

$$= \frac{1}{2}(0.1 - 0.05)(320 - 200) = \mathbf{3 \ kJ}$$

Problem 3.32: An imaginary engine receives heat and does work on a slowly moving piston at such a rate that the cycle of operation of 1 kg of working fluid is represented as a circle 100 mm in diameter on the p–v plane on which 1 mm = 30 kPa and 1 mm = 0.01 m^3/kg (Fig. 3.29).

(a)   How much work is done by each kg of working of an ideal gas for each cycle of operation?

(b)   If the heat rejected by an engine in a cycle is 1000 kJ/kg, what would be its thermal efficiency?

Solution: Given data:
   Diameter of cycle:

Fig. 3.29 p-V diagram for Problem 3.32

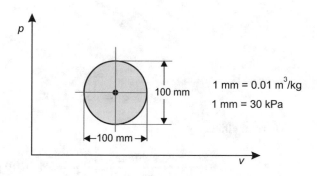

$$d = 100 \text{ mm}$$

Scales:

$$1 \text{ mm} = 30 \text{ kPa},$$

$$1 \text{ mm} = 0.01 \text{ m}^3/\text{kg}$$

(a) Work done:

$$w = \text{area under the cycle}$$
$$= \frac{\pi}{4}d^2 = \frac{3.14}{4} \times d \times d$$
$$= \frac{3.14}{4} \times 100 \text{ mm}_{(x)} \times 100 \text{ mm}_{(y)}$$
$$= \frac{3.14}{4} \times 100 \times 0.01 \times 100 \times 30$$
$$= \mathbf{2355 \text{ kJ/kg}}$$

(b) Heat rejection:

$$q_2 = 1000 \text{ kJ/kg}$$

Work output:

$$w = \text{heat addition} - \text{heat rejection}$$

$$2355 = q_1 - q_2$$

$$2355 = q_1 - 1000$$

or

$$q_1 = 3355 \text{ kJ/kg}$$

Thermal efficiency:

$$\eta_{th} = \frac{\text{Work output}}{\text{Heat supplied}} = \frac{2355}{3355} = 0.7019 = \textbf{70.19\%}$$

Problem 3.33: An insulated rigid tank is divided into two equal parts by a partition. Initially, one part contains nitrogen gas at 2 bar pressure and 20 °C temperature, and the other part contains nitrogen gas at 3.5 bar pressure and 35 °C temperature. The partition is now removed. Analyse the system on the basis of the first law of thermodynamics (Fig. 3.30).

Solution: According to the first law of thermodynamic,

$$Q = \Delta U + W$$

where

$$Q = 0 \quad \text{because of insulated tank,}$$

$$W = 0 \quad \text{because of free expansion,}$$

∴

$$0 = \Delta U + 0$$

Fig. 3.30 Schematic for Problem 3.33

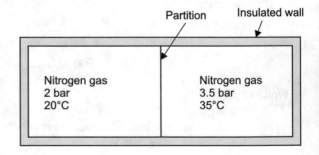

or

$$\Delta U = 0,$$

$$U_2 - U_1 = 0$$

$$U_1 = U_2.$$

Thus, the **internal energy of nitrogen is conserved**.

Problem 3.34: An insulated rigid tank is divided into two equal parts by a partition. Initially, one part contains 4 kg of an ideal gas at 800 kPa and 50 °C, and the other part is evacuated. The partition is now removed, and the gas expands in the entire tank. Determine the final temperature and pressure in the tank (Fig. 3.31).

Solution: Let V be the volume of each part.

$\therefore$

$$\text{Total volume of the tank} = 2V.$$

Given data for one part:

$$m = 4 \text{ kg},$$

$$p_1 = 800 \text{ kPa},$$

$$T_1 = 50 \,^\circ\text{C, and}$$

$$V_1 = V.$$

According to the first law of thermodynamics,

$$Q = dU + W$$

Fig. 3.31 Schematic for Problem 3.34

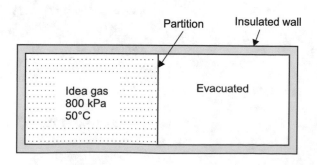

where

$$Q = 0 \quad \text{because of insulated tank,}$$

$$W = 0 \quad \text{because of free expansion,}$$

$\therefore$

$$dU = 0,$$

$$U_2 - U_1 = 0$$

or

$$U_1 = U_2, \text{ and}$$

$$mc_v T_1 = mc_v T_2$$

or

$$T_2 = T_1 = \mathbf{50\,^\circ C}.$$

From the ideal gas equation for state,

$$pV = mRT$$

$$\frac{pV}{T} = \text{constant}$$

For process 1–2,

$$\frac{p_1 V_1}{T_1} = \frac{p_2 V_2}{T_2}$$

$$p_1 V_1 = p_2 V_2 \quad |\because T_1 = T_2$$

$$800 \times V = p_2 \times 2V$$

or

$$p_2 = \mathbf{400\ kPa}$$

Problem 3.35: A metal block of 5 kg and temperature 200 °C is submerged into water whose mass is 8 kg and temperature is 30 °C. If the specific heat of the metal is 0.2 kJ/kgK, what will be the final temperature of the system?

Solution: Given data:
   Mass of metal block:

$$m_M = 5 \text{ kg}$$

Initial temperature of metal:

$$T_{Mi} = 200\,°C$$

Specific heat of metal:

$$c_M = 0.2 \text{ kJ/kgK}$$

Mass of water:

$$m_W = 8 \text{ kg}$$

Initial temperature of water:

$$T_{Wi} = 30\,°C$$

Let

$$T_f = \text{final temperature of the system and metal block in} °C$$

Now applying energy balance equation,

heat lost by the metal block = heat gained by the water

$$m_M c_M (T_{Mi} - T_f) = m_W c_W (T_f - T_{Wi})$$

$$5 \times 0.2 \times (200 - T_f) = 8 \times 4.2 \times (T_f - 30)$$

$$200 - T_f = 33.6(T_f - 30)$$

$$200 - T_f = 33.6 T_f - 1008$$

or

$$33.6T_f + T_f = 200 + 1008$$

$$34.6T_f = 1208$$

or

$$T_f = \mathbf{34.91\,^{\circ}C}$$

Problem 3.36: The specific heats of a gas are of the form $c_p = a + kT$ and $c_v = b + kT$, where a, b, and k are constants, and T is in K. Derive the formula $T^b v^{a-b} e^{kT} = $ constant, for adiabatic expansion of the gas.

Solution:

$$c_p = a + kT$$

$$c_v = b + kT,$$

Gas constant:

$$R = c_p - c_v = a + kT - b - kT = a - b$$

The first law for the process,

$$\delta q = du + \delta W$$

where

$$\delta q = 0 \text{ for adiabatic expansion,}$$

$$du = c_v dT,$$

$$\delta w = pdv,$$

$\therefore$

$$0 = c_v dT + pdv$$

$$0 = (b + kT)dT + RT\frac{dv}{v} \quad \left| \begin{array}{l} pv = RT \\ \text{or } p = \dfrac{RT}{v} \end{array} \right.$$

or

$$(b + kT)dT + RT\,RT\frac{dv}{v} = 0.$$

Dividing by T, we get

$$\left(\frac{b}{T} + k\right)dT + R\frac{dv}{v} = 0$$

$$\frac{bdT}{T} + kdT + (a - b)\frac{dv}{v} = 0$$

On integration, we get

$$b \log_e T + kT + (a - b) \log_e v = C$$

$$\log_e T^b + \log_e e^{kT} + \log_e v^{(a-b)} = C$$

or

$$\log_e T^b e^{kT} v^{(a-b)} = C$$

or

$$T^b e^{kT} v^{a-b} = e^C$$

or

$$T^b v^{a-b} e^{kT} = \text{constant}$$

Summary

1. **First Law of Thermodynamics is called the Law of Conservation of Energy**. According to this law, energy can neither be created nor destroyed, but can be converted from one form to another. It introduces the two properties of the system, the internal energy U and the enthalpy H.

2. **First Law for Process**:

$$\delta Q = dU + dW$$

For process 1–2,

$$Q_{1-2} = (U_2 - U_1) + W_{1-2}$$

For unit mass,

$$q_{1-2} = (u_2 - u_1) + w_{1-2}$$

3.   **First Law for Cyclic Process**:

$$\Sigma \delta Q = \Sigma \delta W$$

OR

$$\oint \delta Q = \oint \delta W$$

and

$$\oint dU = 0$$

4.   A perpetual motion machine of the first kind is an imaginary machine which produces work continuously without any input (no energy is supplied to it). PMM I violates the first law of thermodynamics. It is an impossible machine.

5.   **Enthalpy**. It is defined as the sum of the internal energy and product of pressure and volume. It is denoted by H.

Mathematically,
Enthalpy:

$$H = U + pV$$

Specific enthalpy:

$$h = u + pv$$

For a perfect gas, the change in enthalpy is defined as the heat addition at constant pressure.

$$dH = mc_p dT$$

6.   **Internal Energy**. The energy stored in the molecular structure of a system is called internal energy or microscopic energy (or hidden energy). For an ideal gas, the internal energy is only a function of absolute temperature, and its value is zero at the absolute zero temperature.

For an ideal gas, the change in internal energy is defined as the heat addition at constant volume.

Mathematically,

$$dU = mc_v dT$$

For unit mass,

Change in specific internal energy:

$$du = c_v dT$$

7. **Specific Heat**. Specific heat is the property of the substance. It is defined as the amount of heat that is required to raise the unit temperature of the unit mass of a substance.

    Mathematically,

    Heat transfer:

    $$Q = mcdT$$

    where

    m = mass of the substance in kg,
    c = specific heat of the substance in kJ/kgK, and
    dT = change in temperature in °C or K.

    Liquids and solids have one specific heat value, but gases have two different types of specific heats:

    (i)   Specific heat at constant pressure: c_p;
    (ii) Specific heat at constant volume: c_v;
    (i)   Specific heat at constant pressure: c_p.

    It is defined as the amount of heat that is required to raise the unit temperature of the unit mass of the gas at constant pressure.

    (ii) Specific heat at constant volume: c_v.

    It is defined as the amount of heat that is required to raise the unit temperature of the unit mass of the gas at constant volume.

8. **Heat or Thermal Capacity: C**

    Heat capacity is defined as the product of mass and specific heat. It is denoted by the capital letter C.

    Mathematically,

    Heat capacity:

    $$C = \text{mass} \times \text{specific heat} = mc$$

9. **Relation among c_p, c_v, and R:**

    $$c_p - c_v = R$$

10.   **Relation among c_p, R, and γ:**

$$c_p = \frac{\gamma R}{\gamma - 1}$$

11.   **Relation among c_v, R, and γ:**

$$c_v = \frac{R}{\gamma - 1}$$

12.   **Non-flow Process.** A process occurring in a closed system in which the energy crosses the boundary of the system in the form of heat and work, but there is no mass flow to the system or from the system.

13.   **Flow Process.** A process occurring in an open system in which both mass and energy are transferred to and from the system.

14.   **Work done during a Non-flow Process:**

$$\delta W = pdV$$

Work done between the two states 1 and 2 is

$$W_{1-2} = \int_{1}^{2} pdV$$

15.   **Constant Volume Process (Isometric or Isochoric Process).** This process is governed by Gay-Lussac's law: $\frac{p}{T} = C$

(a) p–v–T relationship:

For process 1–2,

$$\frac{T_2}{T_1} = \frac{p_2}{p_1}$$

(b)   Work done:

$$W_{1-2} = 0$$

(c)   Change in internal energy:

$$dU = mc_v dT$$

(d) Heat transfer:

$$Q_{1-2} = mc_v(T_2 - T_1)$$

(e) Change in enthalpy:

$$dH = mc_p dT$$

16. **Constant Pressure Process (Isobaric Process)**. This process is governed by Charles' law, $\frac{v}{T} = C$.

(a) *p–v–T* relationship: For process 1–2,

$$\frac{T_2}{T_1} = \frac{v_2}{v_1}$$

(b) Work done:

$$W_{1-2} = p(V_2 - V_1) = mR(T_2 - T_1)$$

(c) Change in internal energy:

$$dU = mc_v dT$$

(d) Heat transfer:

$$Q_{1-2} = mc_p(T_2 - T_1)$$

(e) Change in enthalpy:

$$dH = mc_p dT$$

17. **Constant Temperature Process (isothermal process).** This process is governed by Boyle's law, $pv = C$.

(a) p–v–T relationship:

For process 1–2,

$$\frac{p_2}{p_1} = \frac{v_1}{v_2}$$

(b) Work done:

$$W_{1-2} = p_1 V_1 \log_e \frac{V_2}{V_1} = mRT \log_e \frac{V_2}{V_1}$$

(c) Change in internal energy:

$$dU = 0$$

i.e., for process 1–2,

$$U_1 = U_2$$

Internal energy:

$$U = f(T) \quad \text{for ideal gas.}$$

(d) Heat transfer:

$$Q_{1-2} = W_{1-2} = p_1 V_1 \log_e \frac{V_2}{V_1} = mRT_1 \log_e \frac{V_2}{V_1}$$

(e) Change in enthalpy:

$$dH = 0$$

That is, for process 1–2,

$$H_1 = H_2$$

Enthalpy:

$$H = f(T) \quad \text{for ideal gas.}$$

18. **Adiabatic Isentropic Process (Frictionless and Adiabatic Process)**
    This process is followed by the law $pv^{\gamma} = C$.

    (a) *p–v–T* relationship:

    $$\frac{T_2}{T_1} = \left(\frac{p_2}{p_1}\right)^{\frac{\gamma-1}{\gamma}} = \left(\frac{v_1}{v_2}\right)^{\gamma-1}$$

    (i) $\frac{T_2}{T_1} = \left(\frac{p_2}{p_1}\right)^{\frac{\gamma-1}{\gamma}}$

    (ii) $\frac{T_2}{T_1} = \left(\frac{v_1}{v_2}\right)^{\gamma-1}$

    (iii) $\frac{p_2}{p_1} = \left(\frac{v_1}{v_2}\right)^{\gamma}$

    (b) Work done:

    $$W_{1-2} = \frac{p_1 V_1 - p_2 V_2}{\gamma - 1} = \frac{mR(T_1 - T_2)}{\gamma - 1}$$

    (c) Change in internal energy:

    $$U_2 - U_1 = mc_v(T_2 - T_1)$$

    (d) Heat transfer:

    $$Q_{1-2} = 0$$

    (e) Change in enthalpy:

    $$dH = mc_p dT$$

19. **Polytropic Process**:
    This process follows the law $pv^n = C$.

    (a) *p–v–T* relationship:

$$\frac{T_2}{T_1} = \left(\frac{p_2}{p_1}\right)^{\frac{n-1}{n}} = \left(\frac{v_1}{v_2}\right)^{n-1}$$

(i) $\frac{T_2}{T_1} = \left(\frac{p_2}{p_1}\right)^{\frac{n-1}{n}}$

(ii) $\frac{T_2}{T_1} = \left(\frac{v_1}{v_2}\right)^{n-1}$

(iii) $\frac{p_2}{p_1} = \left(\frac{v_1}{v_2}\right)^{n}$

(b)   Work done:

$$W_{1-2} = \frac{p_1 V_1 - p_2 V_2}{n-1} = \frac{mR(T_1 - T_2)}{n-1}$$

(c)   Change in internal energy:

$$U_2 - U_1 = mc_v(T_2 - T_1)$$
$$= \left(\frac{1-n}{\gamma-1}\right) \times W_{1-2}$$

(d)   Heat transfer: Q_{1-2}

$$Q_{1-2} = \left(\frac{\gamma-n}{\gamma-1}\right) \times W_{1-2}$$
$$= \left(\frac{\gamma-n}{1-n}\right) \times \text{change in internal energy}$$

(e)   Change in enthalpy:

$$H_2 - H_1 = mc_p(T_2 - T_1)$$

20.   **Free expansion process**:
      The following are some characteristics of the free expansion process observed:

      (i)   Heat transfer is zero, i.e., $Q_{1-2} = 0$, because the system is insulated.
      (ii)   Work done is zero, i.e., $W_{1-2} = 0$, because of free expansion.

(iii)   The temperature of the free expansion process remains constant,

i.e.,

$$T_1 = T_2$$

Assignment–1

1. Define the first law of thermodynamics.
2. State the first law of thermodynamics for a cycle and a process.
3. Define internal energy. Why does the internal energy of an ideal gas depend only on temperature?
4. Name and state the property by the first law of thermodynamics.
5. What is a PMM I? Why it is impossible?
6. Under what condition is the work done is equal to $\int_1^2 p\,dV$?
7. Show that the reversible adiabatic process is represented by $pV^\gamma = $ constant.
8. Show that internal energy is a property of a system.
9. Derive an expansion for heat transfer in

    (a)   Polytropic process,
    (b)   Isobaric process, and
    (c)   Isothermal process.

10. Show that for a polytropic process

$$Q = \left(\frac{\gamma - n}{\gamma - 1}\right) \times W$$

where

    Q and W are heat and work interactions.
    γ is adiabatic index.
    n is polytropic index.

11. **Explain the following**:

    (a)   Polytropic process,
    (b)   Isometric process, and
    (c)   Isothermal process.
          Also obtain the expression for work in each case.

12. Show that when an ideal gas is compressed in a piston-cylinder arrangement according to the polytropic law $pV^n = $ constant, the heat rejected, increase in internal energy, and work done in the ratio.

$$(\gamma - n) : (1 - n) : (\gamma - 1)$$

Assignment–2

1. In a cyclic process, heat transfers are $+16.7$, -25.2, -3.56, and $+31.5$ kJ. What is the net work for this cyclic process? [**Ans.** 1944 kJ]

2. A domestic refrigerator is loaded with food and the door is closed. During a certain period, the machine consumes 1 kWh of energy, and the internal energy of the system drops by 6000 kJ. Find the net heat transferred in the system. [**Ans.** -9600 kJ]

3. A piston-cylinder device contains 0.2 kg of air at 100 kPa and 30 °C. The piston is moved by compressing air until the pressure becomes 1 MPa and the temperature becomes 150 °C. The work done during the process is 20 kJ. Determine the heat transferred from the air to the surrounding. Take $c_v = 0.718$ kJ/kg K. [**Ans.** -2.768 kJ]

4. A tank containing air is stirred by a paddle wheel. The work input to the paddle wheel is 6000 kJ, and the heat transferred to the surrounding from the tank is 2000 kJ. Determine the work done and change in the internal energy of the system. [**Ans.** 0, 4000 kJ]

5. A rigid tank contains air at 600 kPa and 200 °C. As a result of heat transfer to the surroundings, the temperature and pressure inside the tank drop to 70 °C and 300 kPa, respectively. Determine the work done during this process. [**Ans.** 0]

6. A frictionless piston-cylinder device contains 0.4 m³ of air at 100 kPa and 80 °C. The air is now compressed to 0.1 m³ in such a way that the temperature inside the cylinder remains constant. Determine the work done during this process. [**Ans.** -55.5 kJ]

7. An air compressor takes in air at 10^5 N/m² and 27 °C having a specific volume of 1.5 m³/kg which is compressed to 4.5×10^5 N/m². Find the specific work done, specific heat transfer, and change in specific internal energy if the compression is isothermal.
    [**Ans.** $w_{1-2} = -225.61$ kJ/kg, $q_{1-2} = -225.61$ kJ/kg, and $du = 0$]

8. When a system is taken from state l to state m along path lqm, 168 kJ of heat flows into the system, and the system does 64 kJ of work:

    (i)   How much will be the heat that flows into the system along path lnm if the work done is 21 kJ?

    (ii) When the system is returned from m to l along the curved path, the work done on the system is 42 kJ. Does the system absorb or liberate heat, and how much of the heat is absorbed or liberated?

    (iii) If $U_l = 0$ and $U_n = 84$ kJ, find the heat absorbed in the processes ln and nm.
          [**Ans.** (i) 125 kJ, (ii) -146 kJ, and (iii) 105 kJ, 20 kJ]

9. A fluid system, contained in a piston and cylinder machine, passes through a complete cycle of four processes. The sum of all heat transferred during a cycle is -340 kJ. The system completes 200 cycles per min.
    Complete the following table showing the method for each item, and compute the net rate of work output in kW (Fig. 3.32).

Fig. 3.32 *p-V* diagram for Q9

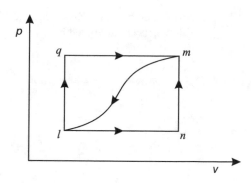

Process	Q (kJ/min)	W (kJ/min)	ΔE (kJ/min)
1–2	0	4340	–
2–3	42000	0	–
3–4	–4200	–	–73200
4–1	–	–	–

[**Ans.** The completed table is given as follows:

Process	Q (kJ/min)	W (kJ/min)	ΔE (kJ/min)
1–2	0	4340	–4340
2–3	42000	0	42000
3–4	–4200	69000	–73200
4–1	–105800	–141340	35540

Net rate of work output $= -1133.33$ kW

10. A closed system constant volume experiences a temperature rise of 25 °C when a certain process occurs. The heat transfer in the process is 30 kJ. The specific heat at constant volume for the pure substance comprising the system is 1.2 kJ/kg°C, and the system contains 2.5 kg of this substance. Determine

   (i) The change in internal energy and
   (ii) The work done. [**Ans.** (i) 75 kJ and (ii) –45 kJ]

11. A system receives 50 kJ of heat while expanding with a volume change of 0.14 m³ against an atmosphere of 1.2×10^5 N/m². A mass of 90 kg in the surroundings is also lifted through a distance of 5.5 m.

   (i) Find the change in energy of the system.
   (ii) The system is returned to its initial volume by an adiabatic process which requires 110 kJ of work. Find the change in energy of the system.
   (iii) For the combined processes of (i) and (ii), determine the change in energy of the system. [**Ans.** (i) 28.35 kJ, (ii) 110 kJ, and (iii) 138.35 kJ]

12. The work and heat per degree change of temperature for a system executing a non-flow process are given by

$$\frac{dW}{dT} = 100 \text{ J/K}$$

and

$$\frac{\delta Q}{dT} = 80 \text{ J/K}$$

Determine the change of internal energy of a system when its temperature increases from 50 to 100 °C. [**Ans.** 1000 J]

13. A fluid system undergoes a non-flow frictionless process following the pressure–volume relation as $p = \frac{5}{V} + 1.5$ where p is in bar and V is in m^3. During the process, the volume changes from 0.15 m^3 to 0.05 m^3, and the system rejects 45 kJ of heat. Determine

   (i)   Change in internal energy and
   (ii) Change in enthalpy. [**Ans.** (i) 519 kJ and (ii) 504 kJ]

14. The properties of a system, during a reversible constant pressure non-flow process at $p = 1.6$ bar, changed from $v_1 = 0.3$ m^3/kg, $T_1 = 20$ °C to $v_2 = 0.55$ m^3/kg, $T_2 = 260$ °C. The specific heat of the fluid is given by

$$c_p = \left(1.5 + \frac{75}{T + 45}\right) \text{kJ/kg\,°C, where } T \text{ is in °C.}$$

   Determine

   (i)   Specific heat added,
   (ii)   Specific work done,
   (iii) Change in specific internal energy, and
   (iv)   Change in specific enthalpy.
          [**Ans.** (i) 475.94 kJ/kg, (ii) 40 kJ/kg, (iii) 435.94 kJ/kg, and (iv) 475.94 kJ/kg]

15. A fluid is contained in a cylinder by a spring-loaded, frictionless piston so that the pressure in the fluid is a linear function of the volume ($p = a + bV$). The internal energy of the fluid is given by the following equation:

$$U = 42 + 3.6pV$$

where U is in kJ, p in kPa, and V in m^3. If the fluid changes from an initial state of 190 kPa, 0.035 m^3 to a final state of 420 kPa, 0.07 m^3, with no work other than that done on the piston, find the direction and magnitude of the work and heat transfer.

[**Ans.** $W_{1-2} = 10.67$ kJ (work done by the system) and

$Q_{1-2} = 92.57$ kJ (heat supplied to the system)]

16.   A cylinder contains 0.45 m^3 of a gas at 1×10^5 N/m^2 and 80 °C. The gas is compressed to a volume of 0.13 m^3, the final pressure being 5×10^5 N/m^2. Determine

   (i)   The mass of gas,
   (ii)   The value of index 'n' for compression,
   (iii) The increase in internal energy of the gas, and
   (iv)   The heat received or rejected by the gas during compression.

   Take $\gamma = 1.4$ and $R = 294.2$ J/kgK.
   [**Ans.** (i) 0.433 kg, (ii) 1.296, (iii) 49.90 kJ, and (iv) -17.54 kJ]

17.   Nitrogen gas at 1 bar and 27 °C is compressed adiabatically up to 10 bar and then expanded isothermally up to initial specific volume and then cooled at constant volume to initial conditions. Find work, heat, and change in internal energy per kg of nitrogen for each process and for the entire processes.
   [**Ans.** $w_{1-2} = -207.2$ kJ/kg, $u_2 - u_1 = 207.6$ kJ/kg, $q_{1-2} = 0$,

   $w_{2-3} = 283.40$ kJ/kg, $u_3 - u_2 = 0$, $q_{2-3} = 283.40$ kJ/kg

   $w_{3-1} = 0$, $u_1 - u_3 = -207.6$ kJ/kg, $q_{3-1} = -207.6$ kJ/kg]

18.   A heat engine cycle is represented by a circle of 4 cm diameter on a T–s plane whose centre lies on a 500 K temperature line. The temperature scale is 1 cm $= 100$ K and entropy scale is 1 cm $= 0.1$ kJ/K. If the engine completes 10 cycles per second, what is the power developed by the engine? Also find out heat to be supplied to the engine and efficiency of the engine?
   [**Ans.** $P = 1256$ kW, $Q_{\text{supplied}} = 262.8$ kJ/cycle, and $\eta = 47.79\%$]

19.   One kg of air at 1 bar and 300 K is compressed adiabatically till its pressure becomes 5 times the original pressure. Then it is expanded at constant pressure and finally cooled at constant volume to return to its original conditions. Calculate (i) heat transfer, (ii) work transfer, and (iii) internal energy for each process and for the cycle.
   Represent the processes on p–v and T–s diagrams.
   [**Ans.** $Q_{1-2} = 0$, $W_{1-2} = -125.93$ kJ, $U_2 - U_1 = 125.93$ kJ

   $Q_{2-3} = 1031.81$ kJ, $W_{2-3} = 294$ kJ, $U_3 - U_2 = -737.81$ kJ,

   $Q_{3-1} = -863.75$ kJ, $W_{2-3} = 0$, $U_1 - U_3 = -863.75$ kJ, and

   $\oint Q = \oint W = 168.06$ kJ]

20.   A system receives 200 kJ of heat at a constant volume process and rejects 220 kJ at constant pressure during which 40 kJ of work is done on the system. The system is brought to its original state by an adiabatic process. Calculate the adiabatic work. If the initial internal energy is 240 kJ, calculate the internal energy at all points.
   [**Ans.** $W_{3-1} = 20$ kJ, $U_2 = 440$ kJ, and $U_3 = 260$ kJ]

Chapter 4
Application of First Law of Thermodynamics to Flow Processes Thermodynamics

Nomenclature

The following list of nomenclature is introduced in this chapter:

V	m/s	Velocity
p	kPa	Pressure
ρ	kg/m^3	Density
T	K	Temperature
$\dot{W}$	N/s	Weight flow rate
m	kg/s	Mass flow rate
Q	m^3/s	Discharge or volume flow rate
A	m^2	Cross-sectional area
v	m^3/kg	Specific volume
SFEE	–	Steady flow energy equation
CV	–	Control volume
CS	–	Control surface
z	m	Potential or datum head
Q	kJ	Heat transfer
q	kJ/kg	Specific heat transfer
W	kJ	Work transfer
w	kJ/kg	Specific work transfer
u	kJ/kg	Specific internal energy
h	kJ/kg	Specific enthalpy
KE	kJ	Kinetic energy
PE	kJ	Potential or datum energy
g	m/s^2	Acceleration due to gravity
H	kJ	Enthalpy

(continued)

© The Author(s) 2022
S. Kumar, *Thermal Engineering Volume 1*,
https://doi.org/10.1007/978-3-030-67274-4_4

E	kJ	Energy
γ	–	Adiabatic index
R	kJ/kgK	Gas constant
n	–	Polytropic index
M	–	Mach number
c_p	kJ/kgK	Specific heat at constant pressure
$UFEE$	–	Unsteady flow energy equation

4.1 Introduction

In the previous chapter, we studied the first law of thermodynamics and its application to non-flow process (closed system). In this chapter, we will derive the steady flow energy equation by application of the first law of thermodynamics to flow process. The derived steady flow energy equation is then applied for analysis of engineering systems. At last, we will discuss the unsteady flow energy equation and its application.

4.2 Steady and Unsteady Flow

A flow is considered to be steady if fluid flow parameters such as mass flow rate (m), velocity (V), pressure (p), density (ρ), and temperature (T) at any point do not change with time. If any one of these parameters changes with time, the flow is said to be unsteady flow.

4.3 Compressible and Incompressible Flow

If the density of the fluid changes from point to point in the direction of fluid flow, it is referred to as compressible flow.

Mathematically,

$$\text{Density} : \rho \neq C$$

For example: Gases (like air, carbon dioxide, etc.) are compressible in nature.

If the density of the fluid remains constant at every point in the fluid flow, it is referred to as an incompressible flow.

Mathematically,

$$\text{Density} : \rho = C$$

For example: Liquids are generally incompressible in nature.

4.4 Rate of Flow

The quantity of the fluid flowing per unit time through a cross-section of flow is called the **rate of flow**. The rate of flow is expressed as the weight of the fluid flowing per second across the section, which is called **weight flow rate**. It is denoted by $\dot{W}$ and its unit is N/s.

The rate of flow is expressed as the mass of the fluid flowing across the section per second, which is called **mass flow rate**. It is denoted by m and its unit is kg/s.

The rate of flow is expressed as the volume of the fluid flowing across the section per second, which is called **volume flow rate** or commonly known as **discharge**. It is denoted by Q and its unit is m³/s.

Mathematically,

$$\text{Discharge} : \boldsymbol{Q = AV = mv}$$

where
$A =$ cross-sectional area; m².
$V =$ average velocity of fluid across the section; m/s.
$m =$ mass flow rate; kg/s.
$v =$ specific volume; m³/kg.

or

$$AV = mv$$

or

$$\boldsymbol{m = \frac{AV}{v}} \tag{4.1}$$

$$\boldsymbol{m = \rho\,AV} \quad \left| \quad \because v = \frac{1}{\rho} \right.$$

4.5 Continuity Equation

The equation based on the **law of conservation of mass** is called **continuity equation**. It means that the mass of fluid can neither be created nor destroyed. If there is no accumulation of mass within the control volume, the mass flow rate entering the system must equal the mass flow rate leaving the system (Fig. 4.1).

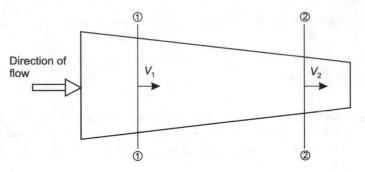

Fig. 4.1 Flow through conical pipe

According to law of conservation of mass,

Mass flow rate at sections (1)–(2) = Mass flow rate at secions (2)–(2)

$$m_1 = m_2$$

From Eq. (4.1),

$$m = \frac{AV}{v}$$

$$m_1 = \frac{A_1 V_1}{v_1} \qquad \text{for sections (1)–(1)}$$

and

$$m_2 = \frac{A_2 V_2}{v_2} \qquad \text{for sections (2)–(2)}$$

$\therefore$

$$\frac{A_1 V_1}{v_1} = \frac{A_2 V_2}{v_2} \tag{4.2}$$

$$\rho_1 A_1 V_1 = \rho_2 A_2 V_2 \qquad \left| \because \rho = \frac{1}{v} \right. \tag{4.3}$$

Equations (4.2) and (4.3) are used for compressible fluid,

where

$$\rho_1 \neq \rho_2 \quad \text{or} \quad v_1 \neq v_2.$$

If fluid is incompressible, then $v_1 = v_2$ or $\rho_1 = \rho_2$, and Eqs. (4.2) and (4.3) are reduced to

$$A_1 V_1 = A_2 V_2$$
$$\rho_1 A_1 V_1 = \rho_2 A_2 V_2 \quad \text{for compressible fluid}$$
$$A_1 V_1 = A_2 V_2 \quad \text{for incompressible fluid}$$

4.6   Steady Flow Energy Equation [SFEE]

Steady flow means the flow in which its properties like $m, p, T, v,$ and V do not change with respect to time at any point in the system.

Consider an open system as shown in Fig. 4.2.

Let p_1, u_1, m_1, v_1, A_1, and V_1 be pressure, specific internal energy, mass flow rate, specific volume, cross-sectional area, and velocity at section (1)–(1) inlet to the system, respectively, and p_2, u_2, m_2, v_2, A_2, and V_2 be pressure, specific internal energy, mass flow rate, specific volume, cross-sectional area, and velocity at section (2)–(2) outlet of the system, respectively.

z_1 and z_2 are datum heads at inlet and exit, respectively.

Q_{1-2} is the rate of heat supplied to the system.

W_{1-2} is the rate of work output from the system.

Note: We assumed both heat and work are positive, i.e., heat supplied to the system is $+ve$ and work output by the system is $+ve$.

The following assumptions are set up to derive the steady flow energy equation:

(i)   Mass flow rate at the inlet and outlet is the same.
(ii)   All the parameters of fluid and flow properties like $p, u, v,$ and V do not change with respect to time at the inlet and outlet of the system.
(iii) Both heat and work do not change with respect to time.

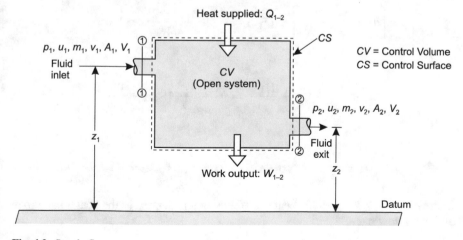

Fig. 4.2 Steady flow system

According to the law of conservation of energy, 'energy can neither be created nor destroyed'.

∴

Net inlet energy to the system = net outlet energy by the system

Energy at section (1)–(1) + heat supplied = energy at section (2)–(2) + work output [Internal energy + flow work + $K.E + P.E$]$_1$ + Q_{1-2} = [Internal energy + flow work + $K.E + P.E$]$_2$ + W_{1-2}.

$$\left[m_1 u_1 + m_1 p_1 v_1 + \frac{1}{2} m_1 V_1^2 + m_1 g z_1 \right] + Q_{1-2} = \left[m_2 u_2 + m_2 p_2 v_2 + \frac{1}{2} m_2 V_2^2 \right.$$
$$\left. + m_2 g z_2 \right] + W_{1-2}$$

$$m_1 \left[u_1 + p_1 v_1 + \frac{1}{2} V_1^2 + g z_1 \right] + Q_{1-2} = m_2 \left[u_2 + p_2 v_2 + \frac{1}{2} V_2^2 + g z_2 \right] + W_1$$

By definition of specific enthalpy,

$$h_1 = u_1 + p_1 v_1 \text{ at inlet}$$
$$h_2 = u_2 + p_2 v_2 \text{ at outlet}$$

and for steady flow, mass flow rate is constant,

i.e.,

$$m_1 = m_2 = m$$

∴

$$m \left[h_1 + \frac{V_1^2}{2} + g z_1 \right] + Q_{1-2} = m \left[h_2 + \frac{V_2^2}{2} + g z_2 \right] + W_{1-2}$$

or

$$h_1 + \frac{V_1^2}{2} + g z_1 + \frac{Q_{1-2}}{m} = h_2 + \frac{V_2^2}{2} + g z_2 + \frac{W_{1-2}}{m}$$

or

$$h_1 + \frac{V_1^2}{2} + g z_2 + q_{1-2} = h_2 + \frac{V_2^2}{2} + g z_2 + w_{1-2} \qquad (4.4)$$

where $q_{1-2} = \frac{Q_{1-2}}{m}$ = specific heat transfer, which is defined as the heat transfer per unit mass flow rate, and $w_{1-2} = \frac{W_{1-2}}{m} =$ specific work, which is defined as the work done per unit of mass flow rate.

Equation (4.4) is called the **steady flow energy equation**. In this equation, the specific enthalpy, kinetic energy, potential energy, specific heat transfer, and specific work done do not change with respect to time in the open system.

4.7 Derive the Steady Flow Energy Equation (SFEE) from First Law of Thermodynamics

From the first law of thermodynamics,
First law for non-flow process,

$$\delta Q = dE + \delta W \tag{4.5}$$

Energy stored in the non-flow process,

$E =$ internal energy + kinetic energy + potential energy
$E = U + K.E. + P.E.$

Total energy stored in the flow process,

$E =$ internal energy + kinetic energy + potential energy + flow energy
$E = U + K.E. + P.E. + pV$
$E = U + pV + K.E. + P.E.$
$E = H + K.E. + P.E.$ $|\because H = U + pV$

Change in energy between the sections 1 and 2 for flow process,

$$dE = dH + d(K.E.) + d(P.E.)$$

We can get the *SFEE* by substituting the value of dE in Eq. (4.5)

$$\delta Q = dH + d(K.E.) + d(P.E.) + \delta W$$

Integrating between states 1 and 2 gives

$$Q_{1-2} = (H_2 - H_1) + \frac{1}{2}m(V_2^2 - V_1^2) + mg(z_2 - z_1) + W_{1-2}$$

For unit mass flow rate,

$$q_{1-2} = (h_2 - h_1) + \frac{1}{2}\left(V_2^2 - V_1^2\right) + g(z_2 - z_1) + w_{1-2}$$

or

$$h_1 + \frac{V_1^2}{2} + gz_1 + q_{1-2} = h_2 + \frac{V_2^2}{2} + gz_2 + w_{1-2}$$

4.8 Derive Euler's Equation and Bernoulli's Equation from Steady Flow Energy Equation

We know that the steady flow energy equation for unit mass

$$h_1 + \frac{V_1^2}{2} + gz_1 + q_{1-2} = h_2 + \frac{V_2^2}{2} + gz_2 + w_{1-2} \tag{4.6}$$

Assumptions of Euler's equation

(i)   fluid is ideal,
(ii) steady and streamline flow,
(iii) incompressible flow, and
(iv) no shaft work and heat interaction.

According to assumption (iv),

$$q_{1-2} = 0, \quad w_{1-2} = 0$$

Equation (4.6) becomes

$$h_1 + \frac{V_1^2}{2} + gz_1 = h_2 + \frac{V_2^2}{2} + gz_2$$

or

$$(h_2 - h_1) + \frac{1}{2}\left(V_2^2 - V_1^2\right) + g(z_2 - z_1) = 0 \tag{4.7}$$

Write the above equation in differential form

$$dh + VdV + gdz = 0$$

By definition of specific enthalpy,

$$h = u + py$$

On differentiating, we get

$$dh = du + pdv + vdp$$
$$dh = \delta q + vdp \qquad |\because \delta q = du + pdv$$
$$dh = vdp \qquad |\because \delta q = 0$$

Substituting $dh = v\,dp$ in Eq. (4.7), we get

$$vdp + VdV + gdz = 0$$

$$\frac{dp}{\rho} + VdV + gdz = 0 \qquad (4.8)$$

Equation (4.8) is known as **Euler's equation**.
Bernoulli's equation is obtained by integration of Euler's equation,

$$\int \frac{dp}{\rho} + \int VdV + \int gdz = \text{constant}$$

$$\frac{p}{\rho} + \frac{V^2}{2} + gz = \text{constant}$$

or

$$\frac{p}{\rho g} + \frac{V^2}{2g} + z = C \qquad (4.9)$$

Equation (4.9) is called **Bernoulli's equation**.

4.9 Flow Work: $W_{1-2} = -\int_1^2 vdp$

We know that the steady flow energy equation

$$m\left[h_1 + \frac{V_1^2}{2} + gz_1\right] + Q_{1\ 2} = m\left[h_2 + \frac{V_2^2}{2} + gz_2\right] + W_{1-2}$$

or

$$Q_{1-2} - W_{1-2} = m[h_2 - h_1] + \frac{m}{2}\left(V_2^2 - V_1^2\right) + mg(z_2 - z_1)$$

For unit mass flow rate,

$$q_{1-2} - w_{1-2} = (h_2 - h_1) + \frac{1}{2}(V_2^2 - V_1^2) + g(z_2 - z_1)$$

Write the above equation in differential form

$$\delta q - \delta w = dh + d(KE) + d(PE) \qquad (4.10)$$

According to the first law of thermodynamics for a closed system,

$$\delta q = du + pdv \qquad (4.11)$$

We know that

$$h = u + pv$$

On differentiating, we get

$$dh = du + pdv + vdp$$

or

$$dh - vdp = du + pdv \qquad (4.12)$$

Equating Eqs. (4.11) and (4.12), we get

$$\delta q = dh - vdp$$

Now substituting the value of δ in Eq. (4.10), we get

$$dh - vdp - \delta w = dh + d(KE) + d(PE)$$

or

$$-vdp - \delta w = d(KE) + d(PE)$$
$$-\delta w = vdp + d(KE) + d(PE)$$
$$\delta w = -vdp - d(KE) - d(PE)$$

If the change in kinetic and potential energies are neglected, i.e., $d(KE) \approx 0$, $d(PE) \approx 0$

$\therefore$

$$\delta w = -vdp$$

Integrating between states 1 and 2 gives (Fig. 4.3)

Fig. 4.3 Steady flow work

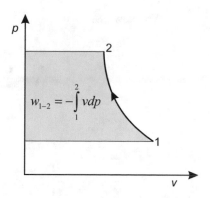

$$\int_1^2 \delta w = -\int_1^2 v dp$$

$$w_{1-2} = -\int_1^2 v dp$$

4.10 Flow Work for Thermodynamics Processes

(i) **Constant volume process (Isochoric): $v = C$**
Flow work:

$$w_{1-2} = -\int_1^2 v dp \text{ neglecting change in kinetic and potential energies}$$
$$= -v[p]_1^2$$
$$= -v(p_2 - p_1)$$
$$= v(p_1 - p_2)$$

(ii) **Constant pressure process (Isobaric): $p = C$**
Flow work:

$$w_{1-2} = -\int_1^2 v dp = 0 \text{ neglecting change in kinetic and potential energies}$$

(iii) **Constant temperature process (Isothermal): $pv = C$**
Flow work:

$$w_{1-2} = -\int_1^2 v dp \text{ neglecting change in kinetic and potential energies}$$

We know that

$$pv = C \quad \text{for isothermal process}$$
$$v = \frac{C}{p}$$

$\therefore$

$$w_{1-2} = -\int_1^2 \frac{C dp}{p}$$
$$= -C\big[\log_e p\big]_1^2$$
$$= -C\big[\log_e p_2 - \log_e p_1\big] = -C \log_e \frac{p_2}{p_1}$$
$$= C \log_e \frac{p_1}{p_2}$$
$$= p_1 v_1 \log_e \frac{p_1}{p_2} = RT_1 \log_e \frac{p_1}{p_2} \quad \big|\because p_1 v_1 = C = RT_1$$
$$= RT_1 \log_e \frac{v_2}{v_1} \quad \left|\begin{array}{l} \because p_1 v_1 = p_2 v_2 \\ \frac{p_1}{p_2} = \frac{v_2}{v_1} \end{array}\right.$$

(iv)   **Adiabatic isentropic process: $pv^\gamma = C$**
Flow work:

$$w_{1-2} = -\int_1^2 v dp \quad \text{neglecting change in kinetic and potential energies}$$

We know that

$$pv^\gamma = C \text{ for adiabatic process}$$
$$v^\gamma = \frac{C}{p}$$
$$v = \frac{C^{1/\gamma}}{p^{1/\gamma}}$$

$\therefore$

$$w_{1-2} = -\int_1^2 \frac{C^{1/\gamma}}{p^{1/\gamma}} dp$$
$$= -C^{1/\gamma} \int_1^2 p^{-1/\gamma} dp$$

$$= -C^{1/\gamma}\left[\frac{p^{-1/\gamma+1}}{-\frac{1}{\gamma}+1}\right]_1^2 = -C^{1/\gamma}\frac{\left[p_2^{-1/\gamma+1} - p_1^{-1/\gamma+1}\right]}{\frac{-1+\gamma}{\gamma}}$$

$$= \frac{-\gamma}{\gamma-1}\left[C^{1/\gamma}p_2^{-\frac{1}{\gamma}+1} - C^{1/\gamma}p_1^{-\frac{1}{\gamma}+1}\right]$$

$$= -\frac{\gamma}{\gamma-1}\left[p_2^{1/\gamma}v_2 p_2^{-\frac{1}{\gamma}+1} - p_1^{1/\gamma}v_1 p_1^{-\frac{1}{\gamma}+1}\right] \quad \left[\because p_1 v_1^{\gamma} = p_2 v_2^{\gamma} = C\right]$$

$$= -\frac{\gamma}{\gamma-1}[p_2 v_2 - p_1 v_1]$$

$$= \frac{\gamma}{\gamma-1}[p_1 v_1 - p_2 v_2]$$

$$= \frac{\gamma}{\gamma-1}[RT_1 - RT_2] \quad \left|\begin{array}{l}\because p_1 v_1 = RT_1 \\ p_2 v_2 = RT_2\end{array}\right.$$

$$= \frac{\gamma}{\gamma-1}R(T_1 - T_2)$$

(v) **Polytropic process: $pv^n = C$**
Flow work:

$$w_{1-2} = -\int_1^2 v\,dp \quad \text{neglecting change in kinetic and potential energies}$$

We calculate the flow work in the polytropic process in a similar manner as the flow work calculation in adiabatic isentropic process. In short, we get flow work for polytropic process by using polytropic index n instead of adiabatic index γ in the flow work for adiabatic isentropic process.

$$\therefore$$

$$w_{1-2} = \frac{n}{n-1}(p_1 v_1 - p_2 v_2) = \frac{n}{n-1}R(T_1 - T_2)$$

Comparison of work done in a non-flow process and flow process is given in Table 4.1.

4.11 Practical Application of Steady Flow Energy Equation (SFEE)

The steady flow energy equation applies to flow processes in many of the engineering systems. Some of the common practical applications of SFEE are discussed in the following devices:

Table 4.1 Comparison of work done in non-flow process and flow process

S. no	Process	Non-flow process $w_{1-2} = \int_1^2 p\,dv$	Flow process $w_{1-2} = -\int_1^2 v\,dp$
1	Constant volume process (Isochoric): $v = C$	0	$v(p_1 - p_2)$
2	Constant pressure process (Isobaric): $p = C$	$p(v_2 - v_1)$	0
3	Constant temperature process (Isothermal): $pv = C$	$p_1 v_1 \log_e \frac{v_2}{v_1}$ OR $p_1 v_1 \log_e \frac{p_1}{p_2}$	$p_1 v_1 \log_e \frac{v_2}{v_1}$ OR $p_1 v_1 \log_e \frac{p_1}{p_2}$
4	Adiabatic isentropic process: $pv^\gamma = C$	$\frac{p_1 v_1 - p_2 v_2}{\gamma - 1}$ OR $\frac{R(T_1 - T_2)}{\gamma}$	$\frac{\gamma}{\gamma - 1}[p_1 v_1 - p_2 v_2]$ OR $\frac{\gamma}{\gamma - 1} R(T_1 - T_2)$
5	Polytropic process: $pv^n = C$	$\frac{p_1 v_1 - p_2 v_2}{n - 1}$ OR $\frac{R(T_1 - T_2)}{n - 1}$	$\frac{n}{n - 1}[p_1 v_1 - p_2 v_2]$ OR $\frac{n}{n - 1} R(T_1 - T_2)$

- (i)   Nozzles
- (ii)   Diffusers
- (iii) Turbines
- (iv)   Compressors
- (v)   Boilers
- (vi)   Heat exchanger
- (vii) Condenser
- (viii) Evaporator
- (ix)   Adiabatic mixing
- (x)   Throttling or wire drawing.

4.11.1   Nozzle

A nozzle is a fluid energy transforming device which increases the velocity (i.e., kinetic energy) of the fluid and simultaneously it decreases the pressure energy of the fluid, i.e., it is used to transfer the pressure energy into kinetic energy. As the fluid flows through the nozzle, it expands to a lower pressure and in this process, pressure falls, and the velocity increases continuously from the entrance to the exit of the nozzle. In case of subsonic flow (Mach number[1] $M < 1$), the nozzle has converging cross-sectional area in the direction of flow whereas in supersonic flow

---

[1] **Mach Number: M**. It is defined as the ratio of the velocity of fluid to the velocity of sound in the same fluid.

Mathematically,

Mach number:

$$M = \frac{V}{a}$$

Fig. 4.4 Converging nozzle

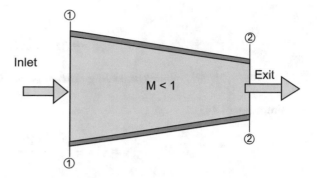

(Mach number $M > 1$) the nozzle has a diverging cross-sectional area in the direction of flow. Let us take a converging nozzle as shown in Fig. 4.4.

We know that the steady flow energy equation

$$m\left[h_1 + \frac{V_1^2}{2} + gz_1\right] + Q_{1-2} = m\left[h_2 + \frac{V_2^2}{2} + gz_2\right] + W_{1-2} \qquad (4.13)$$

Nozzle can be analysed with the following assumptions:

(a) Heat interaction:

$$Q_{1-2} = 0 \qquad \text{for adiabatic flow only}$$

(b) Work interaction:

$$W_{1-2} = 0 \qquad \text{always for nozzle}$$

(c) Datum head:

$$z_1 = z_2 \qquad \text{for horizontal nozzle}$$

Equation (4.13) is reduced to

$$m\left[h_1 + \frac{V_1^2}{2}\right] = m\left[h_2 + \frac{V_2^2}{2}\right]$$

or

$$\begin{array}{cccc} h_1 + & \frac{V_1^2}{2} & = h_2 + & \frac{V_2^2}{2} \\ \downarrow & \downarrow & \downarrow & \downarrow \\ \text{J/kg} & \text{J/kg} & \text{J/kg} & \text{J/kg} \end{array} \qquad (4.14)$$

---

where V = Velocity of fluid.
  a = Velocity of sound.

or

$$h_1 + \frac{V_1^2}{2000} = h_2 + \frac{V_2^2}{2000}$$
$$\downarrow \quad \downarrow \qquad \downarrow \quad \downarrow$$
$$\text{kJ/kg kJ/kg kJ/kg kJ/kg}$$

(4.15)

From Eq. (4.14),

$$\frac{V_2^2}{2} = (h_1 - h_2) + \frac{V_1^2}{2}$$

or

$$V_2^2 = 2(h_1 - h_2) + V_1^2$$
$$V_2 = \sqrt{2(h_1 - h_2) + V_1^2}$$
$$V_2 = \sqrt{2c_p(T_1 - T_2) + V_1^2}\ \text{m/s}$$

(4.16)

where

$$c_p = \text{specific heat at constant pressure in J/kgK}$$
$$= 1005\ \text{J/kgK for air}$$

$$T_1, T_2 \text{ are in } °\text{C or K}$$
$$V_1 \text{ is the velocity at the inlet in m/s}$$

If the inlet velocity V_1 is very small as compared to the exit velocity V_2, i.e.,

$$V_1 \ll V_2, \text{ Eq. (4.16) becomes}$$

$$\boldsymbol{V_2 = \sqrt{2c_p(T_1 - T_2)}\ \textbf{m/s}}$$

From Eq. (4.15),

$$\frac{V_2^2}{2000} = (h_1 - h_2) + \frac{V_1^2}{2000}$$

or

$$V_2^2 = 2000(h_1 - h_2) + V_1^2$$
$$V_2 = \sqrt{2000(h_1 - h_2) + V_1^2}$$

$$= \sqrt{2000c_p(T_1 - T_2) + V_1^2}\, m/s \qquad (4.17)$$

where

$$c_p = \text{specific heat at constant pressure in kJ/kgK}$$
$$= 1.005 \text{ kJ/kgK for air}$$

$$T_1, T_2 \text{ are in } °C \text{ or K}$$
$$V_1 \text{ is the velocity at the inlet in m/s}$$

If the inlet velocity V_1 is very small as compared to the exit velocity V_2, i.e.,

$$V_1 \ll V_2, \text{ Eq. (4.17) becomes}$$

$$V_2 = \sqrt{2000c_p(T_1 - T_2)}\, m/s$$

For numerical problems:

1. **Converging nozzle**

   (i)   Mach number: $M < 1$
   (ii) $A_1 > A_2$
   (iii) $V_2 > V_1$.

2. **Diverging nozzle**

   (i)   Mach number: $M > 1$
   (ii) $A_1 < A_2$
   (iii) $V_2 > V_1$.

3. **Converging–diverging nozzle**

   (i) $M < 1$ for converging portion
   (ii) $M = 1$ at throat
   (iii) $M > 1$ diverging portion
   (iv) $V_2 > V_1$ (Fig 4.5).

4.11.2 Diffuser

A diffuser is also a fluid energy transforming device like a nozzle. A diffuser increases the pressure energy of the fluid and simultaneously it decreases the kinetic energy of the fluid, i.e., it is used to transfer the kinetic energy into pressure energy. As the fluid flows through the diffuser, it compresses to a higher pressure and in this process,

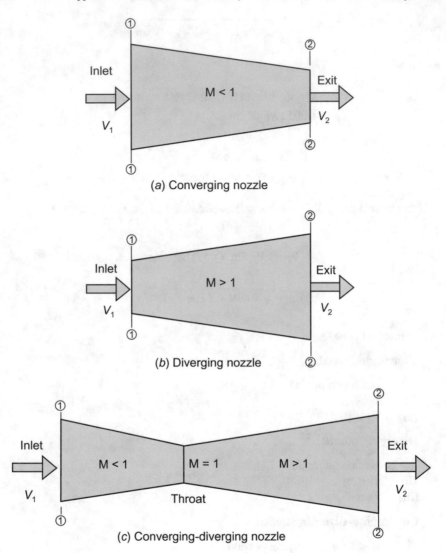

Fig. 4.5 Nozzles

velocity falls, and the pressure increases continuously from the entrance to the exit of the diffuser. The function of the diffuser is reversed than that of the function of the nozzle. In the case of subsonic flow (Mach no. $M < 1$), the diffuser has a diverging cross-sectional area in the direction of flow whereas in supersonic flow (Mach no. $M > 1$), the diffuser has a converging cross-sectional area in the direction of flow.

Let us take a converging diffuser as shown in Fig. 4.6.

The equations for fluid flow through the diffuser are also given by Eqs. (4.14) and (4.15).

Fig. 4.6 Converging diffuser

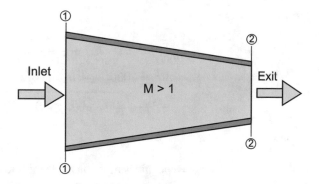

Now starting from Eq. (4.14),

$$h_1 + \frac{V_1^2}{2} = h_2 + \frac{V_2^2}{2}$$

or

$$\frac{V_1^2}{2} = (h_2 - h_1) + \frac{V_2^2}{2}$$
$$V_1^2 = 2(h_2 - h_1) + V_2^2$$
$$V_1 = \sqrt{2(h_2 - h_1) + V_2^2}$$
$$= \sqrt{2c_p(T_2 - T_1) + V_2^2} \text{ m/s} \quad (4.18)$$

where

$$c_p = \text{specific heat at constant pressure in J/kgK}$$
$$= 1005 \text{ J/kgK for air}$$

$$T_1 \text{ and } T_2 \text{ are in } °C \text{ or K}$$
$$V_1 \text{ is the velocity at the exit in m/s}$$

If the exit velocity V_2 is very small as compared to the inlet velocity V_1, i.e., $V_2 \ll V_1$,

Equation (4.18) becomes

$$V_1 = \sqrt{2c_p(T_2 - T_1)} \text{m/s}$$

From Eq. (4.15),

$$h_1 + \frac{V_1^2}{2000} = h_2 + \frac{V_2^2}{2000}$$

or

$$\frac{V_1^2}{2000} = (h_2 - h_1) + \frac{V_2^2}{2000}$$
$$V_1^2 = 2000(h_2 - h_1) + V_2^2$$
$$V_1 = \sqrt{2000c_p(T_2 - T_1) + V_2^2} \, \text{m/s} \qquad (4.19)$$

where

c_p = specific heat at constant pressure in J/kgK

= 1.005 kJ/kgK for air

T_1 and T_2 are in °C or K

V_1 is the velocity at the exit in m/s

If the exit velocity V_2 is very small as compared to the inlet velocity V_1, i.e., $V_2 \ll V_1$,

Equation (4.19) becomes

$$\boldsymbol{V_1 = \sqrt{2000c_p(T_2 - T_1)} \, \text{m/s}}$$

4.11.3   Turbine

A turbine is an energy converting device that converts the fluid energy into work. If the working fluid is a steam in which the heat energy of the steam is converted into work, it is called a steam turbine. In other, if the working fluid is a gas in which the heat energy of the gas is converted into work, it is called a gas turbine (Fig. 4.7).

Fig. 4.7 Turbine

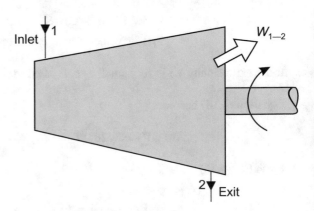

We know that the steady flow energy equation

$$m\left[h_1 + \frac{V_1^2}{2} + gz_1\right] + Q_{1-2} = m\left[h_2 + \frac{V_2^2}{2} + gz_2\right] + W_{1-2} \qquad (4.20)$$

A Turbine can be analysed with the following assumptions:

(a)   Heat transfer: $Q_{1-2} = 0$ for adiabatic flow only.
(b)   Change in kinetic energy is neglected.
(c)   Change in potential or datum energy is neglected, i.e., $z_1 = z_2$.

Equation (4.20) is reduced to

$$mh_1 = mh_2 + W_{1-2}$$

or

$$W_{1-2} = m(h_1 - h_2) = mc_p(T_1 - T_2)\,\text{kW}$$

where

> m is mass flow rate in kg/s
>
> T_1 and T_2 are temperatures at inlet and outlet in °C or K
>
> c_p is specific heat at constant pressure in kJ/kgK

For numerical problems, if assumption (a) is not maintained, i.e., heat transfer has occurred.
   $(Q_{1-2} \neq 0)$, then Eq. (4.20) is reduced to

$$mh_1 + Q_{1-2} = mh_2 + W_{1-2}$$

or

$$W_{1-2} = m(h_1 - h_2) + Q_{1-2} = mc_p(T_1 - T_2)\,\text{kW}$$

where

> m is mass flow rate in kg/s
>
> c_p = specific heat at constant pressure in J/kgK
>
>   = 1.005 kJ/kgK for air
>
> T_1 and T_2 are temperatures at inlet and exit in °C or K
>
> Q_{1-2} is heat transfer in kW

For unit mass flow rate,

$$w_{1-2} = c_p(T_1 - T_2) + q_{1-2} \text{ kJ/kg}$$

where

$$c_p \text{ is in kJ/kgK}$$
$$T_1 \text{ and } T_2 \text{ are in } °C \text{ or K}$$
$$q_{1-2} \text{ is in kJ/kg}$$

w_{1-2} is always $+ve$ which shows that work is done by the system, i.e., the turbine is a work producing device.

4.11.4 Compressor

A compressor is an energy converting device which converts the work into pressure energy of air, i.e., it is used to increase the pressure of air by the need of work (Fig. 4.8).

We know that the steady flow energy equation

$$m\left[h_1 + \frac{V_1^2}{2} + gz_1\right] + Q_{1-2} = m\left[h_2 + \frac{V_2^2}{2} + gz_2\right] + W_{1-2} \qquad (4.21)$$

Compressor can be analysed with the following assumptions:

(a)   Heat transfer: $Q_{1-2} = 0$ for adiabatic flow only.
(b)   Change in kinetic energy is neglected.
(c)   Change in potential or datum energy is neglected, i.e., $z_1 = z_2$.

Equation (4.21) is reduced to

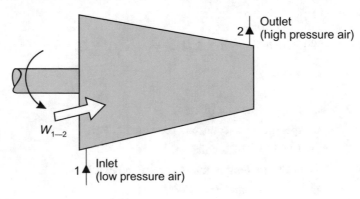

Fig. 4.8 Compressor

$$mh_1 = mh_2 + W_{1-2}$$

or

$$W_{1-2} = m(h_1 - h_2) = mc_p(T_1 - T_2) \, \text{kW}$$

where

m is mass flow rate in kg/s

T_1 and T_2 are temperatures at inlet and outlet in °C or K

c_p = specific heat at constant pressure in kJ/kgK

= 1.005 kJ/kgK for air

For numerical problems, if assumption (a) is not maintained, i.e., heat transfer has occurred.

($Q_{1-2} \neq 0$), then Eq. (4.21) is reduced to

$$mh_1 + Q_{1-2} = mh_2 + W_{1-2}$$

or

$$W_{1\ 2} = m(h_1 - h_2) + Q_{1-2} = mc_p(T_1 - T_2) + Q_{1-2} \, \text{kW}$$

where

m is in kg/s

c_p is in kJ/kgK

T_1 and T_2 are in °C or K

and

Q_{1-2} is in kW

For unit mass flow rate,

$$w_{1-2} = c_p(T_1 - T_2) + q_{1-2} \, \text{kJ/kg}$$

where

c_p is in kJ/kgK

T_1 and T_2 are in °C or K

q_{1-2} is in kJ/kg

w_{1-2} is always $-ve$ which shows that work is done on the system, i.e., the compressor is a work consuming device.

4.11.5 Boiler

A boiler is a metallic vessel used for the generation of steam at constant pressure from the water by application of heat (Fig. 4.9).

We know that the steady flow energy equation

$$m\left[h_1 + \frac{V_1^2}{2} + gz_1\right] + Q_{1-2} = m\left[h_2 + \frac{V_2^2}{2} + gz_2\right] + W_{1-2} \qquad (4.22)$$

A boiler can be analysed with the following assumptions:

(a)   Work interaction: $W_{1-2} = 0$ always for boiler.
(b)   Change in kinetic energy is neglected.
(c)   Change in potential or datum energy is neglected, i.e., $z_1 = z_2$

Equation (4.22) is reduced to

$$mh_1 + Q_{1-2} = mh_2$$

or

$$Q_{1-2} = m(h_2 - h_1)$$

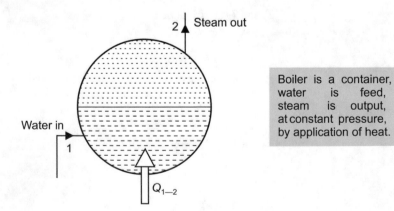

Boiler is a container, water is feed, steam is output, at constant pressure, by application of heat.

Fig. 4.9 Boiler

4.11.6 Heat Exchanger

A heat exchanger is a device in which sensible heat is transferred between two fluids. In the heat exchanger, two fluids are used, one is hot and the other is cold. The heat is lost by the hot fluid and simultaneously the heat is gained by the cold fluid (Fig. 4.10).

We know that the steady flow energy equation

$$
m_h \left[h_{h1} + \frac{V_{h1}^2}{2} + g z_{h1} \right] + m_c \left[h_{c1} + \frac{V_{c1}^2}{2} + g z_{c1} \right] + Q
$$
$$
= m_h \left[h_{h2} + \frac{V_{h2}^2}{2} + g z_{h2} \right] + m_c \left[h_{c2} + \frac{V_{c2}^2}{2} + g z_{c2} \right] + W
$$
$$(4.23)$$

Heat exchanger can be analysed with the following assumptions:

(a) Work interaction: $W = 0$ always for heat exchange.
(b) No heat interaction between the system and surroundings, i.e., $Q = 0$.
(c) Change in kinetic energies is neglected for both the hot and cold fluids.
(d) Change in potential or datum energies is neglected for both the hot and cold fluids, i.e., $z_{c1} = z_{c2}, z_{h1} = z_{h2}$.

Equation (4.23) is reduced to

$$
m_h h_{h1} + m_c h_{c1} = m_h h_{h2} + m_c h_{c2}
$$

or

$$
m_h (h_{h1} - h_{h2}) = m_c (h_{c2} - h_{c1})
$$
$$
m_h c_h (T_{h1} - T_{h2}) = m_c c_c (T_{c2} - T_{c1})
$$

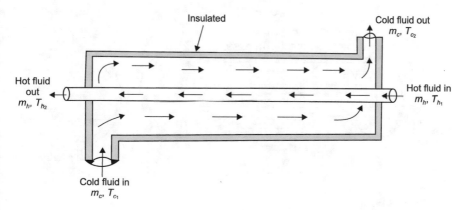

Fig. 4.10 Heat exchanger

Heat lost by hot fluid = Heat gained by cold fluid

where

m_h = mass flow rate of the hot fluid.

m_c = mass flow rate of the cold fluid.

c_h = specific heat of the hot fluid.

c_c = specific heat of the cold fluid.

T_{h1} and T_{h2} = temperatures of the hot fluid at inlet and outlet, respectively.

T_{c1} and T_{c2} = temperatures of the cold fluid at inlet and outlet, respectively.

4.11.7 Condenser

A condenser is a specific type of heat exchanger. It is used to change the phase of vapour to liquid by removing the latent heat of the vapour (Fig. 4.11).

We know that the steady flow energy equation

$$m_v\left[h_{h1} + \frac{V_{h1}^2}{2} + gz_{h1}\right] + m_w\left[h_{c1} + \frac{V_{c1}^2}{2} + gz_{c1}\right] + Q$$
$$= m_v\left[h_{h2} + \frac{V_{h2}^2}{2} + gz_{h2}\right] + m_w\left[h_{c2} + \frac{V_{c2}^2}{2} + gz_{c2}\right] + W$$
$$(4.24)$$

Condenser can be analysed with the following assumptions:

(a)   Work interaction: $W = 0$ always for condenser.

(b)   No heat interaction between the system and surroundings, i.e., $Q = 0$.

(c)   Change in kinetic energies is neglected for both the vapour and cold water.

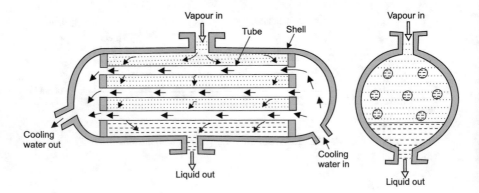

Fig. 4.11 Condenser

(d) Change in potential or datum energies is neglected for both the vapour and cold water, i.e., $z_{h1} = z_{h2}$, $z_{c1} = z_{c2}$.

Equation (4.24) is reduced to

$$m_v h_{h1} + m_w h_{c1} = m_v h_{h2} + m_w h_{c2}$$

or

$$m_v(h_{h1} - h_{h2}) = m_w(h_{c2} - h_{c1})$$

Heat lost by the vapour = heat gained by the cold water

where
m_v = mass flow rate of vapour.
m_w = mass flow rate of cold water.

4.11.8 Evaporator

An evaporator is a specific type of heat exchanger. It is used to change the phase of liquid to vapour by absorbing the latent heat of the liquid (Fig. 4.12).
We know that the steady flow energy equation

$$m_V \left[h_{h1} + \frac{V_{h1}^2}{2} + g z_{h1} \right] + m_W \left[h_{c1} + \frac{V_{cl}^2}{2} + g z_{c1} \right] + Q$$

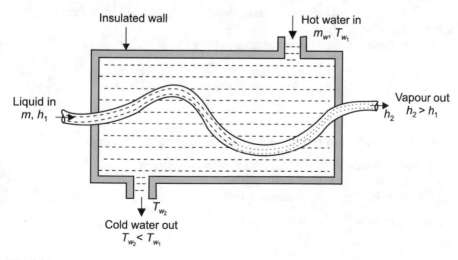

Fig. 4.12 Evaporator

$$= m_V \left[h_{h2} + \frac{V_{h2}^2}{2} + g z_{h2} \right] + m_w \left[h_{c2} + \frac{V_{c2}^2}{2} + g z_{c2} \right] + W$$

$$(4.25)$$

Evaporator can be analysed with the following assumptions:

(a)   Work interaction: $W = 0$ always for evaporator.
(b)   No heat interaction between the system and surroundings, i.e., $Q = 0$.
(c)   Change in kinetic energies is neglected for both the fluids.
(d)   Change in potential or datum energies is neglected for both the fluids, i.e., $z_1 = z_2$, $z_{w1} = z_{w2}$.

Equation (4.25) is reduced to

$$mh_1 + m_w h_{w1} = mh_2 + m_w h_{w2}$$

or

$$mh_1 + m_w h_{w1} = mh_2 + m_w h_{w2}$$

Latent heat gained by the liquid = Sensible heat lost by the water

4.11.9   Adiabatic Mixing

Adiabatic mixing refers to the mixing of two or more streams of fluid under adiabatic condition.

Let two streams of fluid with mass flow rate m_1 and m_2 get mixed together adiabatically (Fig. 4.13).

We know that the steady flow energy equation

$$m_1 \left[h_1 + \frac{V_1^2}{2} + g z_1 \right] + m_2 \left[h_2 + \frac{V_2^2}{2} + g z_2 \right] + Q = m_3 \left[h_3 + \frac{V_3^2}{2} + g z_3 \right] + W$$

$$(4.26)$$

Adiabatic mixing can be analysed with the following assumptions:

(a)   Work interaction: $W = 0$ always for adiabatic mixing.
(b)   Heat transfer: $Q = 0$ adiabatic mixing.
(c)   Change in kinetic and potential energies is neglected.

Equation (4.26) is reduced to

$$m_1 h_1 + m_2 h_2 = m_3 h_3$$

or

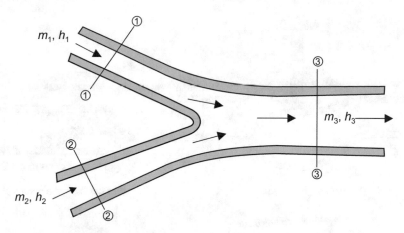

Fig. 4.13 Adiabatic mixing

$$m_1 c_{p1} T_1 + m c_{p2} T_2 = m_3 c_{p3} T_3$$

or

$$m_3 c_{p3} T_3 = m_1 c_{p1} T_1 + m_2 c_{p2} T_2$$

or

$$T_3 = \frac{m_1 c_{p1} T_1 + m_2 c_{p2} T_2}{m_3 c_{p3}} \quad \text{for different gases}$$

$$T_3 = \frac{m_1 T_1 + m_2 T_2}{m_3} \quad \text{for same gases}$$

$$T_3 = \frac{m_1 T_1 + m_2 T_2}{m_1 + m_2} \quad \because m_3 = m_1 + m_2$$

4.11.10 Throttling Process/Wire Drawing Process/Isenthalpic Process

The throttling process is the expansion of a fluid from high pressure to low pressure when fluid flows through a constricted passage, like a partially opened valve, a porous plug, an orifice, or a capillary tube.

Throttling process is shown in Fig. 4.14a by a partially opened value, Fig. 4.14b by a porous plug, Fig. 4.14c by an orifice, and Fig. 4.14d by a capillary tube on a fluid flowing in an insulated pipe.

We know that the steady flow energy equation

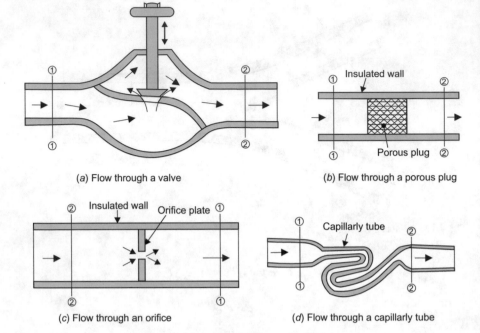

(a) Flow through a valve

(b) Flow through a porous plug

(c) Flow through an orifice

(d) Flow through a capillarly tube

Fig. 4.14 Throttling process

$$m\left[h_1 + \frac{V_1^2}{2} + gz_1\right] + Q_{1-2} = m\left[h_2 + \frac{V_2^2}{2} + gz_2\right] + W_{1-2} \tag{4.27}$$

Throttling process can be analysed with the following assumptions:

(a)   Work interaction: $W_{1-2} = 0$ always for throttling process.
(b)   Heat interaction: $Q_{1-2} = 0$ always for throttling process.
(c)   Change in kinetic and potential energies is neglected.

Equation (4.27) is reduced to

$$m_1 h_1 = m_2 h_2$$

or

$$\boldsymbol{h_1 = h_2}$$

It means that the enthalpy of fluid remains constant during the throttling process (enthalpy of the fluid before throttling is equal to the enthalpy of the fluid after throttling). Thus, a throttling process is an isenthalpic process and it is also called a wire drawing process. Throttling is an irreversible process and involves the degradation of energy.

Uses:

1.   Throttling process is used to determine the dryness fraction of wet steam.
2.   Throttling process is used for a large drop in temperature of the refrigerant in refrigeration and air-conditioning.

For ideal gas,

$$h = c_p T_2$$

∴

$$c_p T_1 = c_p T_2$$

or

$$T_1 = T_2$$

Multiplying c_v on both sides,

$$c_v T_1 = c_v T_2$$

$$u_1 = u_2 \qquad \because u = c_v T$$

The above analysis shows that the enthalpy, temperature, and internal energy of the ideal gas in the throttling process remain constant.

For real gas:

For ideal gases, enthalpy is a function of temperature only. But for real gases, enthalpy is a function of both temperature and pressure. The enthalpy of real gases remains constant during the throttling process but the temperature does no remain constant. The experiment can be conducted by keeping constant temperature and pressure on the upstream condition (i.e., before throttling) but varying the pressure on the down stream condition (i.e., after throttling) and measuring the corresponding temperature. This is achieved by changing different sizes of porous plugs. Both the temperature and pressure of the real gas are measured after throttling. Since the upstream pressure and temperature conditions are kept constant, the enthalpy of the real gas for all measured conditions on down stream pressure and temperature would be constant enthalpy line on T–p diagram. Repeating the experiment for different sets of inlet temperature and pressure and plotting the result, we can construct a T–p diagram for a real gas with several constant enthalpy lines, as shown in Fig. 4.15.

The temperature behaviour of a real gas during a throttling process is described by the **Joule–Thomson coefficient**, which is defined as the slope of an isenthalpic curve at any point in T–p diagram.

Mathematically,

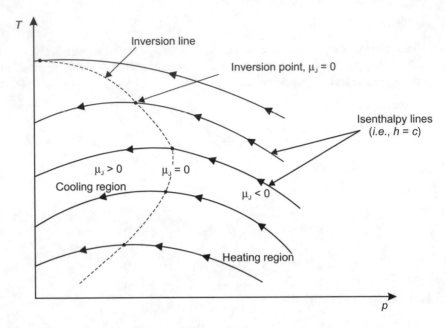

Fig. 4.15 Isenthalpy and inversion lines for a real gas

Joule–Thomson coefficient:

$$\mu_J = \left(\frac{\partial T}{\partial p}\right)_{h=\,\text{constant}}$$

where

μ_J is +ve, −ve, or zero.

The point on the isenthalpic line where $\mu_J = 0$ is called the **inversion point**. Thus, the inversion point denotes the maximum value of temperature on the isenthalpic line in T–p diagram. The line that passes through the inversion points is called the **inversion line**. The temperature at a point where an isenthalpic line intersects the inversion line is called the inversion temperature. The Joule–Thomson coefficient is positive ($\mu_J > 0$) to the left side of the inversion line, is zero at the inversion point and negative ($\mu_J < 0$) to the right side of the inversion line.

A throttling process always proceeds along a constant enthalpy line in the direction of decreasing pressure, i.e., from right to left. Therefore, the temperature of a real gas increases during a throttling process that takes place on the right-hand side of the inversion line, called heating region. However, for a real gas, temperature decreases during a throttling process that takes place on the left-hand side of the inversion line, called the cooling region. It is clear from Fig. 4.15 that a cooling effect cannot be achieved by the throttling process unless the real gas is below its maximum inversion temperature. For example, the maximum inversion temperature of hydrogen gas is

–68 °C. Thus hydrogen gas must be cooled below this temperature if any further cooling is to be achieved by the throttling process.

Remember:

For ideal gas:

Temperature remains constant

For real gas:

Temperature may be increase, decrease or constant

If	$\mu_J < 0$ i.e., μ_J is $- ve$	(i) temperature increases
		(ii) pressure decreases always
		(iii) enthalpy constant always
	$\mu_J = 0$	(i) temperature remains constant
		(ii) pressure decreases always
		(iii) enthalpy constant always
	$\mu_J > 0$ i.e., μ_J is $+ ve$	(i) temperature decreases
		(ii) pressure decreases always
		(iii) enthalpy constant always

4.12 Unsteady Flow Energy Equation (UFEE)

Unsteady flow means the flow in which its properties like $m, p, T, v,$ and V change with respect to time at any point in the system.

Consider a variable flow process as shown in Fig. 4.16.

Let $p_i, u_i, m_i, v_i,$ and V_i be pressure, specific internal energy, mass, specific volume, and velocity at sections (i)–(i) inlet to the system, respectively.

Let $p_e, u_e, m_e, v_e,$ and V_e be pressure, specific internal energy, mass, specific volume, and velocity at sections (e)–(e) exit from the system, respectively.

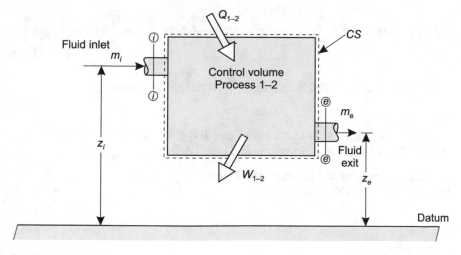

Fig. 4.16 Variable flow process

Let p_1, u_1, m_1, v_1, and V_1 be pressure, specific internal energy, mass, specific volume, and velocity at initial state in the system, respectively.

p_2, u_2, m_2, v_2 and V_2 are pressure, specific internal energy, mass, specific volume and velocity at final state in the system, respectively.

Q_{1-2} is the heat supplied to the system.

W_{1-2} is work output from the system.

According to the first law of thermodynamic for process 1–2 in the system (control volume),

$$Q_{1-2} = dE + W_{1-2}$$
$$Q_{1-2} = (E_2 - E_1) + W_{1-2}$$

For unsteady flow process,

$$Q_{1-2} = (E_2 - E_1) + W_{1-2} + m_e \left(h_e + \frac{V_e^2}{2} + gz_e \right) - m_t \left(h_i + \frac{V_i^2}{2} + gz_i \right)$$

where $m_e \left(h_e + \frac{V_e^2}{2} + gz_e \right)$ is net energy exit to the system

and $m_i \left(h_i + \frac{V_i^2}{2} + gz_i \right)$ is net energy inlet to the system.

$$Q_{1-2} = m_2(u_2 + KE_2 + PE_2) - m_1(u_1 + KE_1 + PE_1)$$
$$+ W_{1-2} + m_i \left(h_i + \frac{V_i^2}{2} + gz_i \right) - m_i \left(h_i + \frac{V_i^2}{2} + gz_i \right) \qquad (4.28)$$

If change in KE and PE is neglected in the control volume, Eq. (4.28) is reduced to

$$Q_{1-2} = m_2 u_2 - m_1 l_1 + W_{1-2} + m_e \left(h_e + \frac{V_e^2}{2} + gz_e \right) - m_i \left(h_i + \frac{V_i^2}{2} + gz_i \right)$$
$$(4.29)$$

Equation (4.29) is called the **unsteady flow energy equation**.

Problem 4.1:

Air enters a nozzle steadily at 2.2 kg/m^3 and 100 m/s and leaves at 1.1 kg/m^3 and 500 m/s. If the inlet area of the nozzle is 50 cm^2, determine.

(a)   the mass flow rate through the nozzle and
(b)   the exit area of the nozzle.

Solution:

Given data:

   At inlet of the nozzle:

$$\rho_1 = 2.2 \, \text{kg/m}^3$$
$$V_1 = 100 \, \text{m/s}$$
$$A_1 = 50 \, \text{cm}^2 = 50 \times 10^{-4} \, \text{m}^2$$

At exit of the nozzle:

$$\rho_2 = 1.1 \, \text{kg/m}^3$$
$$V_2 = 500 \, \text{m/s}$$

(a)   Mass flow rate:

$$m = \rho_1 A_1 V_1 = 2.2 \times 50 \times 10^{-4} \times 100 = \mathbf{1.1 \, kg/s}$$

(b)   Exit area of the nozzle: A_2

$$m = \rho_2 A_2 V_2$$
$$1.1 = 1.1 \times A_2 \times 500$$

or

$$A_2 = \mathbf{2 \times 10^{-3} \, m^2}$$

Problem 4.2:

Air enters steadily at 300 kPa, 350 K, and 50 m/s and leaves at 100 kPa and 320 m/s. The heat loss from the nozzle is estimated to be 3.2 kJ/kg of air flowing. The inlet area of the nozzle is 100 cm^2. Calculate (a) the exit temperature and (b) the exit area of nozzle.

Solution:
Given data:
   At inlet of nozzle:
   Pressure:

$$p_1 = 300 \, \text{kPa}$$

Temperature:

$$T_1 = 350 \, \text{K}$$

Velocity:

$$V_1 = 50 \, \text{m/s}$$

Cross-sectional area:

$$A_1 = 100\,\text{cm}^2$$

At exit of nozzle:
Pressure:

$$p_2 = 100\,\text{kPa}$$

Velocity:

$$V_2 = 320\,\text{m/s}$$

Heat loss:

$$q = -3.2\,\text{kJ/kg}$$

According to the steady flow energy equation,

$$h_1 + \frac{V_1^2}{2} + q = h_2 + \frac{V_2^2}{2}$$
$$c_p T_1 + \frac{V_1^2}{2} + q = c_p T_2 + \frac{V_2^2}{2}$$

where

$$c_p = 1005\,\text{J/kgK}$$
$$T_1 = 350\,\text{K}$$
$$V_1 = 50\,\text{m/s}$$
$$V_2 = 320\,\text{m/s}$$
$$q = -3200\,\text{J/kg}$$

$\therefore$

$$1005.0 \times 350 + \frac{(50)^2}{2} - 3200 = 1005 \times T_2 + \frac{(320)^2}{2}$$
$$351750 + 1250 - 3200 = 1005.0\,T_2 + 51200$$

or

$$1005\,T_2 = 298600$$

or

$$T_2 = 297.11 \, \text{K}$$

Applying the equation of state at inlet,

$$p_1 v_1 = RT_1$$
$$300 \times v_1 = 0.287 \times 350$$

or

$$v_1 = 0.3348 \, \text{m}^3/\text{kg}$$

and at exit,

$$p_2 v_2 = RT_2$$
$$100 \times v_2 = 0.287 \times 297.11$$

or

$$v_2 = 0.8527 \, \text{m}^3/\text{kg}$$

Now applying the continuity equation at inlet and exit for compressible flow,

$$\frac{A_1 V_1}{v_1} = \frac{A_2 V_2}{v_2}$$
$$\frac{100 \times 50}{0.3347} = \frac{A_2 \times 320}{0.8527}$$

or

$$A_2 = 39.80 \, \text{cm}^2$$

Problem 4.3:

Air enters an adiabatic nozzle steadily at 300 kPa, 200 °C, and 30 m/s and leaves at 100 kPa and 180 m/s. The inlet area of the nozzle is 80 cm². Determine (Fig. 4.17).

(a)   the mass flow rate through the nozzle,
(b)   the exit temperature of the air, and
(c)   the exit area of the nozzle.

Solution:

Given data:
   At inlet:

$$p_1 = 300 \, \text{kPa}$$

Fig. 4.17 Converging nozzle

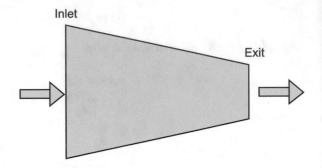

$$T_1 = 200\,°C$$
$$= 200 + 273$$
$$= 473\,K$$
$$V_1 = 30\,m/s$$

At exit:

$$p_2 = 100\,kPa$$
$$V_2 = 180\,m/s$$

At inlet:

$$A_1 = 80\,cm^2 = 80 \times 10^{-4}\,m^2$$
$$p_1 v_1 = RT_1$$
$$300 \times v_1 = 0.287 \times 473$$
$$v_1 = 0.4525\,m^3/kg$$

(a)   Mass flow rate:

$$m = \frac{A_1 V_1}{v_1} = \frac{80 \times 10^{-4} \times 30}{0.4525} = \mathbf{0.5303\,kg/s}$$

(b)   Exit temperature: T_2
Applying the steady flow energy equation at inlet and exit of the nozzle, we have

$$h_1 + \frac{V_1^2}{2000} = h_2 + \frac{V_2^2}{2000}$$

where

$$q_{1-2} = 0 \quad \text{for adiabatic flow}$$
$$w_{1-2} = 0 \quad \text{always for nozzle}$$
$$z_1 = z_2 \quad \text{for horizontal nozzle}$$

$$c_p T_1 + \frac{V_1^2}{2000} = c_p T_2 + \frac{V_2^2}{2000}$$

$$1.005 \times 473 + \frac{(30)^2}{2000} = 1.005 \times T_2 + \frac{(180)^2}{2000}$$

$$475.365 + 0.45 = 1.005\,T_2 + 16.2$$

or

$$T_2 = 457.32 \text{ K} = \mathbf{184.32\,{}^\circ C}$$

(c) Exit area: A_2

$$p_2 v_2 = R T_2$$
$$100 \times v_2 = 0.287 \times 457.32$$

or

$$v_2 = 1.3125 \text{ m}^3/\text{kg}$$

$$\frac{A_1 V_1}{v_1} = \frac{A_2 V_2}{v_2}$$

$$\frac{80 \times 10^{-4} \times 30}{0.4525} = \frac{A_2 \times 180}{1.3125}$$

or

$$A_2 = 38.67 \times 10^{-4} \text{ m}^2 = \mathbf{38.67\,cm^2}$$

Problem 4.4:

A perfect gas ($R = 0.190$ kJ/kgK, $\gamma = 1.35$) flows through a reversible adiabatic nozzle. The properties at the inlet are 22 bar, 500 °C, and 38 m/s. At the exit, the pressure is 2 bar. Determine the exit velocity and exit area for a flow rate of 4 kg/s.

Solution:

Given data:

$$R = 0.190 \text{ kJ/kgK}$$
$$\gamma = 1.35$$

We know that

$$c_p = \frac{\gamma R}{\gamma - 1} = \frac{1.35 \times 0.190}{1.35 - 1} = 0.74 \, \text{kJ/kgK} = 740 \, \text{J/kgK}$$

At inlet:

$$p_1 = 22 \, \text{bar}$$
$$T_1 = 500 \, ^\circ\text{C} = (273 + 500) \, \text{K} = 773 \, \text{K}$$
$$V_1 = 38 \, \text{m/s}$$

At exit:

$$p_2 = 2 \, \text{bar}$$

Flow rate:

$$m = 4 \, \text{kg/s}$$

Flow through a nozzle is a reversible adiabatic,
For process 1–2,

$$\frac{T_2}{T_1} = \left(\frac{p_2}{p_1}\right)^{\frac{\gamma-1}{\gamma}}$$

$$\frac{T_2}{773} = \left(\frac{2}{22}\right)^{\frac{1.35-1}{1.35}}$$

$$\frac{T_2}{773} = (0.09)^{0.26} = 0.5346$$

or

$$T_2 = 0.53446 \times 773 = 413.25 \, \text{K}$$

Applying the steady flow energy equation,

$$h_1 + \frac{V_1^2}{2} + gz_1 + q_{1-2} = h_2 + \frac{V_2^2}{2} + gz_2 + w_{1-2}$$

where

$$z_1 = z_2 \quad \text{for horizontal nozzle}$$
$$q_{1-2} = 0 \quad \text{flow is reversible adiabatic}$$
$$w_{1-2} = 0 \quad \text{always for nozzle}$$

$$h_1 + \frac{V_1^2}{2} = h_2 + \frac{V_2^2}{2}$$

$$(h_1 + h_2) + \frac{V_1^2}{2} = \frac{V_2^2}{2}$$

$$c_p(T_1 - T_2) + \frac{V_1^2}{2} = \frac{V_2^2}{2}$$

or

$$V_2^2 = 2c_p(T_1 - T_2) + V_1^2$$

or

$$V_2 = \sqrt{2c_p\left(T_1 - T_2 + V_1^2\right)}\, \text{m/s}$$

where

c_p is in J/kgK

T_1 and T_2 are in °C or K

V_1 is in m/s

$\therefore$

$$V_2 - \sqrt{2 \times 740(73 - 413.25) + (38)^2}$$
$$V_2 = \sqrt{532430 + 1444} = \sqrt{533874} = \mathbf{730.66\,m/s}$$

Applying the equation of state at exit state 2,

$$p_2 v_2 = RT_2$$

where

p_2 is in kPa

v_2 is in m^3/kg

R is in kJ/kgK

T_2 is in K

$\therefore$

$$200 \times v_2 = 0.190 \times 413.25$$

or

$$v_2 = 0.3925 \, \text{m}^3/\text{kg}$$

From continuity equation,
Mass flow rate:

$$m = \rho_2 A_2 V_2$$

$$m = \frac{A_2 V_2}{v_2}$$

$$4 = \frac{A_2 \times 730.66}{0.3925}$$

or

$$A_2 = 0.002148 \, \text{m}^2 = \mathbf{21.48 \, cm^2}$$

Problem 4.5:
The inlet and exit condition of the fluid passing through a nozzle are $h_1 = 2880$ kJ/kg, $V_1 = 60$ m/s, and $h_2 = 2770$ kJ/kg. Neglecting the heat loss from the nozzle and assuming the nozzle is horizontal, find.

(i)   Exit velocity of the fluid.
(ii)   Mass flow rate through the nozzle if its inlet area is 1000 cm² and specific volume at inlet is 0.187 m³/kg.
(iii) Exit area, if the specific volume at exit is 0.5 m³/kg.

Solution:
Given data:
   At inlet condition:

$$h_1 = 2880 \, \text{kJ/kg}$$

$$V_1 = 60 \, \text{m/s}$$

$$A_1 = 1000 \, \text{cm}^2 = 0.1 \, \text{m}^2$$

$$v_1 = 0.187 \, \text{m}^3/\text{kg}$$

   At exit condition:

$$h_2 = 2770 \, \text{kJ/kg}$$

$$v_2 = 0.5 \, \text{m}^3/\text{kg}$$

(i)   Applying the steady flow energy equation,

$$\left(h_1 + \frac{V_1^2}{2} + g z_1 \right) + q_{1-2} = \left(h_2 + \frac{V_2^2}{2} + g z_2 \right) + w_{1-2}$$

$z_1 = z_2$, nozzle is horizontal (given)
$q_{1-2} = 0$, neglecting heat loss (given)
$w_{1-2} = 0$, work done is zero (always for nozzle).
The above equation will become

$$h_1 + \frac{V_1^2}{2} = h_2 + \frac{V_2^2}{2}$$

[**Note**: If we substitute the values of h_1 and h_2 in kJ/kg, then $\frac{V_1^2}{2}$ and $\frac{V_2^2}{2}$ also to be converted in kJ/kg by dividing 1000].
i.e.,

$$h_1 + \frac{V_1^2}{2000} = h_2 + \frac{V_2^2}{2000}$$

$$2880 + \frac{(60)^2}{2000} = 2770 + \frac{V_2^2}{2000}$$

$$2880 + 1.8 = 2770 + \frac{V_2^2}{2000}$$

or

$$\frac{V_2^2}{2000} = 111.8$$

or

$$V_2^2 = 111.8 \times 2000 = 223600$$

or

$$V_2 = \mathbf{472.86\,m/s}$$

(ii)   Mass flow rate: m

$$m = \frac{A_1 V_1}{v_1} = \frac{0.1 \times 60}{0.187}\,\text{kg/s} = \mathbf{32.08\,kg/s}$$

(iii)   Exit area: A_2
Mass flow rate:

$$m = \frac{A_2 V_2}{v_2}$$

$$32.08 = \frac{A_2 \times 472.86}{0.5}$$

or

$$A_2 = 0.03392 \, \text{m}^2$$

Problem 4.6:

The enthalpy of the fluid at inlet of the nozzle is 3000 kJ/kg and the velocity is 60 m/s, at the discharge the enthalpy is 2762 kJ/kg, and the nozzle is horizontal and there is negligible heat loss.

(a)   Determine the velocity at the exit of the nozzle.
(b)   The inlet area is 0.1 m² and the specific volume at 0.187 m³/kg. Determine the mass flow rate of the fluid through the nozzle.
(c)   The specific volume at the nozzle exit is 0.498 m³/kg. Determine the exit area of the nozzle.

Solution:
Given data:
    At inlet:

$$h_1 = 3000 \, \text{kJ/kg}$$
$$V_1 = 60 \, \text{m/s}$$

    At exit:

$$h_2 = 2762 \, \text{kJ/kg}$$

Applying the steady flow energy equation at inlet and exit of a nozzle, we have (Fig. 4.18)

$$h_1 + \frac{V_1^2}{2} + gz_1 + q_{1-2} = h_2 + \frac{V_2^2}{2} + gz_2 + w_{1-2}$$

Assumptions:

$$q_{1-2} = 0 \quad \text{adiabatic flow}$$

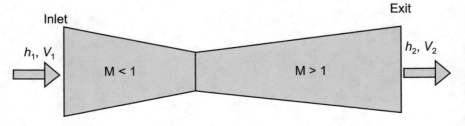

Fig. 4.18 Converging–diverging nozzle

$$w_{1-2} = 0 \quad \text{always for nozzle}$$
$$Z_1 = Z_2 \quad \text{horizontal nozzle}$$

SFEE becomes

$$h_1 + \frac{V_1^2}{2} = h_2 + \frac{V_2^2}{2}$$

(a)   Velocity at the exit of the nozzle: V_2

$$h_2 + \frac{V_2^2}{2000} = h_1 + \frac{V_1^2}{2000}$$
$$\frac{V_2^2}{2000} = (h_1 - h_2) + \frac{V_1^2}{2000}$$
$$V_2^2 = 2000(h_1 - h_2) + V_1^2$$

or

$$V_2 = \sqrt{2000(h_1 - h_2) + V_1^2}$$
$$= \sqrt{2000(3000 - 2762) + (60)^2} = \mathbf{692.53\,m/s}$$

(b)   At inlet:

$$A_1 = 0.1\,\text{m}^2$$
$$v_1 = 0.187\,\text{m}^3/\text{kg}$$

Mass flow rate:

$$m = \frac{A_1 V_1}{v_1} = \frac{0.1 \times 60}{0.187} = \mathbf{32.085\,kg/s}$$

(c)   At exit:

$$v_2 = 0.498\,\text{m}^3/\text{kg}$$
$$m v_2 = A_2 V_2$$
$$32.085 \times 0.498 = A_2 692.53$$

or

$$A_2 = \mathbf{0.02307\,m^2}$$

Problem 4.7:

In a nozzle, the ideal gas expands from a pressure of 20 bar to 3 bar and the process is reversible adiabatic. The inlet conditions are 500 °C and 35 m/s. Determine the area and velocity at the inlet section of the nozzle, if the flow rate is 5 kg/s.

Take $R = 190$ J/kgK and $\gamma = 1.35$.

Solution:

Given data:

At inlet:

$$p_1 = 20 \, \text{bar} = 20 \times 10^5 \, \text{Pa}$$
$$T_1 = 500 \, ^\circ\text{C} = (500 + 273) \, \text{K} = 773 \, \text{K}$$
$$V_1 = 35 \, \text{m/s}$$

At exit:

$$p_2 = 3 \, \text{bar}$$
$$q_{1-2} = 0 \quad \text{flow is reversible adiabatic}$$
$$m = 5 \, \text{kg/s}$$
$$R = 190 \, \text{J/kgK}$$
$$\gamma = 1.35$$

Applying the equation of state at inlet condition,

$$p_1 v_1 = RT_1$$

If

$$R = 190 \, \text{J/kgK}$$
$$p_1 = 20 \times 10^5 \, \text{Pa}$$
$$T_1 = 773 \, \text{K}$$

then, we get v_1 is in m³/kg.

$$\therefore$$

$$20 \times 10^5 \times v_1 = 190 \times 773$$

or

$$v_1 = 0.0734 \, \text{m}^3/\text{kg}$$

The mass flow rate:

$$m = \frac{A_1 V_1}{v_1}$$

$$5 = \frac{A_1 \times 35}{0.0734}$$

or

$$A_1 = 0.010485 \, \text{m}^2 = \mathbf{104.85 \, cm^2}$$

Applying the steady flow energy equation,

$$h_1 + \frac{V_1^2}{2} + gz_1 + q_{1-2} = h_2 + \frac{V_2^2}{2} + gz_2 + w_{1-2}$$

where

$$z_1 = z_2 \quad \text{for horizontal nozzle}$$
$$q_{1-2} = 0 \quad \text{flow is reversible adiabatic}$$
$$w_{1-2} = 0 \quad \text{lways for nozzle}$$

$$h_1 + \frac{V_1^2}{2} = h_2 + \frac{V_2^2}{2}$$
$$c_p T_1 + \frac{V_1^2}{2} = c_p T_2 + \frac{V_2^2}{2}$$

where

$$c_p = \frac{\gamma R}{\gamma - 1} = \frac{1.35 \times 190}{1.35 - 1} = 732.85 \, \text{J/kgK}$$

and

$$\frac{T_2}{T_1} = \left(\frac{p_2}{p_1}\right)^{\frac{\gamma - 7}{\gamma}}$$

$$T_2 = T_1 \left(\frac{p_2}{p_1}\right)^{\frac{\gamma - 1}{\gamma}} = 773 \left(\frac{3}{20}\right)^{\frac{1.35 - 1}{1.35}} = 472.92 \, \text{K}$$

$\therefore$

$$732.85 \times 773 + \frac{(35)^2}{2} = 732.85 \times 472.92 + \frac{V_2^2}{2}$$

$$567105.55 = 346579.42 + \frac{V_2^2}{2}$$

or

$$\frac{V_2^2}{2} = 220526.13$$

or

$$V_2^2 = 441052.26$$

or

$$V_2 = \mathbf{664.11\,m/s}$$

Problem 4.8:
Air passes through a gas turbine at the rate of 5 kg/s. It enters with a velocity of 200 m/s and temperature of 1200 °C. At the exit, the velocity is 150 m/s and the temperature is 400 °C. The air has a heat loss of 15 kJ/s as it passes through the turbine. Determine the power developed (Fig. 4.19).

Solution:
Given data for a gas turbine:
    Mass flow rate:

$$m = 5\,kg/s$$

    At inlet:
    Velocity:

Fig. 4.19 Gas turbine

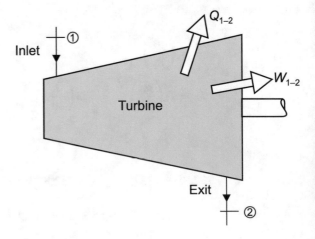

$$V_1 = 200 \, \text{m/s}$$

Temperature:

$$T_1 = 1200 \,°\text{C}$$
$$= (1200 + 273)\text{K}$$
$$= 1473 \, \text{K}$$

At exit:
Velocity:

$$V_2 = 150 \, \text{m/s}$$

Temperature:

$$T_2 = 400 \,°\text{C}$$
$$= (400 + 273)\text{K}$$
$$= 673 \, \text{K}$$

Heat loss:

$$Q_{1-2} = -15 \, \text{kJ/s}$$

Applying the steady flow energy equation at inlet and exit, we have

$$m\left[h_1 + \frac{V_1^2}{2} + gz_1\right] + Q_{1-2} = m\left[h_2 + \frac{V_2^2}{2} + gz_2\right] + W_{1-2}$$

Neglecting change in potential energy

$$m_{\text{kg/s}}\left[\underset{\text{kJ/kg}}{h_1} + \underset{\text{J/kg}}{\frac{V_1^2}{2}}\right] + \underset{\text{kJ/s}}{Q_{1-2}} = m_{\text{kg/s}}\left[\underset{\text{kJ/kg}}{h_2} + \underset{\text{J/kg}}{\frac{V_2^2}{2}}\right] + \underset{\text{kJ/s}}{W_{1-2}}$$

On dividing $\frac{V_1^2}{2}$ and $\frac{V_2^2}{2}$ by 1000 in the above equation,
we have

$$m\left[c_p T_1 + \frac{V_1^2}{2000}\right] + Q_{1-2} = m\left[c_p T_2 + \frac{V_2^2}{2000}\right] + W_{1-2} \quad \left| \begin{array}{l} \text{division by 1000 is} \\ \text{performed in order to} \\ \text{balance units on both} \\ \text{sides of equation} \end{array} \right.$$

$$\left[1.005 \times 1473 + \frac{(200)^2}{2000}\right] - 15 = 5\left[1.005 \times 673 + \frac{(150)^2}{2000}\right] + W_{1-2}$$

$$5[1480.36 + 20] - 15 = 5[676.36 + 11.25] + W_{1-2}$$
$$7501.8 - 15 = 3438.05 + W_{1-2}$$

or

$$W_{1-2} = 4048.75 \, \text{kJ/s}$$

Power (P): Power is defined as the time rate of doing work or work done per unit time.

So, Power developed:

$$P = 4048.75 \, \text{kJ/s} = \textbf{4048.75 kW}$$

Problem 4.9:
Air enters a compressor operating at a steady state at a pressure of 1 bar, a temperature of 290 K, and a velocity of 6 m/s through an inlet with an area of 0.1 m². At the exit, the pressure is 7 bar, the temperature is 450 K, and the velocity is 2 m/s. Heat transfer from the compressor to its surroundings occurs at a rate of 180 kJ/min. Employing the ideal gas model, calculate the power input to the compressor, in kW.

Solution:
Given data:

At inlet of compressor:

$$p_1 = 1 \, \text{bar} = 1 \times 10^2 \, \text{kPa}$$
$$T_1 = 290 \, \text{K}$$
$$V_1 = 6 \, \text{m/s}$$
$$A_1 = 0.1 \, \text{m}^2$$

At exit of compressor

$$p_2 = 7 \, \text{bar} = 7 \times 10^2 \, \text{kPa}$$
$$T_2 = 450 \, \text{K}$$
$$V_2 = 2 \, \text{m/s}$$

Heat transfer:

$$Q_{1-2} = -180 \, \text{kJ/min} = -\frac{180}{60} \, \text{kJ/s} = -3 \, \text{kW}$$

Applying equation of state at inlet condition,

$$p_1 v_1 = R T_1$$

$$\frac{p_1}{\rho_1} = RT_1$$

or

$$\rho_1 = \frac{p_1}{RT_1} \tag{4.30}$$

By continuity equation,

$$m = \rho_1 A_1 V_1 \tag{4.31}$$

Substituting the value of ρ_1 from Eq. (4.30) in Eq. (4.31), we get

$$m_1 = \frac{p_1 A_1 V_1}{RT_1} = \frac{1 \times 10^2 \times 0.1 \times 6}{0.287 \times 290} \, \text{kg/s} = 0.7208 \, \text{kg/s}$$

Applying the steady flow energy equation,

$$m\left(h_1 + \frac{V_1^2}{2} + gz_1\right) + Q_{1-2} = m\left(h_2 + \frac{V_2^2}{2} + gz_2\right) + W_{1-2}$$

Let

$$z_1 = z_2$$

∴

$$m\left(h_1 + \frac{V_1^2}{2}\right) + Q_{1-2} = m\left(h_2 + \frac{V_2^2}{2}\right) + W_{1-2}$$

$$m\left(c_p T_1 + \frac{V_1^2}{2}\right) + Q_{1-2} = m\left(c_p T_2 + \frac{V_2^2}{2}\right) + W_{1-2}$$

$$0.7208\left[1.005 \times 290 + \frac{(6)^2}{2000}\right] - 180 = 0.7208\left[1.005 \times 450 + \frac{(2)^2}{2000}\right] + W_{1-2}$$

$$210.09 - 180 = 325.98 + W_{1-2}$$

or

$$W_{1-2} = -295.89 \, \text{kW}$$

The $-ve$ sign shows the work done on the system.

Problem 4.10:

An air passes through a gas turbine system at 4.5 kg/s. It enters at a velocity of 150 m/s with a specific enthalpy of 3000 kJ/kg. It exits at a velocity of 120 m/s and specific enthalpy 2300 kJ/kg. There is a heat loss of 25 kJ/kg in its surrounding as it passes through turbine. Determine the power developed by the turbine (Fig. 4.20).

Solution:
Given data:
  Mass flow rate:

$$m = 4.5\,\text{kg/s}$$

Inlet velocity:

$$V_1 = 150\,\text{m/s}$$

Specific enthalpy at inlet:

$$h_1 = 3000\,\text{kJ/kg}$$

Exit velocity:

$$V_2 = 120\,\text{m/s}$$

Specific enthalpy at exit:

$$h_2 = 2300\,\text{kJ/kg}$$

Heat loss:

$$q_{1-2} = -25\,\text{kJ/kg}$$

Applying the steady flow energy equation at inlet and exit of a turbine, we have

$$h_1 + \frac{V_1^2}{2} + gz_1 + q_{1-2} = h_2 + \frac{V_2^2}{2} + gz_2 + w_{1-2}$$

Fig. 4.20 Gas-turbine

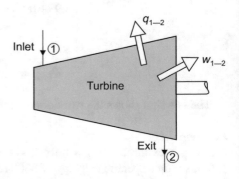

As $z_1 = z_2$ for horizontal turbine,

$$h_1 + \frac{V_1^2}{2} + q_{1-2} = h_2 + \frac{V_2^2}{2} + w_{1-2}$$

Notice that h_1, q_{1-2}, h_2, and w_{1-2} are in kJ/kg but $\frac{V_1^2}{2}$ and $\frac{V_2^2}{2}$ are in J/kg.

So, either you take all terms in kJ/kg or J/kg. In the present case, we will take all terms in kJ/kg on dividing $\frac{V_1^2}{2}$ and $\frac{V_2^2}{2}$ by 1000.

$$h_1 + \frac{V_1^2}{2000} + q_{1-2} = h_2 + \frac{V_2^2}{2000} + w_{1-2}$$
$$3000 + \frac{(150)^2}{2000} - 25 = 2300 + \frac{(120)^2}{2000} + w_{1-2}$$
$$3000 + 11.25 - 25 = 2300 + 7.2 + w_{1-2}$$

or

$$w_{1-2} - 679.05 \text{ kJ/kg}$$

Power developed:

$$P = mw_{1-2} = 4.5 \times 679.05$$
$$= 3055.725 \text{ kJ/s} = \mathbf{3055.725 \, kW}$$

Problem 4.11:
Air flows steadily at the rate of 22.68 kg/min through an air compressor entering at 6.1 m/s velocity, 1.02 bar pressure, and 0.845 m³/kg and leaving at 4.6 m/s, 7.03 bar pressure, and 0.163 m³/kg. The internal energy of the fluid leaving is 88.2 kJ/kg greater than that of the air entering. Cooling water in the compressor jackets absorbs heat from the air at the rate of 3693 kJ/min.

(a) Compute the rate of shaft work input to the air in kW.
(b) Find the ratio of the inlet pipe diameter to the outlet pipe diameter.

Solution:
Given data for compressor:
   Mass flow rate:

$$m = 22.68 \text{ kg/min} = 0.378 \text{ kg/s}$$

At inlet:
Velocity:

$$V_1 = 6.1 \, \text{m/s}$$

Pressure:

$$p_1 = 1.02 \, \text{bar} = 1.02 \times 10^2 \, \text{kPa}$$

Specific volume:

$$v_1 = 0.805 \, \text{m}^3/\text{kg}$$

At outlet:
Velocity:

$$V_2 = 4.6 \, \text{m/s}$$

Pressure:

$$p_2 = 7.03 \, \text{bar} = 7.03 \times 10^2 \, \text{kPa}$$

Specific volume:

$$v_2 = 0.163 \, \text{m}^3/\text{kg}$$

Specific internal energy:

$$u_2 = (u_1 + 88.2) \, \text{kJ/kg}$$

Change in specific internal energy:

$$u_2 - u_1 = 88.2 \, \text{kJ/kg}$$

Heat lost:

$$Q_{1-2} = -3693 \, \text{kJ/min}$$
$$= -\frac{3693}{60} \, \text{kJ/s} = -61.55 \, \text{kW}$$

Now, the change in specific enthalpy:

$$h_2 - h_1 = (u_2 + p_2 v_2) - (u_1 + p_1 v_1) \qquad [\because h = u + pv]$$
$$= (u_2 - u_1) + p_2 v_2 - p_1 v_1$$
$$= 88.2 + 7.03 \times 10^2 \times 0.163 - 1.02 \times 10^2 \times 0.845$$
$$= 88.2 + 114.589 - 86.19 = 116.599 \, \text{kJ/kg}$$

(a) Compute the rate of shaft work input to the air in kW
Applying the steady flow energy equation, we have

$$m\left(h_1 + \frac{V_1^2}{2} + g z_1\right) + Q_{1-2} = m\left(h_2 + \frac{V_2^2}{2} + g z_2\right) + W_{1-2}$$

Neglecting change in potential energy,

$$mh_1 + \frac{m V_1^2}{2} + Q_{1-2} = mh_2 + \frac{m V_2^2}{2} + W_{1-2}$$

Now after simplification of the steady flow energy equation according to the given problem, the next step is to make all the terms involved in the above equation in the same units as

$$\left.\begin{array}{l} m \text{ is in kg/s} \\ h \text{ is in kJ/kg} \end{array}\right\} mh_1, mh_2 \text{ are in kW}$$

$$V \text{ is in m/s} \left\{\begin{array}{l} \frac{m V_1^2}{2}, \frac{m V_2^2}{2} \text{ are in W} \\ \text{units balance:} \\ \frac{m V_1^2}{2000}, \frac{m V_1^2}{2000} \text{ are in kW} \end{array}\right.$$

Q_{1-2} and W_{1-2} are in kW.
The above equation is written as

$$W_{1-2} = m(h_1 - h_2) + \frac{m V_1^2}{2000} - \frac{m V_2^2}{2000} + Q_{1-2}$$

or

$$W_{1-2} = -m(h_2 - h_1) + \frac{m V_1^2}{2000} - \frac{m V_2^2}{2000} + Q_{1-2}$$

$$= -0.378 \times 116.599 + \frac{0.378 \times (6.1)^2}{2000} - \frac{0.378 \times (4.6)^2}{2000} - 61.55$$

$$= -44.07 + 0.007 - 0.004 - 61.55$$

$$= -105.617 \, \text{kW}$$

The −ve sign indicates that the work is done on the system.

(b) Ratio of inlet pipe diameter to the outlet diameter: $\frac{d_1}{d_2}$
Let d_1 be the diameter of the inlet pipe and
d_2 be diameter of the outlet pipe.
For steady flow:

$$m = \frac{AV}{v} = \text{constant}$$

i.e.,

$$\frac{A_1 V_1}{v_1} = \frac{A_2 V_2}{v_2}$$

$$\frac{\pi}{4} \frac{d_1^2 V_1}{v_1} = \frac{\pi}{4} \frac{d_2^2 V_2}{v_2}$$

or

$$\frac{d_1^2 V_1}{v_1} = \frac{d_2^2 V_2}{v_2}$$

or

$$\left(\frac{d_1}{d_2}\right)^2 = \frac{V_2}{V_1} \times \frac{v_1}{v_2}$$

$$\left(\frac{d_1}{d_2}\right)^2 = \frac{4.6}{6.1} \times \frac{0.845}{0.163}$$

$$\left(\frac{d_1}{d_2}\right)^2 = 3.909$$

or

$$\frac{d_1}{d_2} = \sqrt{3.909} = \mathbf{1.977}.$$

Problem 4.12:

In a steady flow apparatus, 135 kJ of work is done by each kg of fluid. The specific volume of the fluid, pressure, and velocity at the inlet are 0.37 m³/kg, 600 kPa, and 16 m/s, respectively. The inlet is 32 m above the floor, and the discharge pipe is at the floor level. The discharge conditions are 0.63 m³/kg, 100 kPa, and 270 m/s. The total heat loss between the inlet and discharge is 9 kJ/kg of fluid. In flowing through this apparatus, does the specific internal energy increase or decrease, and by how much (Fig. 4.21)?

Solution:

Given data:

Work done by the system:

$$w = 135 \, \text{kJ/kg}$$

At inlet:

Specific volume:

$$v_1 = 0.37 \, \text{m}^3/\text{kg}$$

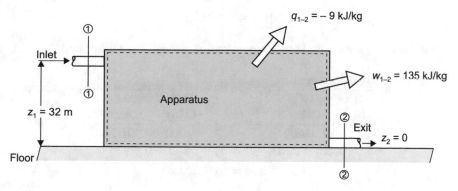

Fig. 4.21 Schematic for Problem 4.12

Pressure:

$$p_1 = 600\,\text{kPa}$$

Velocity:

$$V_1 = 16\,\text{m/s}$$

Datum head:

$$z_1 = 32\,\text{m}$$

At exit:
Datum head:

$$z_2 = 0$$

Specific volume:

$$v_2 = 0.62\,\text{m}^3/\text{kg}$$

Pressure:

$$p_2 = 100\,\text{kPa}$$

Velocity:

$$V_2 = 270\,\text{m/s}$$

Total heat loss:

$$q_{1-2} = -9 \, \text{kJ/kg}$$

According to the steady flow energy equation per unit mass flow rate,

$$h_1 + \frac{V_1^2}{2000} + \frac{gz_1}{1000} + q_{1-2} = h_2 + \frac{V_2^2}{2000} + \frac{gz_2}{1000} + w_{1-2}$$

$$h_1 + \frac{(16)^2}{2000} + \frac{9.81 \times 32}{1000} - 9 = h_2 + \frac{(270)^2}{2000} + \frac{9.81 \times 0}{1000} + 135$$

$$h_1 + 0.128 + 0.314 - 9 = h_2 + 36.45 + 0 + 135$$

$$h_1 - 8.558 = h_2 + 171.45$$

or

$$h_2 - h_1 = -180$$

where

$$h_2 = u_2 + p_2 v_2$$
$$h_1 = u_1 + p_1 v_1$$

∴

$$u_2 + p_2 v_2 - u_1 - p_1 v_1 = -180$$

$$u_2 + 100 \times 0.62 - u_1 - 600 \times 0.37 = -180$$

$$u_2 + 62 - u_1 - 222 = -180$$

$$u_2 - u_1 = \mathbf{-20 \, kJ/kg}$$

The −ve sign shows that the specific internal energy decreases by 20 kJ/kg.

Problem 4.13:

In a certain steady flow process, 12 kg of fluid per minute enters at a pressure of 1.4 bar, density 25 kg/m³, velocity 120 m/s, and internal energy 920 kJ/kg. The fluid properties at exit are pressure 5.6 bar, density 5 kg/m³, velocity 180 m/s, and internal energy 720 kJ/kg. During the process, the fluid rejects 60 kW of heat and rises through 60 m. Determine work done during the process in kW (Fig. 4.22).

Solution:

Given data:

Mass flow rate:

$$m = 12 \, \text{kg/min} = \frac{12}{60} \, \text{kg/s} = 0.2 \, \text{kg/s}$$

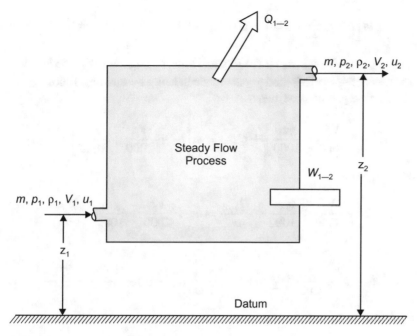

Fig. 4.22 Schematic for Problem 4.13

At inlet:

$$p_1 = 1.4\,\text{bar} = 140\,\text{kPa}$$
$$\rho_1 = 25\,\text{kg/m}^3$$
$$V_1 = 120\,\text{m/s}$$
$$u_1 = 920\,\text{kJ/kg}$$

At exit:

$$p_2 = 5.6\,\text{bar} = 560\,\text{kPa}$$
$$\rho_2 = 5\,\text{kg/m}^3$$
$$V_2 = 180\,\text{m/s}$$
$$u_2 = 720\,\text{kJ/kg}$$
$$z_2 - z_1 = 60\,\text{m}$$

Heat rejection:

$$Q_{1-2} = -60\,\text{kW}$$

Applying the steady flow energy equation between sections 1 and 2, we get

$$m\left[h_1 + \frac{V_1^2}{2} + gz_1\right] + Q_{1-2} = m\left[h_2 + \frac{V_2^2}{2} + gz_2\right] + W_{1-2}$$

Let the units of h_1 and h_2 be in kJ/kg and units of Q_{1-2} and W_{1-2} be in kW. The units of $\frac{V_1^2}{2}$, gz_1, $\frac{V_2^2}{2}$ and gz_2 are made in kJ/kg by dividing 1000. The above equation is written as

$$m\left[h_1 + \frac{V_1^2}{2000} + \frac{gz_1}{1000}\right] + Q_{1-2} = m\left[h_1 + \frac{V_2^2}{2000} + \frac{gz_2}{1000}\right] + W_{1-2}$$

or

$$h_1 + \frac{V_1^2}{2000} + \frac{gz_1}{1000} + \frac{Q_{1-2}}{m} = h_2 + \frac{V_2^2}{2000} + \frac{gz_2}{1000} + \frac{W_{1-2}}{m}$$

or

$$h_2 - h_1 + \frac{1}{2000}\left(V_2^2 - V_1^2\right) + \frac{g}{1000}(z_2 - z_1) + \frac{w_{1-2}}{m} = \frac{Q_{1-2}}{m}$$

By definition of specific enthalpy,

$$h = u + pv$$

$$h = u + \frac{p}{\rho} \quad \therefore v = \frac{1}{\rho}$$

At inlet:

$$h_1 = u_1 + \frac{p_1}{\rho_1}$$

$$= 920 + \frac{140}{25}$$

$$= 920 + 5.6 = 925.6\,\text{kJ/kg}$$

At exit:

$$h_2 = u_2 + \frac{p_2}{\rho_2}$$

$$= 720 + \frac{560}{5} = 720 + 112 = 832\,\text{kJ/kg}$$

$\therefore$

$$832 - 925.6 + \frac{1}{2000}\left[(180)^2 - (120)^2\right] + \frac{9.81}{1000} \times 60 + \frac{W_{1-2}}{0.2} = -\frac{60}{0.2}$$

$$-93.6 + 9 + 0.5886 + \frac{W_{1-2}}{0.2} = -300$$

$$-84.01 + \frac{W_{1-2}}{0.2} = -300$$

or

$$\frac{W_{1-2}}{0.2} = -215.99$$

or

$$W_{1-2} = -215.99 \times 0.2$$
$$= \mathbf{-43.198\,kW}$$

The $-ve$ sign indicates that the work done on the process is 43.198 kW.

Problem 4.14:
During the flight, the airspeed of a turbojet engine is 250 m/s. Ambient air temperature -14 °C. The gas temperature at the nozzle outlet is 610 °C. The enthalpy of air at entry is 250 kJ/kg and the enthalpy of gases at exit is 900 kJ/kg. The fuel–air ratio is 0.0180: 1. The chemical energy of fuel is 45000 kJ/kg. Heat loss from the engine is 21 kJ/kg of air. Calculate the velocity of gases at the exit (Fig. 4.23).

Solution:
Given data:
   Airspeed of turbojet:

$$V_a = 250\,\text{m/s}$$

Ambient temperature:

$$T_a = -14\,°\text{C}$$

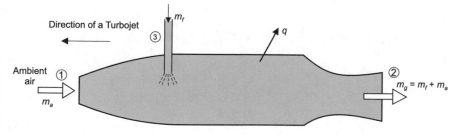

Fig. 4.23 Turbojet

Gas temperature at the nozzle outlet:

$$T_g = 610\,°C$$

Enthalpy of air at inlet:

$$h_a = 250\,kJ/kg$$

Enthalpy of gases at exit:

$$h_g = 900\,kJ/kg$$

Fuel–air ratio:

$$m_f : m_a = 0.0180 : 1$$

i.e.,

$$\frac{m_f}{m_a} = 0.0180$$

Chemical energy of fuel:

$$C.V. = 45000\,kJ/kg$$

Heat loss from the engine:

$$q = -21\,kJ/kg\,of\,air$$

Velocity of gases at exit:

$$V_g = ?$$

According to the steady flow energy equation,

$$m_a\left(h_a + \frac{V_a^2}{2} + gz_1\right) + Q + m_f C.V. = m_g\left(h_g + \frac{V_g^2}{2} + gz_2\right) + W$$

where

$$z_1 = z_2$$

Work interaction is zero, i.e.,

$$W = 0$$

Mass of gases:

$$m_g = m_a + m_f$$

$\therefore$

$$m_a\left(h_a + \frac{V_a^2}{2}\right) + Q + m_f C.V. = \left(m_a + m_f\right)\left[h_g + \frac{V_g^2}{2}\right]$$

$$h_a + \frac{V_a^2}{2} + \frac{Q}{m_a} + \frac{m_f}{m_a}C.V. = \left(1 + \frac{m_f}{m_a}\right)\left[h_g + \frac{V_g^2}{2}\right]$$

$$h_a + \frac{V_a^2}{2} + q + \frac{m_f}{m_a}C.V. = \left(1 + \frac{m_f}{m_a}\right)\left[h_g + \frac{V_g^2}{2}\right]$$

If specific enthalpies, heat loss, and calorific value of fuel are in kJ/kg. Then, terms $\frac{V_a^2}{2}$ and $\frac{V_g^2}{2}$ are also made in kJ/kg by dividing 1000.

$$h_a + \frac{V_a^2}{2000} + q + \frac{m_f}{m_a}C.V. = \left(1 + \frac{m_f}{m_a}\right)\left[h_g + \frac{V_g^2}{2000}\right]$$

$$250 + \frac{(250)^2}{2000} - 21 + 0.0180 \times 45000 = (1 + 0.0180)\left[900 + \frac{V_g^2}{2000}\right]$$

$$250 + 31.25 - 21 + 810 = 916.2 + \frac{1.018}{2000}V_g^2$$

$$1070.25 = 916.2 + \frac{1.018}{2000}V_g^2$$

or

$$\frac{1.018}{2000}V_g^2 = 150.05$$

or

$$V_g^2 = 150.05 \times \frac{2000}{1.018} = 302652.259$$

or

$$V_g = 550.13\,\text{m/s}$$

Problem 4.15:
Reconsider Problem 4.14. If 5% of the chemical energy is not released due to incomplete combustion. Calculate the velocity of the exhaust jet.

Solution:

Given data:

5% of the chemical energy is not released due to incomplete combustion,

i.e., 95% of the chemical energy is released due to fuel combustion $= 0.95\, m_f C.V.$

According to the steady flow energy equation, by neglecting potential energy and work interaction,

$$m_a\left(h_a + \frac{V_a^2}{2}\right) + Q + 0.95 m_f C.V. = \left(m_a + m_f\right)\left[h_g + \frac{V_g^2}{2}\right]$$

$$h_a + \frac{V_a^2}{2} + \frac{Q}{m_a} + 0.95\frac{m_f}{m_a}C.V. = \left(1 + \frac{m_f}{m_a}\right)\left(h_g + \frac{V_g^2}{2}\right)$$

$$h_a + \frac{V_a^2}{2} + q + 0.95\frac{m_f}{m_a}C.V. = \left(1 + \frac{m_f}{m_a}\right)\left(h_g + \frac{V_g^2}{2}\right)$$

If specific enthalpies, heat loss, and calorific value of fuel are in kJ/kg. Then, terms $\frac{V_a^2}{2}$ and $\frac{V_g^2}{2}$ are also taken in kJ/kg on dividing $\frac{V_a^2}{2}$ and by 1000.

$$h_a + \frac{V_a^2}{2000} + q + 0.95\frac{m_f}{m_a}C.V. = \left(1 + \frac{m_f}{m_a}\right)\left[h_g + \frac{V_g^2}{2000}\right]$$

$$250 + \left(\frac{250}{2000}\right)^2 - 21 + 0.95 \times 0.0180 \times 45000 = (1 + 0.0180)\left(900 + \frac{V_g^2}{2000}\right)$$

$$250 + 31.25 - 21 + 76.95 = 916.2 + \frac{1.018}{2000}V_g^2$$

$$1029.75 = 916.2 + \frac{1.018}{2000}V_g^2$$

or

$$\frac{1.018}{2000}V_g^2 = 113.05$$

$$V_g^2 = \frac{113.05 \times 2000}{1.018}$$

or

$$V_g^2 = 222102.16$$

or

$$V_g = \mathbf{471.27\,m/s}$$

Problem 4.16:
Air at a temperature of 15 °C passes through a heat exchange at a velocity of 30 m/s where its temperature is raised to 800 °C. It then enters a turbine with the same velocity of 30 m/s and expands until the temperature falls to 650 °C. On leaving the turbine, the air is taken at a 60 m/s to a nozzle where it expands until the temperature has fallen to 500 °C. If the airflow rate is 2 kg/s, determine.

(a) The rate of heat transfer to the air in the heat exchanger.
(b) The power output from the turbine, assuming no heat loss.
(c) The velocity at the exit from the nozzle, assuming no heat loss.

Take the specific enthalpy of air as $h = c_p T$, where c_p is the specific heat at constant pressure and is equal to 1.005 kJ/kgK. T is the temperature (Fig. 4.24).

Solution:
Given data:
   At inlet of the heat exchanges:
   Temperature of air:

$$T_1 = 15\,°C$$

Velocity of air:

$$V_1 = 30\,\text{m/s}$$

At exit of the heat exchanges or inlet of the turbine:
Temperature of air:

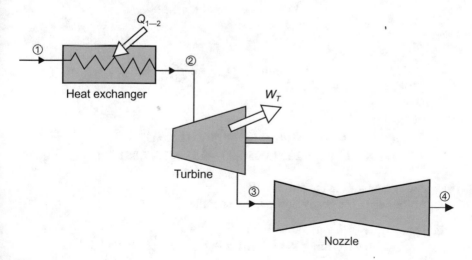

Fig. 4.24 Schematic for Problem 4.16

$$T_2 = 800\,^\circ\text{C}$$

Velocity:

$$V_2 = 30\,\text{m/s} = V_1$$

At turbine exit or nozzle inlet:

$$T_3 = 650\,^\circ\text{C}$$
$$V_3 = 60\,\text{m/s}$$

At nozzle exit:

$$T_4 = 500\,^\circ\text{C}$$

Mass flow rate of air:

$$m = 2\,\text{kg/s}$$

(a)   Rate of heat transfer to the air in the heat exchanges: Q_{1-2}
According to the steady flow energy equation,

$$m\left[h_2 + \frac{V_2^2}{2} + gz_2\right] + Q_{2-3} = m\left[h_3 + \frac{V_3^2}{2} + gz_3\right] + W_T$$

where

$$V_1 = V_2, z_1 = z_2, W_{1-2} = 0$$

$\therefore$

$$mh_1 + Q_{1-2} = mh_2$$

or

$$Q_{1-2} = m(h_2 - h_1) = mc_p(T_2 - T_1)$$
$$= 2 \times 1.005(800 - 15) = \mathbf{1577.85\,kJ/s}$$

(b)   Power output from the turbine: W_T
According to the steady flow energy equation,

$$m\left[h_2 + \frac{V_2^2}{2} + gz_2\right] + Q_{2-3} = m\left[h_3 + \frac{V_3^2}{2} + gz_3\right] + W_T$$

where

$$z_2 = z + 3, \quad Q_{2-3} = 0$$

∴

$$mh_2 + \frac{mV_2^2}{2} = mh_3 + \frac{mV_3^2}{2} + W_T$$

or

$$mh_2 + \frac{mV_2^2}{2000} = mh_3 + \frac{mV_3^2}{2000} + W_T \quad \text{by units similarity}$$

or

$$
\begin{aligned}
W_T &= m(h_2 - h_3) + \frac{m}{2000}\left(V_2^2 - V_3^2\right) \\
&= mc_p(T_2 - T_3) + \frac{m}{2000}\left(V_2^2 - V_3^2\right) \\
&= 2 \times 1.005(800 - 650) + \frac{2}{2000}\left[(30)^2 - (60)^2\right] \\
&= 301.5 - 2.7 = \mathbf{298.8\,kW}
\end{aligned}
$$

(c)   Velocity at the exit from the nozzle: V_4
According to the steady flow energy equation,

$$m\left[h_3 + \frac{V_3^2}{2} + gz_3\right] + Q_{3-4} = m\left[h_4 + \frac{V_4^2}{2} + gz_4\right] + W_{3-4}$$

where

$$z_3 = z_4, \quad Q_{3-4} = 0, \quad W_{3-4} = 0$$

∴

$$mh_3 + \frac{mV_3^2}{2} = mh_4 + \frac{mV_4^2}{2}$$

or

$$h_3 + \frac{V_3^2}{2} = h_4 + \frac{V_4^2}{2}$$

$$h_3 + \frac{V_3^2}{2000} = h_4 + \frac{V_4^2}{2000} \quad \text{by units similarity}$$

or

$$\frac{V_4^3 - V_3^2}{2000} = h_3 - h_4$$

$$V_4^2 - V_3^2 = 2000c_p(T_3 - T_4)$$

$$V_4^2 = 2000c_p(T_3 - T_4) + V_3^2$$

or

$$V_4 = \sqrt{2000c_p(T_3 - T_4) + V_3^2}$$
$$= \sqrt{2000 \times 1.005(650 - 500) + (60)^2}$$
$$= \sqrt{301500 + 3600} = \sqrt{305100} = \mathbf{552.35\,m/s}$$

Problem 4.17:

The following data were collected to design an air-conditioning system for a restaurant in Delhi:

$$\text{Solar heat gain through walls and roof} = 15000\,\text{kJ/h}$$
$$\text{Solar heat gain through glass} = 14400\,\text{kJ/h}$$
$$\text{Occupants} = 25$$
$$\text{Heat gain per person} = 660\,\text{kJ/h}$$

Internal lighting load = 15 lamps of 100 W capacity each and 10 fluorescent tubes of 80 W each. Determine the rate at which heat to removed by a restaurant air-conditioning system, so that a steady state is maintained in the restaurant.

Solution:

Given data:

Solar heat gain through walls and roof:

$$Q_{SW} = 15000\,\text{kJ/h} = \frac{15000}{3600}\,\text{kW} = 4.16\,\text{kW}$$

Solar heat gain through glass:

$$Q_{SG} = 14500\,\text{kJ/h} = \frac{14400}{3600}\,\text{kW} = 4\,\text{kW}$$

Heat gain from 25 occupants:

$$Q_0 = 25 \times 660\,\text{kJ/h}$$
$$= 16500\,\text{kJ/h} = \frac{16500}{3600}\,\text{kW} = 4.58\,\text{kW}$$

Heat gain from 10 lamps:

$$Q_L = 15 \times 100W = 1500\,W = 1.5\,kW$$

Heat gain from 10 fluorescent

$$Q_F = 10 \times 80\,W = 800\,W = 0.8\,kW$$

According to energy balance equation,
Rate of heat removed by a restaurant air-conditioning system = sum of heat gain by a restaurant

$$= Q_{SW} + S_{WG} + Q_O + Q_L + Q_F$$
$$= 4.16 + 4 + 4.58 + 1.5 + 0.8 = 15.04\,kW$$

Problem 4.18:
The following data is given for water turbine at inlet and exit.

Variable	Inlet	Exit
Pressure	1.15 MPa	0.05 MPa
Velocity	30 m/s	15.5 m/s
Height above datum	10 m	2 m

If the discharge of the water 40 m³/s, determine the net hydraulic energy that is converted into work.

Solution:
Given data:
   At inlet:
   Pressure:

$$p_1 = 1.15\,MPa = 1150\,kPa$$

Velocity:

$$V_1 = 30\,m/s$$

Datum head:

$$z_1 = 10\,m$$

At exit;
Pressure:

$$p_1 = 0.05\,\text{MPa} = 50\,\text{kPa}$$

Velocity:

$$V_2 = 15.5\,\text{m/s}$$

Datum head:

$$z_2 = 2\,\text{m}$$

Discharge:

$$Q = 40\,\text{m}^3/\text{s}$$

also

$$Q = mv$$

$$Q = \frac{m}{\rho} \quad \Big| \quad \because \text{ Specific volume: } v = \frac{1}{\rho}$$

$$40 = \frac{m}{1000} \qquad \because \rho = 1000\,\text{kg/m}^3 \text{ for water}$$

or

$$m = 40 \times 10^3\,\text{kg/s}$$

We know that the steady flow energy equation

$$m\left[h_1 + \frac{V_1^2}{2} + gz_1 \right] + Q_{1-2} = m\left[h_2 + \frac{V_2^2}{2} + gz_2 \right] + W_{1-2}$$

$$m\left[u_1 + p_1 v_1 + \frac{V_1^2}{2} + gz_1 \right] + Q_{1-2} = m\left[u_2 + p_2 v_2 + \frac{V_2^2}{2} + gz_2 \right] + W_{1-2}$$

where

$$u_1 = u_2 \quad \therefore T_1 = T_2 \text{ for water turbine}$$

$$v_1 = v_2 = \frac{1}{\rho}$$

$$Q_{1-2} = 0$$

$\therefore$

$$m\left[\frac{p_1}{\rho} + \frac{V_1^2}{2} + gz_1 \right] = m\left[\frac{p_2}{\rho} + \frac{V_2^2}{2} + gz_2 \right] + W_{1-2}$$

Note: Make all terms in kJ/kg

$$m\left[\frac{p_1}{\rho} + \frac{V_1^2}{2000} + \frac{gz_1}{1000}\right] = m\left[\frac{p_2}{\rho} + \frac{V_2^2}{2000} + \frac{gz_2}{1000}\right] + W_{1-2}$$

$$40 \times 10^3\left[\frac{1150}{1000} + \frac{(30)^2}{2000} + \frac{9.81 \times 10}{1000}\right] = 40 \times 10^3\left[\frac{50}{1000} + \frac{(15.5)^2}{2000}\right.$$

$$\left. + \frac{9.81 \times 2}{1000}\right] + W_{1-2}$$

$$40 \times 10^3[1.15 + 0.45 + 0.0981] = 40 \times 10^3[0.05 + 0.120 + 0.0196] + W_{1-2}$$

$$67.92 \times 10^3 = 7.58 \times 10^3 + W_{1-2}$$

or

$$W_{1-2} = 60.34 \times 10^3 \text{ kJ/s or kW} = \mathbf{60.34\,MW}$$

Problem 4.19:

A turbine operates under steady flow condition, receiving steam at the following state:

$$\text{Pressure} = 12\,\text{bar}$$
$$\text{Temperature} = 188\,°\text{C}$$
$$\text{Specific enthalpy} = 2785\,\text{kJ/kg}$$
$$\text{Velocity} = 33.3\,\text{m/s}$$
$$\text{Elevation} = 3\,\text{m}$$

The steam leaves the turbine at the following state:

$$\text{Pressure} = 0\,\text{bar}$$
$$\text{Specific enthalpy} = 2512\,\text{kJ/kg}$$
$$\text{Velocity} = 100\,\text{m/s}$$
$$\text{Elevation is zero}$$

A heat loss of 0.29 kJ/s occurs during the expansion of steam in the turbine. If the rate of steam flow through the turbine is 0.42 kg/s, determine the power output of the turbine.

Solution:
Given data:
At inlet:

$$p_1 = 12\,\text{bar}$$

$$T_1 = 188\,^{\circ}\text{C}$$
$$h_1 = 2785\,\text{kJ/kg}$$
$$V_1 = 33.3\,\text{m/s}$$
$$z_1 = 3\,\text{m}$$

At exit:

$$p_2 = 0.2\,\text{bar}$$
$$h_2 = 2512\,\text{kJ/kg}$$
$$V_2 = 100\,\text{m/s}$$
$$z_2 = 0$$

Heat loss:

$$Q_{1-2} = -0.29\,\text{kJ/s}$$

Mass flow rate:

$$m = 0.42\,\text{kg/s}$$

We know that the steady flow energy equation

$$m\left[h_1 + \frac{V_1^2}{2} + gz_1\right] + Q_{1-2} = m\left[h_2 + \frac{V_2^2}{2} + gz_2\right] + W_{1-2}$$

Note: Make all terms in kJ/s

$$m\left[h_1 + \frac{V_1^2}{2000} + \frac{gz_1}{1000}\right] + Q_{1-2} = m\left[h_2 + \frac{V_2^2}{2000} + \frac{gz_2}{1000}\right] + W_{1-2}$$

$$0.42\left[2785 + \frac{(33.3)^2}{2000} + \frac{9.81 \times 3}{1000}\right] - 0.29 = 0.42\left[2512 + \frac{(100)^2}{2000} + \frac{9.81 \times 0}{1000}\right] + W$$

$$0.42[2785 + 0.5544 + 0.0294] - 0.29 = 0.42[2512 + 5 + 0] + W_{1-2}$$

$$1169.94 - 0.29 = 1057.14 + W_{1-2}$$

$$1169.65 = 1057.14 + W_{1-2}$$

or

$$W_{1-2} = 112.51\text{kJ/s} = \mathbf{112.51\,kW}$$

Problem 4.20:

A turbocompressor delivers 2.33 m³/s at 276 kPa, 43 °C which is heated at this pressure to 430 °C and finally expanded in a turbine which delivers 1860 kW. During the expansion, there is a heat transfer of 90 kJ/s to the surroundings. Determine the turbine exhaust temperature if changes in kinetic and potential energies are negligible (Fig. 4.25).

Solution:

Given data:

At exit of a compressor:
Discharge:

$$Q = 2.33\,\text{m}^3/\text{s}$$

Pressure:

$$p_2 = 276\,\text{kPa}$$

Temperature:

$$T_2 = 43\,^\circ\text{C} = (43 + 273)\,\text{K} = 316\,\text{K}$$

For turbine inlet,
Temperature:

$$T_3 = 430\,^\circ\text{C} = (430 + 273)\,\text{K} = 703\,\text{K}$$

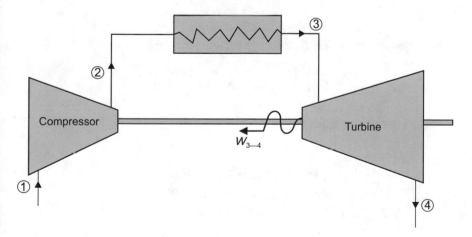

Fig. 4.25 Schematic for Problem 4.20

Work deliver:

$$W_{3-4} = 1860\,\text{kW}$$

Heat transfer:

$$Q_{3-4} = -90\,\text{kJ/s}$$

Now applying the gas equation at state 2,

$$p_2 v_2 = RT_2$$
$$276 \times v_2 = 0.287 \times 316$$

or

$$v_2 = 0.3285\,\text{m}^3/\text{kg} \quad (\because\ R = 0.287\,\text{kJ/kgK})$$

We know that discharge

$$Q = mv$$

$\therefore$

$$2.33 = m \times 0.3285$$

or

$$m = 7.09\,\text{kg/s}$$

Now applying the steady flow energy equation for turbine,

$$m\left[h_3 + \frac{V_3^2}{2} + gz_3\right] + Q_{3-4} = m\left[h_4 + \frac{V_4^2}{2} + gz_4\right] + W_{1-2}$$

Given condition: Changes in kinetic and potential energies are negligible
$\therefore$

$$mh_3 + Q_{3-4} = mh_4 + W_{1-4}$$
$$mc_p T_3 + Q_{3-4} = mc_p T_4 + W_{1-4}$$
$$7.09 \times 1.005 \times 703 - 90 = 7.09 \times 1.005 \times T_4 + 1860$$
$$5009.19 - 90 = 7.12 T_4 + 1860$$

or

$$7.12 T_4 = 3059.19$$

or

$$T_4 = 429.66K = (429.66 - 273)\,°C = \mathbf{156.66\,°C}$$

Problem 4.21:

A gas enters an adiabatic rotary compressor at a temperature of 16 °C, a pressure of 100 kPa, and the specific enthalpy of 391.2 kJ/kg. The gas leaves the compressor at a temperature of 245 °C, a pressure of 6 bar, and the specific enthalpy of 534.5 kJ/kg.
   Determine

(a)   The external work done per unit mass of the gas. Assuming the change in kinetic energy of the gas to be negligible.
(b)   The external work done per unit mass of the gas when the gas velocity at inlet is 80 m/s and at exit is 160 m/s.

Solution:
Given data:
   Heat transfer:

$$q_{1-2} = 0 \qquad \because \text{ Adiabatic compression}$$

At inlet:
Temperature:

$$T_1 = 16\,°C$$

Pressure:

$$p_1 = 100\,\text{kPa}$$

Specific enthalpy:

$$h_1 = 391.2\,\text{kJ/kg}$$

At exit:
Temperature:

$$T_2 = 245\,°C$$

Pressure:

$$p_2 = 6\,\text{bar} = 600\,\text{kPa}$$

Specific enthalpy:

$$h_2 = 534.5 \, \text{kJ/kg}$$

(a)   We know that the steady flow energy equation for unit mass

$$h_1 + \frac{V_1^2}{2} + gz_1 + q_{1-2} = h_2 + \frac{V_2^2}{2} + gz_2 + w_{1-2}$$

where

$$z_1 = z_2$$
$$q_{1-2} = 0$$
$$\Delta KE = 0$$

$\therefore$

$$h_1 = h_2 + w_{1-2}$$
$$391.2 = 534.5 + w_{1-2}$$

or

$$w_{1-2} = 391.2 - 534.5 = \mathbf{-143.3 \, kJ/kg}$$

The $-ve$ sign shows that work done on the system, i.e., 143.3 kJ/kg work required to drive the compressor.

(b)   Velocity at inlet:

$$V_1 = 80 \, \text{m/s}$$

Velocity at exit:

$$V_2 = 160 \, \text{m/s}$$

We know that the steady flow energy equation for unit mass

$$h_1 + \frac{V_1^2}{2} + gz_1 + q_{1-2} = h_2 + \frac{V_2^2}{2} + gz_2 + w_{1-2}$$

where

$$z_1 = z_2$$
$$q_{1-2} = 0$$

$\therefore$

$$h_1 + \frac{V_1^2}{2} = h_2 + \frac{V_2^2}{2} + w_{1-2}$$

Note: Make all terms in kJ/kg

$$h_1 + \frac{V_1^2}{2000} = h_2 + \frac{V_2^2}{2000} + w_{1-2}$$
$$391.2 + \frac{(80)^2}{2000} = 534.5 + \frac{(160)^2}{2000} + w_{1-2}$$
$$391.2 + 3.2 = 534.5 + 12.8 + w_{1-2}$$
$$394.4 = 547.3 + w_{1-2}$$

or

$$w_{1-2} = -\mathbf{152.9\,kJ/kg}$$

The $-ve$ sign shows that work done on the system, i.e., 152.9 kJ/kg work required to drive the compressor.

Problem 4.22:
The centrifugal pump delivers 50 kg of water per second. The inlet and outlet pressure are 1 bar and 4.2 bar, respectively. The suction is 2.2 m below the centre of the pump and delivery is 8.5 m above the centre of the pump. The suction and delivery pipe diameter are 20 cm and 10 cm, respectively. Determine the power required to drive the pump (Fig. 4.26).

Solution:
Given data:

$$m = 50\,kg/s$$

Pressure at inlet:

$$p_1 = 1\,bar = 100\,kPa$$

Pressure at outled:

$$p_1 = 1\,bar = 100\,kPa$$

Suction below the centre of the pump $= 2.2$ m.
Delivery above the centre of the pump $= 8.5$ m.
Diameter of suction pipe:

$$d_1 = 20\,cm = 0.2\,m$$

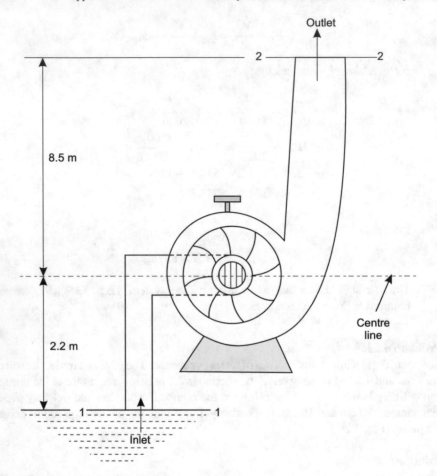

Fig. 4.26 Schematic for Problem 4.22

∴ Cross-sectional area:

$$A_1 = \frac{\pi}{4}d_2^2 = \frac{3.14}{4} \times (0.2)^2 = 0.0314 \text{m}^2$$

Diameter of suction pipe:

$$d_2 = 10 \text{ cm} = 0.1 \text{ m}$$

∴ Cross-sectional area:

$$A_2 = \frac{\pi}{4}d_2^2 = \frac{3.14}{4} \times (0.1)^2 = 0.00785 \text{m}^2$$

Let the datum at inlet state 1. Applying the steady flow energy equation at sections 1 and 2, we get

$$m\left[h_1 + \frac{V_1^2}{2} + gz_1\right] + Q = m\left[h_2 + \frac{V_2^2}{2} + gz_2\right] + W$$

By definition of specific enthalpy,

$$h = u + pv \quad \therefore v = \frac{p}{\rho}$$

$$h = u + \frac{p}{\rho}$$

At inlet:

$$h_1 = u_1 + \frac{p_1}{\rho} \quad \therefore \rho = c$$

At outlet:

$$h_2 = u_2 + \frac{p_2}{\rho}$$

$$z_1 = 0, z_2 = 2.2 + 8.5 = 10.7\,\mathrm{m}$$

$$Q = 0$$

$$m\left[u_1 + \frac{p_1}{\rho} + \frac{V_1^2}{2}\right] = m\left[u_2 + \frac{p_2}{\rho} + \frac{V_2^2}{2} + 9.81 \times 10.7\right] + W$$

For units balancing,

$$m\left[u_1 + \frac{p_1}{\rho} + \frac{V_1^2}{2000}\right] = m\left[u_2 + \frac{p_2}{\rho} + \frac{V_2^2}{2000} + \frac{9.81 \times 10.7}{1000}\right] + W$$

or

$$W = m(u_1 - u_2) + m\left(\frac{p_1 - p_2}{\rho}\right) + m\left(\frac{V_1^2 - V_2^2}{2000}\right) - \frac{981 \times 10.7}{1000}$$

where

$$u_1 = u_2 \quad \because T = C$$

$$m = \rho A_1 V_1$$

$$50 = 1000 \times 0.0314 \times V_1$$

or

$$V_1 = 1.59\,\text{m/s}$$

and also

$$m = \rho A_2 V_2 \qquad \because \rho = C, \text{ water is incompressible}$$
$$50 = 1000 \times 0.00785 \times V_2$$

$m = \rho A_2\, V_2 \therefore \rho = C$, water is incompressible.
$50 = 1000 \times 0.00785 \times V_2$.
or

$$V_2 = 6.37\,\text{m/s}$$
$$W = 0 + 50\left(\frac{1000 - 420}{1000}\right) + 50\left[\frac{(1.59)^2 - (6.37)^2}{2000}\right] - \frac{9.81 \times 10.7 \times 50}{1000}$$
$$= -16 - 0.95 - 5.25 = \mathbf{-22.2\,kW}$$

Problem 4.23:

A rigid tank contains 20 m³ of air at 10 MPa and 25 °C. If the air is allowed to escape with no heat transfer to 200 kPa, determine the mass of air leaving from the tank and the final temperature of the air in the tank (Fig. 4.27).

Solution:

Given data:
   Initial volume of air:

$$V_i = 20\,\text{m}^3$$

   Initial pressure of air:

$$p_i = 10\,\text{MPa} = 10 \times 10^3\,\text{kPa}$$

Fig. 4.27 Schematic for Problem 4.23

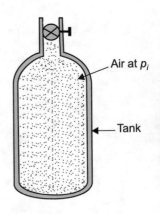

Air at p_i

Tank

Initial temperature of air:

$$T_i = 25\,°C = (273 + 25)\,K = 298\,K$$

Final pressure of air:

$$p_f = 200\,kPa$$

Final volume of air:

$$V_f = V_i = 20\,m^3$$

The initial mass of air in the tank is found by using the equation of state

$$p_i V_i = m_i R T_i$$
$$10 \times 10^3 \times 20 = m_i \times 0.287 \times 298$$

or

$$m_i = 2338.47\,kg$$

Specific volume at initial state,

$$v_i = \frac{V_i}{m_i} = \frac{20}{2338.47}$$
$$= 8.55 \times 10^{-3} m^3/kg$$

For a process with no heat transfer, i.e., adiabatic,

$$\frac{p_f}{p_i} = \left(\frac{v_i}{v_f}\right)^\gamma$$
$$\frac{200}{10 \times 10^3} = \left(\frac{8.55 \times 10^{-3}}{v_f}\right)^{1.4}$$
$$0.02 = \left(\frac{8.55 \times 10^{-3}}{v_f}\right)^{1.4}$$

or

$$v_f = \frac{8.55 \times 10^{-3}}{(0.02)^{1/1.4}} = 0.1396\,m^3/kg$$

also

$$v_f = \frac{V_f}{m_f}$$

∴

$$0.1396 = \frac{20}{m_f} \quad \because V_f = V_i$$

or

$$m_f = 143.26 \, \text{kg}$$

Mass of air leaving from the tank:

$$\Delta m = m_i - m_f = 2338.47 - 143.26 = \mathbf{2195.21 \, kg}$$

$$\frac{T_f}{T_i} = \left(\frac{p_f}{p_i}\right)^{\frac{\gamma-1}{\gamma}}$$

$$T_f = 298 \times \left(\frac{200}{10 \times 10^3}\right)^{\frac{1.4-1}{1.4}}$$

$$T_f = 97.72 \, \text{K} = (97.72 - 273)\,°\text{C} = \mathbf{-175.28\,°C}$$

Problem 4.24:
An insulated rigid tank is initially evacuated. A valve is opened, and atmospheric air at 95 kPa and 17 °C enters the tank until the pressure in the tank reaches 95 kPa, at which point the valve is closed. Find the final temperature of the air in the tank (Fig. 4.28).

Fig. 4.28 Schematic for Problem 4.24

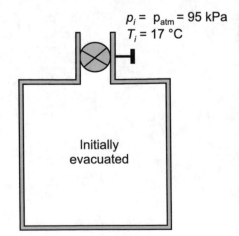

$p_i = p_{atm} = 95 \, \text{kPa}$
$T_i = 17 \,°\text{C}$

Initially evacuated

Solution:
Given data:
   At inlet:

$$p_i = 95 \, \text{kPa}$$
$$T_i = 17\,^\circ\text{C} = (17 + 273)\text{K}$$
$$= 290 \, \text{K}$$

Initial condition in the tank:

$$m_1 = 0$$
$$p_1 = 0$$
$$T_1 = 0$$

Final condition:

$$p_2 = 95 \, \text{kPa}$$
$$T_2 = ?$$

Applying unsteady flow energy equation, we have

$$m_i \left[h_i + \frac{V_i^2}{2} + gz_i \right] + Q_{1-2} = m_2 \left[u_2 + \frac{V_2^2}{2} \right] - m_1 \left[u_1 + \frac{V_1^2}{2} \right]$$
$$+ m_e \left[h_e + \frac{V_e^2}{2} + gz_e \right] + W_{1-2}$$

where

$$m_1 = 0 \quad \text{initially evacuated}$$
$$m_e = 0 \quad \text{no exit}$$
$$Q_{1-2} = 0 \quad \text{insulated tank}$$
$$W_{1-2} = 0$$
$$z_i = z_e$$

V_i and V_2 are neglected.
$\therefore$

$$m_i h_i = m_2 u_2$$
$$h_i = u_2 \qquad \because m_i = m_2$$
$$c_p T_i = c_v T_2$$

$$\frac{c_p}{c_v} T_i = T_2$$

$$\gamma T_i = T_2 \qquad \because \gamma = c_p/c_v$$

$$1.4 \times 290 = T_2$$

or

$$T_2 = 406\,\text{K} = \mathbf{133\,°C}$$

Problem 4.25:

A 2 m³ rigid tank initially contains air at 100 kPa and 22 °C. The tank is connected to a supply line through a valve. Air is flowing in the supplying line at 600 kPa and 22 °C. The valve is opened, and the air is allowed to enter the tank until the pressure in the tank reaches the line pressure, at which the valve is closed. At that instant, the air in the tank is at 77 °C. Determine (Fig. 4.29).

(a)   the mass of air that entered the tank and
(b)   the heat transfer during the process.

Solution:

Given data:

   Initial condition in the tank:

$$V = 2\,\text{m}^3$$

$$p_1 = 100\,\text{kPa}$$

$$T_1 = 22\,°\text{C}$$

$$= 22 + 273$$

$$= 295\,\text{K}$$

Fig. 4.29 Schematic for Problem 4.25

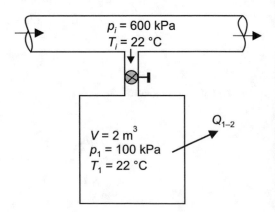

$$p_1 V = m_1 R T_1$$
$$100 \times 2 = m_1 \times 0.287 \times 295$$

or

$$m_1 = 2.36 \, \text{kg}$$

At inlet:

$$p_i = 600 \, \text{kPa}$$
$$T_1 = 22 \, ^\circ\text{C} = 295 \, \text{K}$$
$$m_i = ?$$

Final condition:

$$p_1 = 600 \, \text{kPa}$$
$$T_2 = 77 \, ^\circ\text{C} = 350 \, \text{K}$$
$$V_2 = V_1 = V = 2 \, \text{m}^3$$
$$p_2 V = m_2 R T_2$$
$$600 \times 2 = m_2 \times 0.287 \times 350$$

or

$$m_2 = 11.94 \, \text{kg}$$

(a)   Mass of air entered the tank: m_i

$$m_i = m_2 - m_1 = 11.94 - 2.36 = \mathbf{9.58 \, kg}$$

(b)   Heat transfer during the process: Q_{1-2}
Applying unsteady flow energy equation, we have

$$m_i \left[h_i + \frac{V_i^2}{2} + g z_i \right] + Q_{1-2} = m_2 \left[u_2 + \frac{V_2^2}{2} \right] - m_1 \left[u_1 + \frac{V_1^2}{2} \right]$$
$$+ m_e \left[h_e + \frac{V_e^2}{2} + g z_e \right] + W_{1-2}$$

where

$$m_e = 0 \quad \text{no exit}$$
$$W_{1-2} = 0$$
$$z_i = 0$$

V_i, V_2 are neglected

$\therefore$

$$m_i h_i + Q_{1-2} = m_2 u_2 - m_1 u_1$$
$$m_i c_p T_i + Q_{1-2} = m_2 c_v T_2 - m_1 c_v T_1$$

where

$$\left. \begin{array}{l} c_p = 1.005 \, \text{kJ/kgK} \\ c_v = 0.718 \, \text{kJ/kgK} \end{array} \right\} \text{for air}$$

$\therefore$

$$9.58 \times 1.005 \times 295 + Q_{1-2} = 11.94 \times 0.718 \times 350 - 2.36 \times 0.718 \times 295$$
$$2840.23 + Q_{1-2} = 3000.52 - 499.87$$

or

$$Q_{1-2} = -339.58 \, \text{kJ}$$

The $-ve$ signs show that the heat transfer from the system, i.e., tank.

Problem 4.26:

The internal energy of air is given, at ordinary temperature,
by

$$u = u_0 + 0.718 \, \text{T}$$

where u is in kJ/kg.
u_0 is any arbitrary value of u at 0 °C, kJ/kg and T is the temperature in °C.
Also for air,

$$pv = 0.287(T + 273)$$

where p is in kPa and
V is in m³/kg.

(i)   An evacuated bottle is fitted with a valve through which air from the atmo-
       sphere, at 760 mm of Hg and 25 °C, is allowed to flow slowly to fill the bottle.
       If no heat is transferred to or from the air in the bottle, what will its temperature
       be when the pressure in the bottle reaches 760 mm of Hg?
(ii)   If the bottle initially contained 0.03 m³ of air at 400 mm Hg and 25 °C, what
       will the temperature be when the pressure in the bottle reaches 760 mm of Hg?
       (Fig. 4.30)

Fig. 4.30 Schematic for
Problem 4.26

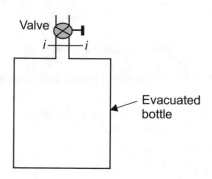

Solution:

Given data:

$$u = u_0 + 0.718\,T$$

Here

$$c_v = 0.718\,\text{kJ/kgK}$$

and

$$pv = 0.287(T + 273)$$

Here

$$R = 0.287\,\text{kJ/kgK}$$

(i)   Atmosphere pressure head:

$$h_i = 760\,\text{mm of Hg}$$
$$= 0.760\,\text{m of Hg}$$

∴ Atmosphere pressure:

$$p_i = \rho g h_i$$
$$= 13600 \times 9.81 \times 0.76 \quad \rho_{Hg} = 13600\,\text{kg/m}^3$$
$$= 101396.16\,\text{N/m}^2 = 101.396\,\text{kN/m}^2$$
$$= \text{Inlet pressure air in evacuate bottle}$$

Atmosphere temperature:

$$T_i = \text{Inlet temperature of air in evacuate bottle}$$

$$= 25\,^\circ\text{C} = 298\,\text{K}$$

Initial condition in the evacuated bottle,

$$m_1 = 0, \quad p_1 = 0$$

Final condition in the bottle,
Pressure head:

$$h_2 = 760\,\text{mm of Hg}$$
$$= 0.760\,\text{m of Hg}$$

$\therefore$ Pressure:

$$p_2 = \rho h_2 g$$
$$= 13600 \times 0.760 \times 9.81\,\text{N/m}^2$$
$$= 101396.16\,\text{N/m}^2$$
$$= 101.396\,\text{kN/m}^2$$

Applying unsteady flow energy equation,

$$m_i\left[h_i + \frac{V_i^2}{2} + gz_i\right] + Q_{1-2} = m_e\left[h_e + \frac{V_e^2}{2} + gz_e\right] + m_2\left[u_2 + \frac{V_2^2}{2}\right]$$
$$- m_1\left[u_1 + \frac{V_1^2}{2}\right] + W_{1-2}$$

For given problem, V_i and V_2 are neglected.

$$z_i = 0$$
$$Q_{1-2} = 0,\ W_{1-2} = 0$$

No exit condition, i.e., m_e, h_e, V_e, z_e are zero.
Evacuated bottle,

$$m_1 = 0$$

$\therefore$

$$m_i h_i = m_2 u_2$$
$$h_i = u_2 \quad \therefore m_i = m_2$$
$$c_p T_i = c_v T_2$$

or

$$\frac{c_p}{c_v} T_i = T_2$$
$$\gamma T_i = T_2$$
$$1.4 \times 298 = T_2$$

or

$$T_2 = 417.2 \, \text{K}$$
$$T_2 = \mathbf{144.2\,^\circ C}$$

(ii)   Initial condition of air inside the bottle,

$$V_1 = 0.03 \, \text{m}^3$$

Pressure head:

$$h_1 = 400 \, \text{mm of HG} = 0.4 \, \text{m of HG}$$

∴ Pressure:

$$p_1 = \rho g h_1 = 13600 \times 9.81 \times 0.4$$
$$= 53366.4 \, \text{N/m}^2 = 53.366 \, \text{kN/m}^2$$

Temperature:

$$T_1 = 25\,^\circ C = (25 + 273) \, \text{K} = 298 \, \text{K}$$

We know that

$$p_1 V_1 = m_1 R T_1$$
$$53.366 \times 0.03 = m_1 \times 0.287 \times 298$$

or

$$m_1 = 0.01871 \, \text{kg}$$

Final condition in the bottle,

$$V_2 = V_1 = 0.03 \, \text{m}^3$$

Pressure head:

$$h_2 = 760 \, \text{mm of HG} = 0.76 \, \text{m of HG}$$

Pressure:

$$p_2 = \rho g h_2 = 13600 \times 9.81 \times 0.76$$
$$= 101396.16 \, \text{N/m}^2 = 101.396 \, \text{kN/m}^2$$
$$p_2 V_2 = m_2 R T_2$$
$$101.396 \times 0.03 = m_2 \times 0.287 \times T_2$$

or

$$m_2 T_2 = 10.598 \tag{4.32}$$

Applying unsteady flow energy equation,

$$m_i \left[h_i + \frac{V_i^2}{2} + g z_i \right] + Q_{1-2} = m_e \left[h_e + \frac{V_e^2}{2} + g z_e \right] + m_2 \left[u_2 + \frac{V_2^2}{2} \right]$$
$$- m_1 \left[u_1 + \frac{V_1^2}{2} \right] + W_{1-2}$$

The above equation is reduced according to the given condition,

$$m_i h_i = m_2 u_2 - m_1 u_1$$
$$(m_2 - m_1) h_i = m_2 u_2 - m_1 u_1 \quad \left| \begin{array}{l} \because m_2 = m_i + m_1 \\ or \ m_i = m_2 - m_1 \end{array} \right.$$

$$(m_2 - m_1) c_p T_i = m_2 c_v T_2 - m_1 c_v T_1$$
$$(m_2 - 0.01871) c_P \times 298 = c_v [m_2 T_2 - m_1 T_1]$$
$$(m_2 - 0.01871) \frac{c_p}{c} \times 298 = 10.598 - 0.01871 \times 298$$
$$(m_2 - 0.01871) \times \gamma \times 298 = 5.022$$
$$(m_2 - 0.01871) \times 1.4 \times 298 = 5.022$$

or

$$(m_2 - 0.01871) = 0.01203$$

or

$$m_2 = 0.01203 + 0.01871 = 0.03074 \, \text{kg}$$

Substituting the value of m in Eq. (4.32), we get

$$0.03074 \times T_2 = 10.598$$

or

$$T_2 = 344.76\,\text{K} = \mathbf{71.76\,^\circ C}$$

Problem 4.27:

A compressed air bottle of 0.3 m³ volume contains air at 35 bar, 40 °C. This air is used to drive a turbogenerator supplying power to a device that consumes 5 kW. Determine the time for which the device can be operated if the actual output of the turbogenerator is 60% of the maximum theoretical output. The ambient pressure to which the tank pressure has fallen is 1 bar. For air, $\gamma = 1.4$, $c_v = 0.718$ kJ/kgK, and $c_p = 1.005$ kJ/kgK (Fig. 4.31).

Solution:

Given data:
   Initial condition of air in bottle:
   Volume of the bottle:

$$V_1 = 0.3\,\text{m}^3$$
$$p_1 = 35\,\text{bar} = 3500\,\text{kPa}$$
$$T_1 = 40\,^\circ\text{C} = (40 + 273)\text{K} = 313\,\text{K}$$

Final condition of air in bottle:

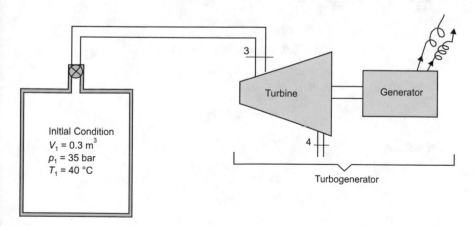

Initial Condition
$V_1 = 0.3\,\text{m}^3$
$p_1 = 35\,\text{bar}$
$T_1 = 40\,^\circ\text{C}$

3

Turbine

4

Generator

Turbogenerator

Fig. 4.31 Schematic for Problem 4.27

$$p_2 = 1\,\text{bar} = 100\,\text{kPa}$$
$$V_2 = V_1 = 0.3\,\text{m}^3$$

For turbine,
Exhaust pressure:

$$p_4 = p_2 = 1\,\text{bar}$$

Now applying an ideal gas equation of state to the initial condition of air in the bottle,

$$p_1 V_1 = m_1 R T_1$$
$$3500 \times 0.3 = m_1 \times 0.287 \times 313$$

or

$$m_1 = 11.688\,\text{kg}$$
$$\frac{T_2}{T_1} = \left(\frac{p_2}{p_1}\right)^{\frac{\gamma-1}{\gamma}} \quad \because \text{adiabatic process follows in the bottle}$$
$$\frac{T_2}{313} = \left(\frac{1}{35}\right)^{\frac{1.4-1}{1.4}}$$

or

$$T_2 = 313 \times (0.02857)^{0.285} = 313 \times 0.363 = 113.62\,\text{K}$$

Final mass of air in the bottle:

$$p_2 V_2 = m_2 R T_2$$
$$100 \times 0.3 = m_2 \times 0.287 \times 113.62$$

or

$$m_2 = 0.92\,\text{kg}$$

For turbine,

$$\frac{T_3}{T_4} = \left(\frac{p_3}{p_4}\right)^{\frac{\gamma-1}{\gamma}}$$

As

$$T_3 = T_1, p_3 = p_1 \quad \left| \begin{array}{l} \because \text{ inital outlet condition of} \\ \text{the bottle} = \text{inlet condition} \\ \text{of the turbine} \end{array} \right.$$

$\therefore$

$$\frac{313}{T_4} = \left(\frac{35}{1}\right)^{\frac{1.4+1}{1.4}}$$

$$\frac{313}{T_4} = (35)^{0.285}$$

or

$$T_4 = 113.62 \, \text{K}$$

also

$$T_3 = T_2, p_3 = p_2 \quad \left| \begin{array}{l} \because \text{ final outlet condition of} \\ \text{the bottle} = \text{inlet condition} \\ \text{of the turbine} \end{array} \right.$$

$$\frac{113.62}{T_4} = \left(\frac{1}{1}\right)^{\frac{\gamma-1}{\gamma}} = 1$$

or

$$T_4 = 113.62 \, \text{K}$$

Work delivered by the turbine: W_T

$$\begin{aligned}
W_T &= \text{energy at turbine inlet} - \text{energy at turbine outlet} \\
&= \text{energy exit of the bottle} - \text{energy at turbine outlet} \\
&= (m_1 u_1 - m_2 u_2) - (m_1 - m_2)h_4 \\
&= m_1 c_v T_1 - m_2 c_v T_2 - (m_1 - m_2)c_p T_4 \\
&= c_v(m_1 T_1 - m_2 T_2) - (m_1 - m_2)c_p T_4 \\
&= 0.718(11.688 \times 313 - 0.92 \times 113.62) \\
&\quad - (11.688 - 0.92) \times 1.005 \times 113.62 \\
&= 2551.63 - 1229.57 = 1322.06 \, \text{kJ}
\end{aligned}$$

Actual work required to drive the generator:

$$\begin{aligned}
W_{\text{act}} &= 60\% \text{ of the maximum theoretical output} \\
&= 0.6 \times W_T = 0.6 \times 1322.06 = 793.236 \, \text{kJ} \quad\quad (4.33)
\end{aligned}$$

Power required to drive the generator:

$$P = 5\,\text{kW} = 5\,\text{kJ/s}$$

Work required to drive the generator:

$$W = 5 \times \Delta t\,\text{kJ} \tag{4.34}$$

where
Δt = time duration for which turbogenerator is run; in seconds.
Equating Eqs. (4.33) and (4.34), we have

$$5\Delta t = 793.236$$

or

$$\Delta t = \textbf{158.64\,s}$$

Problem 4.28:
A compressed air bottle of volume 0.15 m³ contains air at 40 bar and 27 °C. It is used to drive a turbine which exhausts to the atmosphere of 1 bar. If the pressure in the bottle is allowed to fall to 2 bar, determine the amount of work that could be delivered by the turbine (Fig. 4.32).

Solution:
Given data:
    Volume of the bottle:

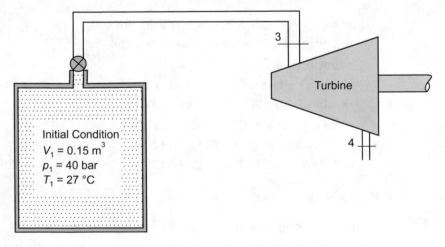

Fig. 4.32 Schematic for Problem 4.28

$$V = 0.15\,\text{m}^3$$

Initial condition of air in bottle:

$$p_1 = 40\,\text{bar} = 40 \times 10^2\,\text{kPa}$$
$$T_1 = 27\,°\text{C} = (27 + 273)\,\text{K} = 300\,\text{K}$$

Final condition of air in bottle:

$$p_2 = 2\,\text{bar} = 2 \times 10^2\,\text{kPa}$$
$$V_2 = V_1 = 0.15\,\text{m}^3$$

For turbine,
Exhaust pressure:

$$p_4 = 1\,\text{bar} = 1 \times 10^2\,\text{kPa}$$

Now applying an ideal gas equation of state to the initial condition of air in the bottle,

$$pV_1 = m_1 R T_1$$
$$40 \times 10^2 \times 0.15 = m_1 \times 0.287 \times 300$$

or

$$m_1 = 6.968\,\text{kg}$$
$$\frac{T_2}{T_1} = \left(\frac{p_2}{p_1}\right)^{\frac{\gamma-1}{\gamma}} \qquad \because \text{adiabatic process follows in the bottle}$$
$$\frac{T_2}{300} = \left(\frac{2}{40}\right)^{\frac{1.4-1}{1.4}}$$
$$T_2 = 300(0.05)^{0.2857} = 127.47\,\text{K}$$

Final mass of air in the bottle:

$$p_2 V_2 = m_2 R T_2$$
$$2 \times 10^2 \times 0.15 = m_2 \times 0.287 \times 127.47$$

or

$$m_2 = 0.82\,\text{kg}$$

For turbine:

$$\frac{T_3}{T_4} = \left(\frac{p_3}{p_4}\right)^{\frac{\gamma-1}{\gamma}}$$

As

$$T_3 = T_1, p_3 = p_1 \quad \left|\begin{array}{l} \because \text{ initial condition of the bottle} = \\ \text{inlet condition of the turbine} \end{array}\right.$$

$\therefore$

$$\frac{300}{T_4} = \left(\frac{40}{1}\right)^{\frac{1.4-1}{1.4}}$$

$$\frac{300}{T} = 2.861$$

or

$$T_4 = 104.85 \,\text{K}$$

also

$$T_3 = T_2, p_3 = p_2 \quad \left|\begin{array}{l} \because \text{ final condition of the bottle} = \\ \text{inlet condition of the turbine} \end{array}\right.$$

$$\frac{127.74}{T_4} = \left(\frac{2}{1}\right)^{\frac{1.4-1}{1.4}}$$

$$\frac{127.74}{T_4} = 1.218$$

or

$$T_4 = 104.85 \,\text{K}$$

Work delivered by the turbine: W_T

$$\begin{aligned} W_T &= \text{energy at turbine inlet} - \text{energy at turbine outlet} \\ &= \text{energy exit of the bottle} - \text{energy at turbine outlet} \\ &= (m_1 u_1 - m_2 u_2) - (m_1 - m_2)h_4 \\ &= m_1 c_v T_1 - m_2 c_v T_2 - (m_1 - m_2)c_p T_4 \\ &= c_v(m_1 T_1 - m_2 T_2) - (m_1 - m_2)c_p T_4 \\ &= 0.718(6.968 \times 300 - 0.82 \times 127.74) \\ &\quad - (6.968 - 0.82) \times 1.005 \times 104.85 \\ &= 1425.69 - 647.84 = \mathbf{777.85\,kJ} \end{aligned}$$

Note: m_2 is also find out by available values of

$$p_1 = 40 \,\text{bar}, \ m_1 = 6.969 \,\text{kg}$$
$$p_2 = 2 \,\text{bar}$$

For adiabatic process,

$$p_1 v_1^\gamma = p_2 v_2^\gamma$$
$$p_1 \left(\frac{V_1}{m_1} \right)^\gamma = p_2 \left(\frac{V_1}{m_2} \right)^\gamma$$
$$p_1 m_2^\gamma = p_2 m_1^\gamma \quad \because V_1 = V_2 = V$$

or

$$m_2^\gamma = \left(\frac{p_2}{p_1} \right) \times m_1^\gamma$$

or

$$m_2 = \left(\frac{p_2}{p_1} \right)^{1/\gamma} \times m_1 = \left(\frac{2}{40} \right)^{1/.4} \times 6.968 = \mathbf{0.82 \,kg}$$

Summary

1. **Steady and Unsteady Flow**. A flow is considered to be steady if fluid flow parameters such as mass flow rate (m), velocity (V), pressure (p), density (ρ), and temperature (T) at any point do not change with time. If any one of these parameters changes with time, the flow is said to be unsteady flow.
2. **Compressible and Incompressible Flow**. If the density of the fluid changes from point to point in the fluid flow, it is referred to as compressible flow. Mathematically,

$$\rho \neq C$$

If the density of the fluid remains constant at every point in the fluid flow, it is referred to as an incompressible flow.
Mathematically,

$$\rho = C$$

3. Discharge:

$$Q = AV = mv$$

or

$$AV = mv$$

or

$$m = \frac{AV}{v}$$
$$m = \rho AV$$

4.   **Continuity Equation**. The equation based on the law of conservation of mass is called the continuity equation. It means that mass can neither be created nor destroyed. If there is no accumulation of mass within the control volume, the mass flow rate entering the system must be equal to the mass flow rate leaving the system.

5.   **Steady Flow Energy Equation (SFEE)**.

$$m\left[h_1 + \frac{V_1^2}{2} + gz_1\right] + Q_{1-2} = m\left[h_2 + \frac{V_2^2}{2} + gz_2\right] + W_{1-2}$$

For unit mass flow rate,

$$h_1 + \frac{V_1^2}{2} + gz_1 + q_{1-2} = h_2 + \frac{V_2^2}{2} + gz_2 + w_{1-2}$$

6.   **Nozzle**. A nozzle is a fluid energy transforming device that increases the velocity of the fluid and simultaneously decreases the pressure energy of the fluid.

Velocity at exit of nozzle:

$$V_2 = \sqrt{2(h_1 - h_2) + V_1^2} \text{ m/s}$$

where
h_1, h_2 are in J/kg, V_1 in m/s

$$V_2 = \sqrt{2c_p(T_1 - T_2) + V_1^2} \text{ m/s}$$

where
c_p is in J/kgK.
T_1 and T_2 are in °C or K.
V_1 is in m/s.
As

$$V_1 \ll V_2$$

$$V_1 = \sqrt{2(h_2 - h_1) + V_2^2}\,\text{m/s}$$

$$= \sqrt{2c_p(T_2 - T_1) + \frac{V_2^2}{2}}\,\text{m/s}$$

$$= \sqrt{2000c_p(T_1 - T_2) + V_1^2}\,\text{m/s}$$

where

h_1 and h_2 are in kJ/kg.

c_p is in kJ/kgK.

T_1 and T_2 are in °C or K.

7. **Diffuser**. A diffuser is also a fluid energy transforming device as a nozzle that increases the pressure energy of the fluid and simultaneously decreases the kinetic energy of the fluid.

Velocity at inlet of diffuser:

$$V_1 = \sqrt{2(h_2 - h_1) + V_2^2}\,\text{m/s}$$

$$= \sqrt{2c_p(T_2 - T_1) + \frac{V_2^2}{2}}\,\text{m/s}$$

where

h_1 and h_2 are in J/kg.

V_2 is in m/s.

c_p is in J/kgK.

T_1 and T_2 arc in °C or K.

As

$$V_2 \ll V_1$$

$\therefore$

$$V_1 = \sqrt{2(h_2 - h_1)}\,\text{m/s}$$
$$= \sqrt{2c_p(T_2 - T_1)}\,\text{m/s}$$

and

$$V_1 = \sqrt{2000(h_2 - h_1)}\,\text{m/s}$$
$$= \sqrt{2000c_p(T_2 - T_1)}\,\text{m/s}$$

where

h_1 and h_2 are in kJ/kg.

c_p is in kJ/kg.

T_1 and T_2 are in °C or K.

8. **Turbine**. It is an energy converting device that converts the fluid energy into work.

Work done:

$$W_{1-2} = m(h_1 - h_2)$$

For unit mass flow rate,

$$w_{1-2} = (h_1 - h_2) = c_p(T_1 - T_2)$$

9. **Compressor**. It is used to increase the pressure of air by need of work.

Work done:

$$W_{1-2} = m(h_1 - h_2)$$

For unit mass flow rate,

$$w_{1-2} = h_1 - h_2 = c_p(T_1 - T_2)$$

10. **Boiler**. It is a metallic vessel that is used for the generation of steam at constant pressure from the water by application of heat.

Heat supplied:

$$Q_{1-2} = m[h_2 - h_1]$$

11. **Heat Exchanger**. It is a device in which sensible heat is transferred between two fluids. In the heat exchanger, two fluids are used, one is hot and the other is cold. The heat is lost by the hot fluid and simultaneously the heat is gained by the cold fluid.

Mathematically,

$$\text{heat lost by hot fluid} = \text{heat gained by cold fluid}$$
$$m_h c_p(T_{h1} - T_{h2}) = m_c c_c(T_{c2} - T_{c1})$$

12. **Condenser**. A condenser is a specific type of heat exchanger. It is used to change the phase of vapour to liquid by removing the latent heat of the vapour.

Mathematically,

$$\text{heat lost by the vapour} = \text{heat gained by cold water}$$
$$m_V(h_{h1} - h_{h2}) = m_w(h_{c2} - h_{c1})$$

13. **Evaporator**. An evaporator is a specific type of heat exchanger. It is used to change the phase of liquid to vapour by absorbing the latent heat of the liquid. Mathematically,

latent heat gained by the liquid $=$ sensible heat lost by the water

$$m(h_2 - h_1) = m_w(h_{w1} - h_{w2})$$

14. **Adiabatic Mixing**. Adiabatic mixing refers to the mixing of two or more streams of fluid under adiabatic condition.
For mixing of two streams,

$$m_1 h_1 + m_2 h_2 = m_3 h_3$$

$$T_3 = \frac{m_1 c_{p1} T_1 + m_2 c_{p2} T_2}{m_3 c_{p3}} \quad \text{for different gases}$$

$$= \frac{m_1 T_1 + m_2 T_2}{m_3} \quad \text{for same gases}$$

15. **Throttling Process**. Throttling process is the expansion of a fluid from high pressure to low pressure when fluid flows through a constricted passage, like a partially opened valve, a porous plug, an orifice, and a capillary tube.
Condition for throttling process,

$$h_1 = h_2$$

It means that the enthalpy of fluid remains constant during the throttling process.

16. **Joule–Thomson Coefficient**:

$$\mu_J = \left(\frac{\partial T}{\partial p}\right)_{h=\text{constant}}$$

where μ_J is +ve, –ve, or zero during a throttling process for a real gas.
That is,

$$\mu_J \begin{cases} < 0 & \text{temperature increases} \\ = 0 & \text{temperature remains constant} \\ > 0 & \text{tempertaure decreases} \end{cases}$$

17. **Unsteady Flow Energy Equation (UFEE)**

$$Q_{1-2} = m_2 u_2 - m_1 u_1 + W_{1-2} + m_e\left(h_e + \frac{V_e^2}{2} + gz_e\right) - m_i\left(h_i + \frac{V_i^2}{2} + gz_i\right)$$

Subscripts 1 and 2 refer to the initial and final conditions in the open system.

Subscripts i and e refer to the inlet and exit conditions of the open system.

Assignment-1

1. What is the difference between steady flow and non-steady flow processes?
2. Derive the steady flow energy equation.
3. Derive the steady flow energy equation from the first law of thermodynamics.
4. Derive Euler's equation and Bernoulli's equation from the steady flow energy equation.
5. Under what conditions does the steady flow energy equation reduce to Euler's equation?
6. Show that the work done in a steady flow process is given by
7. Show that for adiabatic isentropic flow process,
    Work done:

$$w_{1-2} = \frac{\gamma R}{\gamma - 1}(T_1 - T_2)$$

    where
    γ = adiabatic index of an ideal gas
    R = gas constant.
8. The kinetic energy of a fluid increases as it is accelerated to an insulated nozzle. Where does this energy come from? How will heat loss from the surface of the nozzle affect the fluid velocity at the nozzle exit?
9. What is the difference between nozzle and diffuser? Also, draw the diagrams of the nozzle and diffuser.
10. What are the assumptions for a steady flow process? Write the general energy equation for a steady flow system.
11. Write down the simplified steady flow energy equation for a unit mass flow rate for

    (a) nozzle
    (b) diffuser
    (c) turbine
    (d) compressor

12. Write down the simplified steady flow energy equation for a unit mass flow rate for

    (a) boiler
    (b) heat exchanger, and
    (c) adiabatic mixing

13. Applying the steady flow energy equation to a nozzle, derive an equation for the velocity at exit.
14. Explain the throttling process. What is the difference between a throttling process and a free expansion process?

15. Show that the enthalpy of a fluid before throttling is equal to that after throttling.
16. Define Joule–Thomson coefficient, inversion point, and inversion line.
17. Sketch the isenthalpic curves for a real gas and indicate thereupon the inversion line.
18. Write the general energy equation for an unsteady flow process.

Assignment 2

1. Air enters an adiabatic nozzle steadily at 300 kPa, 250 °C, and 40 m/s and leaves at 100 kPa and 200 m/s. The inlet area of the nozzle in 75 cm². Determine

    (a) the mass flow rate through the nozzle,
    (b) the exit temperature of the air, and
    (c) the exit area of the nozzle.

    [**Ans.** (*a*) 0.6 kg/s (*b*) 230.89 °C (*c*) 43.38 cm²]

2. The velocity and enthalpy of fluid at the inlet of a certain nozzle are 50 m/s and 2800 kJ/kg, respectively. The enthalpy at the exit of the nozzle is 2600 kJ/kg. The nozzle is horizontal and insulated so that no heat transfer takes place from it. Find

    (a) Velocity of the fluid at exit of the nozzle.
    (b) Mass flow rate, if the area at the inlet of the nozzle is 0.09 m² and the specific volume is 0.185 m³/kg.
    (c) Exit area of the nozzle, if the specific volume at the exit of the nozzle is 0.495 m³/kg.

    [**Ans.** (*a*) 634.42 m/s (*b*) 24.32 kg/s (*c*) 0.0189 m²]

3. Air at 10 °C and 80 kPa enters the diffuser of a jet engine steadily with a velocity of 200 m/s. The inlet area of the diffuser is 0.4 m². The air leaves the diffuser with a velocity that is very small compared with the inlet velocity. Determine

    (a) the mass flow rate of the air and
    (b) the temperature of the air leaving the diffuser.

    [**Ans.** (*a*) 78.81 kg/s (*b*) 30 °C]

4. A turbine, operating under steady flow conditions, receives 5400 kg of steam per hour. The steam enters the turbine at a velocity of 50 m/s, an elevation of 5.5 m, and a specific enthalpy of 2800 kJ/kg. It leaves the turbine at a velocity of 90 m/s, an elevation of 1.5 m, and a specific enthalpy of 2300 kJ/kg. Heat losses from the turbine to the surroundings at the rate of 4.5 kJ/s. Determine the power output of the turbine.
    [**Ans.** 741.35 kW]

5. A gas at 7.5 bar, 750 °C, and 140 m/s is passed through a turbine of a jet engine. The gas leaves the turbine at 2 bar, 550 °C, and 280 m/s. The process may be assumed adiabatic. The enthalpies of gas at the entry and exit of the turbine

are 950 kJ/kg and 650 kJ/kg of gas, respectively. Determine the capacity of the turbine if the gas flow rate is 3.5 kg/s.
[**Ans**. 947.1 kW]

6.   In an air compressor, air flows steadily at the rate of 15 kg/min. The air enters the compressor at 5 m/s with a pressure of 1 bar and a specific volume of 0.5 m^3/kg. It leaves the compressor at 7.5 m/s with a pressure 7 bar and a specific volume of 0.15 m^3/kg. The internal energy of the air leaving the compressor is 165 kJ/kg greater than that of the air entering. The cooling water in the compressor jackets absorbs heat from the air at the rate of 125 kJ/s. Find

(a)   Power required to drive the compressor.
(b)   Ratio of the inlet pipe diameter to outlet diameter.

[**Ans**. (a) 180 kW (b) 2.236]

7.   In a centrifugal compressor, the suction and delivery pressures are 100 kPa and 550 kPa, respectively. The compressor draws 15 m^3/min of air which has a specific volume of 0.77 m^3/kg. At the delivery point, the specific volume is 0.20 m^3/kg. The compressor is driven by a 40 kW motor and during the passage of air through the compressor, the heat lost to the surroundings is 30 kJ/kg of air. Neglecting changes in the potential and kinetic energy, make calculations for the increase in internal energy per kg of air.
[**Ans**. 60.23 kJ/kg]

8.   In a conference hall comfortable temperature conditions are maintained in winter by circulating hot water through a piping system. The water enters the piping system at 3 bar pressure and 50 °C temperature (specific enthalpy = 240 kJ/kg) and leaves at 2.5 bar pressure and 30 °C temperature (specific enthalpy = 195 kJ/kg). The exit from the piping system is 15 m above the entry. If 30 MJ/hr of heat needs to be supplied to the hall, make calculations for the quantity of water circulated through the pipe per minute. Assume that there are no pumps in the system and that the change in kinetic energy is negligible.
[**Ans**. 11.148 kg/min]

9.   A hot water stream at 80 °C enters a mixing chamber with a mass flow rate of 0.5 kg/s where it is mixed with a stream of cold water at 20 °C. If it is desired that the mixture leaves the chamber at 42 °C, determine the mass flow rate of the cold water stream. Assume all the streams are at a pressure of 250 kPa.
[**Ans**. 0.865 kg/s]

10.   The airspeed of a turbojet engine in flight is 270 m/s. The ambient air temperature is –15 °C. The gas temperature at the outlet of the nozzle is 600 °C. Corresponding specific enthalpy values for air and gas are, respectively, 260 and 912 kJ/kg. The fuel–air ratio is 0.0190. The chemical energy of the fuel is 44.5 MJ/kg. Owing to incomplete combustion, 5% of the chemical energy is not released in the reaction. Heat loss from the engine is 21 kJ/kg of air. Calculate the velocity of the exhaust jet.
[**Ans**. 541.40 m/s]

11.   Air at a temperature of 20 °C passes through a heat exchanger at a velocity of 40 m/s where its temperature is raised to 820 °C. It then enters a turbine with

the same velocity of 40 m/s and expands till the temperature falls to 620 °C. On leaving the turbine, the air is taken at a velocity of 55 m/s to a nozzle where it expands until the temperature has fallen to 510 °C. If the airflow rate is 2.5 kg/s, calculate

(a)   Rate of heat transfer to the air in the heat exchanger.
(b)   The power output from the turbine assuming no heat loss.
(c)   The velocity at the exit from the nozzle, assuming no heat loss.

Take the specific enthalpy of air as $h = c_p T$, where c_p is the specific heat at constant pressure and is equal to 1.005 kJ/kg°C. T is the temperature.
[**Ans.** (a) 2010 kJ/s (b) 523.21 kW (c) 473.41 m/s]

12.   An air receiver of volume 5.5 m³ contains air at 16 bar and 42 °C. A valve is opened and some air is allowed to blow out to the atmosphere. The pressure of the air in the receiver drops rapidly to 12 bar when the valve is then closed. Determine the mass of air that has left the receiver.
[**Ans.** 18.04 kg]

13.   After the completion of the exhaust stroke of an IC engine, the piston-cylinder device remains filled up with 1×10^{-4} kg of combustion products at 527 °C. During the subsequent suction stroke, the piston moves outward and 16×10^{-4} kg of air at 17 °C is sucked inside the cylinder. The suction process occurs at constant pressure and heat interaction is negligible. Determine the temperature of gases at the end of the suction stroke. For air and gases, take $c_p = 1$ kJ/kgK.
[**Ans.** 48.87 °C]

Chapter 5
Second Law of Thermodynamics

Nomenclature

The following is the nomenclature introduced in this chapter:

Q	kJ	Heat transfer
η	–	Efficiency
T	K	Temperature
PMM II	–	Perpetual-motion machine of the second kind
W	kJ	Work
R	–	Refrigerator
HP	–	Heat pump
$(COP)_R$	–	Coefficient of performance of a refrigerator
$(COP)_{HP}$	–	Coefficient of performance of a heat pump
η_{carnot}	–	Efficiency of the Carnot engine
HE_R	–	Reversible heat engine
HE_I	–	Irreversible heat engine
$\eta_{th,rev}$	–	Thermal efficiency of a reversible heat engine

5.1 Introduction

In this chapter, we will discuss the limitations of the first law of thermodynamics that had given birth to the second law of thermodynamics. The second law of thermodynamics is used to check the feasibility of the process and direction of the process. This law also determines how much part of supplied heat is converted into work. It introduces a state function called entropy which is a very useful tool to determine the degree of irreversibility of a process.

© The Author(s) 2022
S. Kumar, *Thermal Engineering Volume 1*,
https://doi.org/10.1007/978-3-030-67274-4_5

Fig. 5.1 Source
(high-temperature reservoir)
and sink (low-temperature
reservoir)

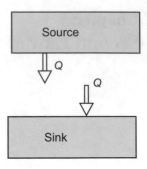

5.2   Thermal Reservoir

A thermal reservoir is a system having a very large heat capacity. A thermal reservoir can supply or absorb finite amounts of heat without any appreciable change in its temperature. It is also called the thermal energy reservoir or characterized by its temperature remaining constant during heat interaction. For example, oceans, large rivers, and the atmospheric air are thermal reservoirs because the interaction of a finite amount of heat cannot make appreciable change in their temperature, i.e., waste energy dumped in oceans/large rivers by thermal power plants do not cause any significant change in temperature of the ocean. In a similar way, the atmosphere does not warm up as a result of heat losses from residential buildings in winter. Thermal reservoirs are classified into two categories depending upon heat interaction, i.e., heat supplying or heat rejecting reservoirs. The reservoir which is at high temperature and supplies energy in the form of heat is called a **source**. For example, sun, nuclear reactor, furnace, etc. The reservoir which is at low temperature and absorbs energy in the form of heat is called a **sink**. For example, oceans, large rivers, atmospheric air, etc. (Fig. 5.1).

5.3   Limitations of First Law of Thermodynamics

The limitations of the first law of thermodynamics are expressed as follows:

1. It is impossible to convert the heat supplied into an equivalent amount of work. That is, a heat engine must reject some heat to a low-temperature reservoir in order to complete the cycle.
2. First law of thermodynamics does not specify the direction of heat flow. Whether the heat flows from low-temperature system to high-temperature system or from high-temperature system to low-temperature system.

5.4 Second Law of Thermodynamics

There are two statements of the second law of thermodynamics (i) Kelvin-Planck statement, which is related to heat engines. (ii) Clausius statement, which is related to heat pumps and refrigerators.

(i) **Kelvin-Planck statement**: *It is impossible to construct a heat engine that works in a cycle and converts net heat supplied into an equivalent amount of work.*

It is also stated that no heat engine can have a thermal efficiency of 100%. It means that only a part of heat transfer at high temperature in a cyclic process can be converted into work, the remaining part has to be rejected to surroundings at a lower temperature. Thus, thermal efficiency of a heat engine is always less than 100% (Fig. 5.2).

(ii) **Clausius statement**: *It is impossible to construct a device that operates in a cycle and produces no effect other than the transfer of heat from a low-temperature body to a high-temperature body without external aid of work.*

The statement implies that heat cannot flow itself from a sink (i.e., reservoir at low temperature) to a source (i.e., reservoir at high temperature). In the case of refrigerator/ heat pump, an external work is required for extracting heat from low-temperature body and rejecting it to high-temperature body (Fig. 5.3).

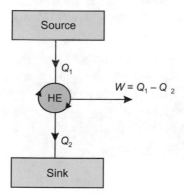

1. Practically possible heat engine.
2. According to second law of thermodynamics—Kelvin-Planck statement
3. $\eta < 100\%$

Fig. 5.2 Kelvin-Planck statement

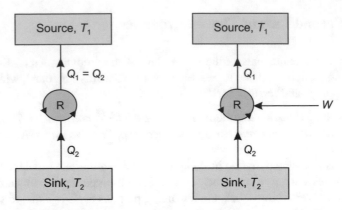

(a) Practically impossible refrigerator   (b) Practically possible refrigerator

Fig. 5.3 Clausius statement

5.5   Perpetual-Motion Machine of the Second Kind—PMM II

Perpetual-Motion machine of the second kind is a hypothetical device that takes heat from a source and converts it completely into work. That is, the thermal efficiency of PMM II is 100%. It violates the second law of thermodynamics. Thus, PMM II is a practically impossible device (Fig. 5.4).

Remember

- PMM II is an imaginary machine.
- PMM II violates the second law of thermodynamics.
- The thermal efficiency of PMM II is 100%.

Fig. 5.4 PMM II

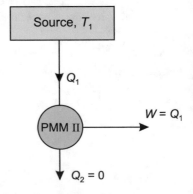

- PMM II is impossible machine.

5.6 Heat Engine

A heat engine is a mechanical work producing device which is used for converting a part of heat into work, while the remaining waste heat is rejected to a sink. It works according to the second law of thermodynamics stated by Kelvin and Planck. It is shown in Fig. 5.5, which is characterized by the following features:

Thermal efficiency: The fraction of the heat supplied that is converted to the net work output is a measure of the performance of a heat engine and is called thermal efficiency (Fig. 5.6).

Mathematically,

$$\text{Thermal efficiency: } \eta_{th} = \frac{\text{network output}}{\text{heat supplied}} = \frac{W}{Q_1}$$

Fig. 5.5 Heat engine

HE: Heat Engine

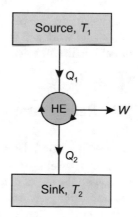

Fig. 5.6 Conservation of energy

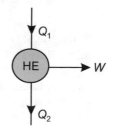

Applying the principle of energy conservation to the heat engine, sum of incoming energies = sum of outgoing energies

$$Q_1 = W + Q_2$$

or

$$W = Q_1 - Q_2$$

$\therefore$

$$\eta_{th} = \frac{Q_1 - Q_2}{Q_1} = 1 - \frac{Q_2}{Q_1}$$

The thermal efficiency of a heat engine is always less than unity since both Q_1 and Q_2 are defined as positive quantities. In the case of IC engines, the values of the thermal efficiencies are relatively low.

That is,

$$\eta_{th} \simeq 25\% \text{ for petrol engines}$$
$$\simeq 40\% \text{ for diesel engines.}$$

It means that the petrol engine converts 25% of the chemical energy of the petrol into useful work and the diesel engine converts 40% of the chemical energy of the diesel into useful work.

5.7   Refrigerator and Heat Pump

Both refrigerator and heat pump work according to the second law of thermodynamics stated by Clausius—heat transfer from a low-temperature body to a high-temperature body with the aid of external work. So, refrigerator and heat pump are mechanical work absorbing devices. The main difference between a refrigerator and a heat pump is their use of different purposes.

Refrigerators are used for cooling purpose while heat pumps are used for heating purpose.

Refrigerator. It is a device that is used to attain and maintain a temperature below that of the surroundings, the aim being to cool some product or space to some required low temperature.

This can be achieved by transferring heat from low-temperature space to high-temperature space by applying input work on it. It is shown in Fig. 5.7a.

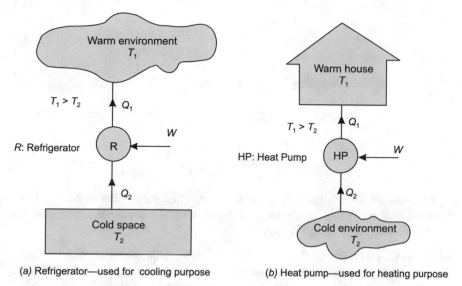

(a) Refrigerator—used for cooling purpose   (b) Heat pump—used for heating purpose

Fig. 5.7 Refrigerator and heat pump

Coefficient of Performance. The performance of a refrigerator is expressed in terms of the coefficient of performance, denoted by $(COP)_R$. It is defined as the ratio of desired effect (i.e., heat removed from the cold space) to work input for achieving the desired effect.

Mathematically,

Coefficient of performance of a refrigerator:

$$(COP)_R = \frac{\text{Desired effect}}{\text{Work input}} = \frac{\text{Cooling effect} : Q_2}{\text{Work input} : W}$$

$$\mathbf{(COP)_R = \frac{Q_2}{W}} \tag{5.1}$$

Applying the principle of energy conservation to a refrigerator:

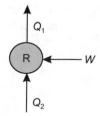

sum of incoming energies = sum of outgoing energies

$$W + Q_2 = Q_1$$

or

$$W = Q_1 - Q_2$$

$\therefore$

$$(COP)_R = \frac{Q_2}{Q_1 - Q_2} \tag{5.2}$$

The value of $(COP)_R$ can be greater than one, when the amount of heat removed from the cold space can be greater than the amount of work input. Otherwise, the value of $(COP)_R$ may be less than one.

Heat Pump. It is a device that is used to attain and maintain a temperature higher than that of the surroundings, the aim being to heat up some space to the required high temperature. This can be achieved by transferring heat from low-temperature space to high-temperature space by applying input work on it. It is shown in Fig. 5.7b.

Coefficient of Performance. The coefficient of performance of heat is expressed as the ratio of the desired effect (i.e., heat supplied to the hot space) to work input for achieving the desired effect. Mathematically,

Coefficient of performance of a heat pump:

$$(COP)_{HP} = \frac{\text{Desired effect}}{\text{Work input}} = \frac{\text{Heating effect: } Q_1}{\text{Work input: } W}$$

$$(COP)_{HP} = \frac{Q_1}{W}$$

$$(COP)_{HP} = \frac{Q_1}{Q_1 - Q_2} \qquad \because W = Q_1 - Q_2$$

$$= \frac{Q_1 - Q_2 + Q_2}{Q_1 - Q_2} = 1 + \frac{Q_2}{Q_1 - Q_2}$$

$$(COP)_{HP} = 1 + (COP)_R \tag{5.3}$$

For the same values of Q_1 and Q_2. Eq. (5.3) shows that the coefficient of performance of a heat pump is equal to one more than the coefficient of performance of a refrigerator.

5.8 Equivalence of Two Statements of the Second Law of Thermodynamics

The Kelvin-Planck and Clausius statements of the second law of thermodynamics are two different statements but their basic facts are equivalent to each other. It means that a device that violates the Kelvin-Planck statement leads to the violation of the Clausius statement, and vice versa. The equivalence of the two statements can be demonstrated as follows.

5.8.1 Violation of the Kelvin-Planck Statement Leads to the Violation of the Clausius Statement

Consider the heat engine and refrigerator are operating between the same two reservoirs as shown in Fig. 5.8a. Let the heat engine take the heat Q_1 from a source and convert it completely into work W. It means that the thermal efficiency of the heat engine is 100% and that it violates the Kelvin-Planck statement. Let the work output of the heat engine be supplied to a refrigerator that takes heat in the amount of Q_2 from a sink and rejects heat in the amount of $Q_1 + Q_2$ to a source. During this process, a source receives a net amount of heat Q_2 [i.e., $(Q_1 + Q_2) - Q_1 = Q_2$]. Thus, the combination of the heat engine and refrigerator is shown in Fig. 5.8b that transfer heat in an amount of Q_2 from a sink to a source without any work input from outside. This indicates the violation of the Clausius statement. Hence, a violation of the Kelvin-Planck statement leads to the violation of the Clausius statement.

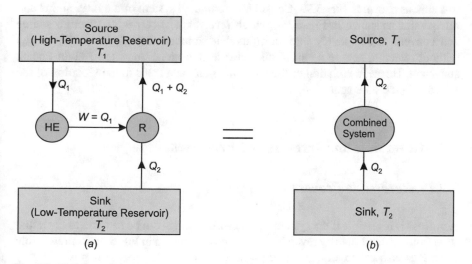

Fig. 5.8 Violation of the Kelvin-Planck statement leads to the violation of the Clausius statement

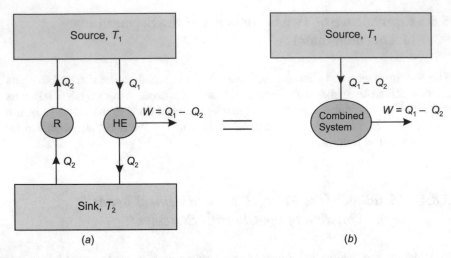

Fig. 5.9 Violation of the Clausius statement leads to the violation of the Kelvin-Planck statement

5.8.2 Violation of the Clausius Statement Leads to the Violation of the Kelvin-Planck Statement

Consider the heat engine and refrigerator are operating between the same two reservoirs as shown in Fig. 5.9a. Let the refrigerator take the heat Q_2 from a sink and transfer the same amount of heat to a source without any work input from outside as shown in Fig. 5.9a. It violates the Clausius statement.

Let the heat engine take heat Q_1 from a source, give work W and reject heat Q_2 to a sink as shown in Fig. 5.9a. Consider a combined system of the heat engine and refrigerator as shown in Fig. 5.9b, which receives heat $(Q_1 - Q_2)$ from a source and converts it completely into an equivalent amount of work $W = (Q_1 - Q_2)$ without rejecting heat to a sink. This indicates the violation of the Kelvin-Planck statement. Hence, a violation of the Clausius statement leads to the violation of the Kelvin-Planck statement.

5.9   Reversible and Irreversible Processes

5.9.1 Reversible Process

A process is reversible if the system and its surroundings can be restored to the initial state from the final state by reversing the process as shown in Fig. 5.10a. A reversible process is also called a quasi-static process. A reversible process is possible only in the absence of fluid friction and heat transfer with finite temperature difference. Since these conditions are impossible to achieve in actual processes, all real flows in the turbine, compressors, nozzle are irreversible. The reversible process is used only as an 'ideal reference process' for comparison with its equivalent actual process.

Following are some examples of the reversible process.

1. Compression and expansion of spring.
2. Frictionless adiabatic compression and expansion of gas.
3. Isothermal compression and expansion of gas.
4. Electrolysis.

5.9.2 Irreversible Process

A process that does not satisfy the above conditions of a reversible process is an irreversible process as shown in Fig. 5.10b.

Following are examples of the irreversible process:

1. Throttling process
2. Free expansion
3. Diffusion
4. Combustion
5. Plastic deformation
6. Heat transfer
7. Flow of electricity
8. Flow of real fluid.

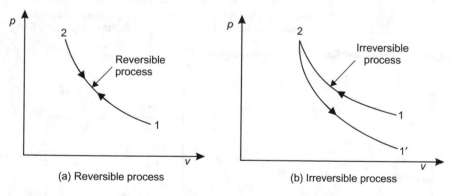

Fig. 5.10 Reversible and irreversible processes

5.10 Carnot Cycle and Carnot Heat Engine

The Carnot cycle is a reversible cycle (or ideal cycle). This cycle was proposed in 1824 by French engineer Sadi Carnot. The theoretical heat engine that operates on the Carnot cycle is called the Carnot heat engine. The Carnot cycle is composed of the following four reversible processes:

1.   Reversible isothermal compression.
2.   Reversible adiabatic compression.
3.   Reversible isothermal expansion.
4.   Reversible adiabatic expansion.

The Carnot heat engine is shown in Fig. 5.11 and the Carnot cycle on p-v and T-s diagrams is shown in Fig. 5.12.

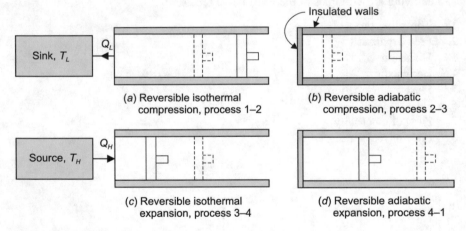

Fig. 5.11 Schematic diagrams of Carnot heat engine

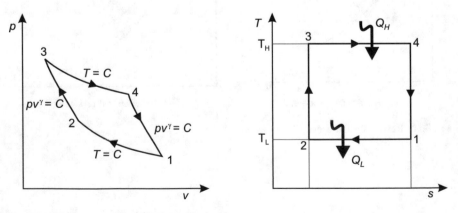

Fig. 5.12 Carnot cycle in p-v and T-s diagrams

Process 1-2, reversible isothermal compression

At state 1, the cylinder head is brought into contact with a sink at temperature T_L. The piston is pushed inward by an external force, doing work on the working fluid. As the working fluid is compressed slowly due to a slow movement of the piston, its temperature tends to rise slightly.

Due to this slight rise in temperature of the working fluid, heat is transferred from the working fluid to the sink, causing the temperature of the working fluid to drop to T_L. Thus, the temperature of the working fluid remains constant T_L.

> Reversible isothermal process is possible when frictionless piston moves very slowly.

Process 2-3, reversible adiabatic compression

At state 2, the sink (i.e., reservoir at low temperature T_L) that was in contact with the cylinder head is removed and replaced by insulation so that the system becomes adiabatic, and the gas is compressed in a reversible adiabatic mode. The temperature rises from T_L to T_H during this reversible adiabatic compression process.

> Reversible adiabatic process is maintained when frictionless piston moves very fast in adiabatic cylinder.

Process 3-4, reversible isothermal expansion

At state 3, the insulation of the cylinder head is removed, the cylinder head is brought into contact with a source at temperature T_H. The working fluid is allowed to expand slowly, doing work on the surroundings. As the working fluid expands, the temperature of the working fluid tends to fall due to a slow movement of the piston. Due to this small fall of temperature of the working fluid, some amount of heat is transferred from the reservoir into the working fluid, causing the temperature of the working fluid to raise to T_H. Thus, the temperature of the working fluid remains constant T_H.

Process 4-1, reversible adiabatic expansion

At state 4, the source (i.e., reservoir at high temperature T_H) that was in contact with the cylinder head is removed and replaced by insulation so that the system becomes adiabatic. The working fluid continues to expand, doing work on the surroundings until its temperature drops from T_H at state 4 to T_L at state 1. The piston is assumed to be frictionless and the cylinder is adiabatic.

For process 3-4,
Heat supplied:

$$Q_H = \text{area under process } 3 - 4$$
$$= T_H \times (S_4 - S_3)$$
$$= T_H(S_4 - S_3)$$

Heat rejection:

$$Q_L = \text{area under process } 1 - 2$$
$$= T_L(S_1 - S_2)$$

Net work output:

$$W = \text{heat supplied} - \text{heat rejection}$$
$$= Q_H - Q_L$$

Thermal efficiency of the Carnot engine,

$$\eta_{\text{Carnot}} = \frac{\text{work output}}{\text{heat supplied}}$$
$$= \frac{W}{Q_H} = \frac{Q_H - Q_L}{Q_H} = 1 - \frac{Q_L}{Q_H}$$
$$= 1 - \frac{T_L(S_1 - S_2)}{T_H(S_4 - S_3)}$$
$$\eta_{\text{Carnot}} = 1 - \frac{T_L}{T_H} \quad \because S_1 - S_2 = S_4 - S_3$$

The Carnot heat engine is also known as a reversible heat engine and Carnot efficiency is equal to the thermal efficiency of a reversible heat engine, i.e.,

$$\eta_{\text{Carnot}} = \eta_{\text{th,rev}} = 1 - \frac{T_L}{T_H} \tag{5.4}$$

Equation (5.4) can give the following information:

(i)   The Carnot efficiency is independent of the working substance and depends upon the temperatures of sink and source only.
(ii) Always T_L and T_H are used in absolute temperature (i.e., in kelvin). Using °C for temperature gives wrong results.
(iii) If $T_L = 0$ K, then $\eta_{\text{Carnot}} = 100\%$. This means that there is no heat rejection to the sink which is a violation of the Kelvin-Planck statement of the second law of thermodynamics.

(iv)   The Carnot efficiency increases with an increase in temperature of a source and a decrease in temperature of a sink.

(v)   If $T_L = T_H$, no work will be done and efficiency will be zero.

5.11   Why the Carnot Cycle is not Practically Feasible?

The Carnot engine is a hypothetical device and it cannot be practically possible due to the following reasons:

1.   All the four processes used in the Carnot cycle have to be reversible. The reversible processes are possible only if the processes have no internal friction among the fluid particles and no mechanical friction between the piston and cylinder walls.
2.   Heat addition and rejection at constant temperature is impossible.
3.   Frequent change of cylinder head (i.e., insulating head for adiabatic process and diathermic head for isothermal process) is not possible.
      OR
      The isothermal process can be achieved only if the piston moves very slowly to allow heat transfer so that the temperature remains constant. The adiabatic process can be achieved only if the piston moves very fast in order to approach a reversible adiabatic process, i.e., heat transfer is negligible due to a very short period of time. Thus, it is impossible to maintain the variations in the speed of the piston during the processes of a cycle.
4.   The isothermals and adiabatic lines in p-v diagram are largely extended both in the horizontal and vertical directions. The cylinder involves high pressure and large displacement volume, and the size of the engine becomes heavy and bulky. The Carnot cycle is an impractical device or hypothetical device. Now the question is why we study the Carnot cycle? The answer is that the Carnot cycle is taken as a standard or reference against which the performance of any practical heat engine is compared.

5.12   Carnot Theorem and Its Corollaries

It states that *the efficiency of an irreversible heat engine is always less than the efficiency of a reversible heat engine operating between the same temperature limits.*

    Consider two heat engines operating between the same temperature limits, as shown in Fig. 5.13a. One engine is reversible and the other is irreversible.

    Let

$Q_1 = $ Heat supplied to both the heat engine from source.
$W_R = $ Work output of the reversible heat engine.
$W_I = $ Work output of the irreversible heat engine.
$Q_{2,R} = $ Heat rejected by the reversible heat engine.

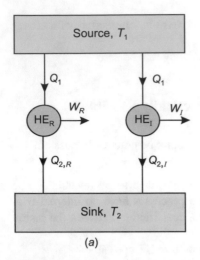 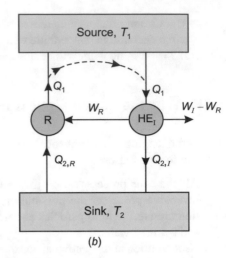

(a)   (b)

Fig. 5.13 Carnot Theorem

$Q_{2,I}$ = Heat rejected by the irreversible heat engine.

In the violation of the Carnot theorem, we assume that the irreversible heat engine is more efficient than the reversible heat engine.

i.e.,

$$\eta_R < \eta_I,\ W_R < W_I\ \text{and}\ Q_{2,R} > Q_{2I}$$

Now let the reversible heat engine be reversed and operate as a refrigerator. The irreversible heat engine continues to operate as a heat engine shown in Fig. 5.13b. The refrigerator will receive a work input of W_R and reject heat Q_1 to the source. The irreversible heat engine is receiving the same amount of heat from the source, and the net heat exchange for the source is zero. Thus, there is no need of the source. The combined system produces a net work ($W_I - W_R$) while exchanging heat with a single reservoir at temperature T_2. This is a violation of the Kelvin-Planck statement of the second law. Therefore, the assumption that we made in the beginning that 'the irreversible heat engine is more efficient than the reversible heat engine' is wrong. Hence, we conclude that no heat engine can be more efficient than a reversible heat engine (*i.e.*, Carnot engine) operating between the same temperature limits.

Corollary of Carnot's Theorem Corollary-1: *The efficiencies of all reversible heat engines operating between the same two reservoirs (i.e., source and sink) are the same.*

Consider both the reversible heat engines operating between the same temperature limits as shown in Fig. 5.14a.

We assume that the efficiency of reversible heat engine $HE_{R,2}$ is more than the efficiency of reversible heat engine $HE_{R,1}$, i.e., $\eta_{R,1} < \eta_{R,2}$.

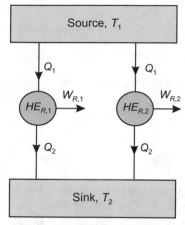

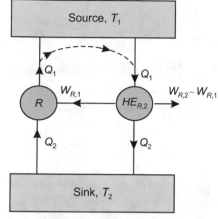

(a) Two reversible heat engines

(b) Refrigerator and reversible heat engine

Fig. 5.14 Corollary-1

Now let the reversible heat engine be reversed and operate as a refrigerator, as shown in Fig. 5.14b.

The refrigerator will receive a work input of $W_{R,1}$ and reject heat Q_1 to the source. The reversible heat engine is receiving the same amount of heat from the source, and the net heat exchange for the source is zero. Thus, there is no need of the source. The combined system produces a net work ($W_{R,2} - W_{R,1}$) while exchanging heat with a single reservoir at temperature T_2. This is a violation of the Kelvin-Planck statement of the second law. Therefore, our assumption made in the beginning that the efficiency of reversible heat engine $HE_{R,2}$ is more than the efficiency of reversible heat engine $HE_{R,1}$ is wrong. Hence, we conclude that the efficiencies of all reversible heat engines operating between the same two reservoirs are the same.

Corollary-2: *The efficiency of any reversible heat engine does not depend on the working fluid in the cycle and depends only on the temperatures of the source and the sink.*

Consider a reversible heat engine and refrigerator operate between the same temperature limits, as shown in Fig. 5.15a. The reversible heat engine receives Q_1 heat from the source at temperature T_1 and rejects heat Q_2 to the sink at temperature T_2. The reversible heat engine output is used to drive the refrigerator since the refrigerator is a reversed Carnot engine, its work and heat transfer will be the same as the reversible heat engine but with directions reversed. It is clear from Fig. 5.15a that the net output of the reversible heat engine and the refrigerator is zero.

Now let the efficiency of the reversible heat engine depend upon the working fluid. The heat engine may deliver more work than that required to drive the refrigerator, as shown in Fig. 5.15b. The refrigerator will receive a work input of W and reject heat Q_1 to the source. The heat engine is receiving the same amount of heat from

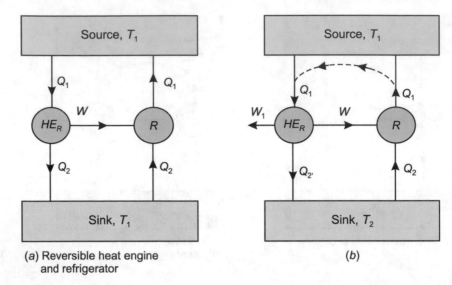

(a) Reversible heat engine
    and refrigerator

(b)

Fig. 5.15 Corollary-2

the source, and the net heat exchange for the source is zero. Thus, there is no need of the source. The combined system produces a net work W_1 while exchanging heat with a single reservoir at temperature T_2. This is a violation of the Kelvin-Planck statement of the second law. Therefore, our assumption made in the beginning that the efficiency of the reversible heat engine depends upon the working fluid and which deliver more work than that required to drive the refrigerator is wrong. Hence, we conclude that the efficiency of any reversible heat engine does not depend on the working fluid in the cycle and depends only on the temperatures of the source and the sink.

5.13   Thermodynamic Temperature Scale

A temperature scale that is independent of the properties of the substance that are used to measure temperature is called a thermodynamic temperature scale. It is established based on the fact that the thermal efficiency of a reversible heat engine is a function of the two reservoir temperatures only.

Mathematically,

$$\eta_{\text{th,rev}} = f(T_1, T_2) \tag{5.5}$$

Consider two reversible heat engines HE_1 and HE_2 operating between two reservoirs of temperatures T_1, T_2 and T_2, T_3. The heat engine HE_1 receives heat Q_1 from the thermal reservoir at T_1 and rejects heat Q_2 to the thermal reservoir at T_2. The

Fig. 5.16 Concept of the thermodynamics temperature scale

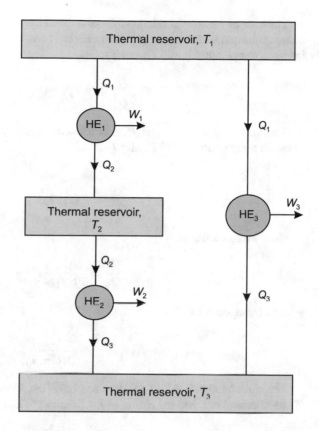

heat engine HE_2 receives heat Q_2 from the thermal reservoir at T_2 and rejects heat Q_3 to the thermal reservoir at T_3 as shown in Fig. 5.16.

For the reversible heat engine HE_1 working between temperature limits T_1 and T_2

Thermal efficiency:

$$\eta_{\text{th},1} = 1 - \frac{Q_2}{Q_1} = 1 - \frac{1}{Q_1/Q_2} \tag{5.6}$$

From Eqs. (5.5) and (5.6), we get

$$1 - \frac{1}{Q_1/Q_2} = f(T_1, T_2)$$

$$\text{or} \quad \frac{Q_1}{Q_2} = f_1(T_1, T_2) \tag{5.7}$$

Thus, some functional relationship as defined by f_1 is established between heat interactions and temperatures. Similarly for the reversible heat engine HE_2 working between temperature limits T_2 and T_3,

$$\frac{Q_2}{Q_3} = f_1(T_2, T_3) \tag{5.8}$$

The heat engines HE_1 and HE_2 constitute another heat engine HE_3 working between temperature limits T_1 and T_3.

∴

$$\frac{Q_1}{Q_3} = f_1(T_1, T_3) \tag{5.9}$$

Now we can write as

$$\frac{Q_1}{Q_2} = \frac{Q_1/Q_3}{Q_2/Q_3}$$

which corresponds to

$$f_1(T_1, T_2) = \frac{f_1(T_1, T_3)}{f_1(T_2, T_3)} \tag{5.10}$$

The right hand side of Eq. (5.10) must be a function of T_1 and T_2 only. Hence, T_3 must cancel out, and the nature of the function f_1 has the following form:

$$f_1(T_1, T_2) = \frac{\phi(T_1)}{\phi(T_2)}$$

From Eq. (5.7), $\frac{Q_1}{Q_2} = f_1(T_1, T_2)$, the above equation becomes

$$\frac{Q_1}{Q_2} = \frac{\phi(T_1)}{\phi(T_2)}$$

Similarly

$$\frac{Q_2}{Q_3} = \frac{\phi(T_2)}{\phi(T_3)}$$

and

$$\frac{Q_1}{Q_3} = \frac{\phi(T_1)}{\phi(T_3)}$$

Lord Kelvin proposed that the function $\phi(T)$ can be arbitrary chosen based on the Kelvin scale or absolute thermodynamic temperature scale as absolute thermodynamic temperature scale as

$$\phi(T) = T, \text{ temperature in Kelvin scale.}$$

Therefore,

$$\frac{Q_1}{Q_2} = \frac{T_1}{T_2}$$

$$\frac{Q_2}{Q_3} = \frac{T_2}{T_3}$$

and

$$\frac{Q_1}{Q_3} = \frac{T_1}{T_3}$$

where T_1, T_2, and T_3 are temperatures in absolute thermodynamic scale (or Kelvin scale). On the Kelvin scale, the temperature ratios depend on the ratios of heat transfer between a reversible heat engine and the reservoirs and are independent of the physical properties of the working substance.

Problem 5.1 Find the COP of a refrigerator which is extracting heat at the rate of 5 kcal/min and the compressor work is 1 kJ/s.

Solution: Given data (Refer Fig. 5.17):

$$Q_2 = 5 \, \text{kcal/min}$$

Fig. 5.17 Refrigerator

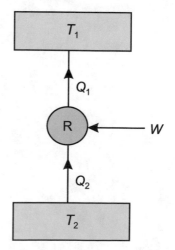

$$= \frac{5 \times 4.2}{60} \text{kJ/s} = 0.35 \text{ kJ/s}$$

$$W = 1 \text{ kJ/s}$$

$$\text{COP} = \frac{Q_2}{W} = \frac{0.35}{1} = \textbf{0.35}$$

Problem 5.2 A Carnot heat engine received 125 J of heat from its higher temperature 200 °C reservoir and rejects 100 J to its sink. Determine the temperature of the sink.

Solution: Given data (Refer Fig. 5.18):

$$Q_1 = 125 \text{ J}$$
$$T_1 = 200 °\text{C} = (200 + 273) \text{ K} = 473 \text{ K}$$
$$Q_2 = 100 \text{ J}$$

Carnot efficiency:

$$\eta = 1 - \frac{T_2}{T_1}$$

also

$$\eta = 1 - \frac{Q_2}{Q_1}$$

Fig. 5.18 Carnot Engine

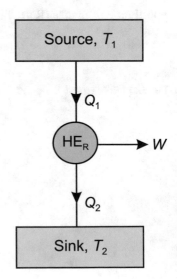

$\therefore$

$$1 - \frac{T_2}{T_1} = 1 - \frac{Q_2}{Q_1}$$

or

$$\frac{T_2}{T_1} = \frac{Q_2}{Q_1}$$

$$\frac{T_2}{473} = \frac{100}{125}$$

or

$$T_2 = 378.4\,\text{K} = (378.4 - 273)\,^{\circ}\text{C}$$
$$= \mathbf{105.4\,^{\circ}C}$$

Problem 5.3 Determine the thermal efficiency of a reversible heat engine in the following cases:

(i)   Reversible heat engine works between NTP and STP.
(ii) Reversible heat engine works between 0 and 100 °C.
(iii) 0 K and 0 °C.

Solution: Refer Fig. 5.19.

(i)   At NTP: Normal temperature and pressure.

$$T_2 = 0\,^{\circ}\text{C} = 273\,\text{K}$$

At STP: Standard temperature and pressure.

$$T_1 = 15\,^{\circ}\text{C} = (15 + 273)\,\text{K} = 288\,\text{K}$$

Fig. 5.19 Reversible heat engine

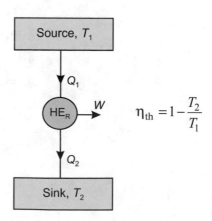

$$\eta_{th} = 1 - \frac{T_2}{T_1}$$

Thermal efficiency:

$$\eta_{th} = 1 - \frac{T_2}{T_1}$$
$$= 1 - \frac{273}{288}$$
$$= 1 - 0.9479$$
$$= 0.0521 = \mathbf{5.21\%}$$

(ii)   Low temperature:

$$T_2 = 0\,°C = 273\,K$$

High temperature:

$$T_1 = 100\,°C$$
$$= (273 + 100)\,K$$
$$= 373\,K$$

Thermal efficiency:

$$\eta_{th} = 1 - \frac{T_2}{T_1} = 1 - \frac{273}{373}$$
$$= 1 - 0.7319$$
$$= 0.268 = \mathbf{26.81\%}$$

(iii) Low temperature:

$$T_2 = 0\,K$$

High temperature:

$$T_1 = 0\,°C = 273\,K$$

Thermal efficiency:

$$\eta_{th} = 1 - \frac{T_2}{T_1} = 1 - \frac{0}{273}$$
$$= 1 - 0 = 1 = \mathbf{100\%}$$

Problem 5.4 A Carnot cycle operates between 0 and 100 °C. Determine

(a)   its thermal efficiency if it operates as a heat engine,
(b)   its COP if it operates as a refrigerator, and
(c)   its COP if it operates as a heat pump.

Solution: Refer Fig. 5.20.
   Low temperature:

$$T_2 = 0\,°C = 273\,K$$

   High temperature:

$$T_1 = 100\,°C = (100 + 273)\,K = 373\,K$$

(a)   Heat engine:

Efficiency:

$$\eta = 1 - \frac{T_2}{T_1} = 1 - \frac{273}{373}$$
$$= 1 - 0.7319 = 0.2681 = \mathbf{26.81\%}$$

(b)   Refrigerator:

$$(COP)_R = \frac{T_2}{T_1 - T_2}$$

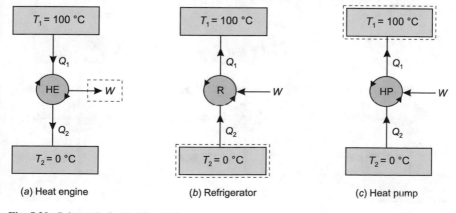

(a) Heat engine   (b) Refrigerator   (c) Heat pump

Fig. 5.20 Schematic for Problem 5.4

$$= \frac{273}{373 - 273} = \frac{273}{100} = \mathbf{2.73}$$

(c)   Heat Pump:

$$(COP)_{HP} = \frac{T_1}{T_1 - T_2} = \frac{373}{373 - 273} = \frac{373}{100} = \mathbf{3.73}$$

Problem 5.5 A heat engine is supplied with 278 kJ/s of heat at a constant fixed temperature of 283 °C and heat rejection takes place at 5 °C. The following results were reported:

(i)   208 kJ/s of heat are rejected.
(ii)   139 kJ/s of heat are rejected.
(iii) 70 kJ/s of heat are rejected.

Classify which of the results report reversible cycle or irreversible cycle or impossible results.

Solution: Given data:
   Heat supplied: $Q_1 = 278$ kJ/s
   Temperature of source: $T_1 = 283$ °C $= 556$ K
   Temperature of sink: $T_2 = 5$ °C $= 278$ K
   Carnot efficiency: $\eta_{Carnot} = 1 - \frac{T_2}{T_1} = 1 - \frac{278}{556} = 1 - 0.5 = 0.50 = 50\%$

(i)   From Fig. 5.21a,
       Heat rejected:

$$Q_2 = 208 \, \text{kJ/s}$$

       Efficiency of a heat engine:

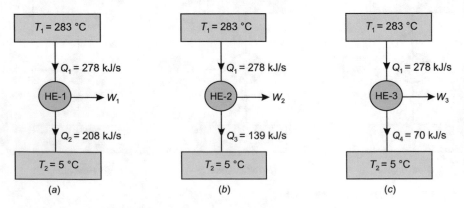

Fig. 5.21 Schematic for Problem 5.5

$$\eta_1 = 1 - \frac{Q_2}{Q_1} = 1 - \frac{208}{278}$$
$$= 1 - 0.7482 = 0.2518 = 25.18\%$$

The Carnot efficiency is more than the actual efficiency of a heat engine. Thus, the heat engine in Fig. 5.21a followed the **irreversible cycle**.

(ii)   From Fig. 5.21b,
       Heat rejected:

$$Q_3 = 139 \, \text{kJ/s}$$

Efficiency of a heat engine:

$$\eta_2 = 1 - \frac{Q_3}{Q_1} = 1 - \frac{139}{278} = 1 - 0.5 = 0.5 = 50\%$$

The Carnot efficiency is equal to the actual efficiency of a heat engine. Thus, the heat engine in Fig. 5.21b followed the **reversible cycle**.

(iii) From Fig. 5.21c,
       Heat rejected:

$$Q_4 = 70 \, \text{kJ/s}$$

Efficiency of a heat engine:

$$\eta_3 = 1 - \frac{Q_4}{Q_1} = 1 - \frac{70}{278} = 1 - 0.2517 = 0.7489 = 74.83\%$$

The Carnot efficiency is less than the actual efficiency of a heat engine. Thus, the heat engine in Fig. 5.21c is **impossible**.

Problem 5.6 A Carnot heat engine operates between a source at 1000 K and a sink at 300 K. If the heat engine is supplied with heat at a rate of 800 kJ/min, determine

(a)   the thermal efficiency and
(b)   the power output of the heat engine.

Solution: Given data.
    Temperature of a source: $T_1 = 1000$ K (Refer Fig. 5.22).
    Temperature of a sink: $T_2 = 300$ K
    Heat supplied: $Q_1 = 800 \, \text{kJ/min} = \frac{800}{60} \, \text{kJ/s} = 13.33 \, \text{kJ}$

(a)   Thermal efficiency:

$$\eta_{\text{th}} = 1 - \frac{T_2}{T_1} = 1 - \frac{300}{1000}$$

Fig. 5.22 Schematic for
Problem 5.6

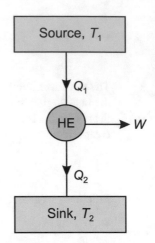

(b)   Power output:

$$W = 1 - 0.3 = 0.70 = \textbf{70\%}$$

Thermal efficiency:

$$\eta_{th} = \frac{W}{Q_1}$$
$$0.7 = \frac{W}{13.33}$$

or

$$W = 0.7 \times 13.33$$
$$= 9.33 \, \text{kJ/s} = \textbf{9.33 kW}$$

Problem 5.7 A heat engine operates between a source at 550 °C and a sink at 25 °C. If heat is supplied to the heat engine at a steady rate of 1200 kJ/min, determine the maximum power output of this heat engine.

Solution: Given data:
   Temperature of source: $T_1 = 550 \, °C = (550 + 273) \, \text{K} = 823 \, \text{K}$
   Temperature of sink: $T_2 = 25 \, °C = (25 + 273) \, \text{K} = 298 \, \text{K}$
   Heat supplied: $Q_1 = 1200 \, \text{kJ/min} = \frac{1200}{60} \, \text{kJ/s} = 20 \, \text{kJ/s} = 20 \, \text{kW}$
   Maximum power output means the heat engine is reversible.
   i.e., Thermal efficiency: $\eta_{th} = 1 - \frac{T_2}{T_1}$
                            $\eta_{th} = 1 - \frac{298}{823} = 1 - 0.3620 = 0.638$
   also

$$\eta_{th} = \frac{W}{Q_1}$$

$$0.638 = \frac{W}{20}$$

or

$$W = 0.638 \times 20 = \mathbf{12.76\,kW}$$

Problem 5.8 An engine manufacturer claims to have developed a heat engine with the following specification:

Power developed = 80 kW
Fuel burnt = 5.5 kg/hr
Calorific value of fuel = 77000 kJ/kg.

Temperature limits 1150 and 400 K. Is the claim of the manufacturer true or false? Give reasons for your answer.

Solution: Given data:
Power developed: $W = 80$ kW
Mass of fuel burnt: $m_f = 5.5$ kg/hr $= \frac{5.5}{3600} = 1.527 \times 10^{-3}$ kg/s
Calorific value of fuel: $CV = 77000$ kJ/kg
Heat supplied: $Q_1 = m_f CV = 1.527 \times 10^{-3} \times 77000\,\text{kJ/s} = 117.579\,\text{kW}$
Thermal efficiency:

$$\eta_{th} = \frac{W}{Q_1} = \frac{80}{117.579}$$
$$= 0.6803 = 68.03\%$$

Thermal efficiency of a Carnot heat engine:

$$\eta_{Carnot} = 1 - \frac{T_2}{T_1} = 1 - \frac{400}{1150}$$
$$= 1 - 0.3478 = 0.6522 = 65.22\%$$

Since, the thermal efficiency of a heat engine is greater than the thermal efficiency of a Carnot heat engine, so the claim is false and the heat engine needs to be redesigned.

Problem 5.9 An inventor claims to have developed a heat engine that receives 700 kJ of heat from a source at 500 K and produces 300 kJ of net work while rejecting the waste heat to a sink at 290 K. Is this a reasonable claim? Why?

Solution: Given data:
Heat received from a source: $Q_1 = 700$ kJ
Temperature of a source: $T_1 = 500$ K
Work output: $W = 300$ kJ
Temperature of a sink: $T_2 = 290$ K

Thermal efficiency of a heat engine:

$$\eta = \frac{W}{Q_1} = \frac{300}{700} = 0.4285 = 42.85\%$$

We know that the Carnot efficiency,

$$\eta_{\text{Carnot}} = 1 - \frac{T_2}{T_1} = 1 - \frac{290}{500} = 1 - 0.58 = 0.42 = 42\%$$

Since, no heat engine can be more efficient than a Carnot heat engine (or reversible heat engine), the claim is false.

Problem 5.10 A domestic refrigerator that has a power input of 450 W and a COP of 2.5 is to cool five large watermelons, 10 kg each, to 8 °C. If the watermelons are initially at 30 °C, how long it will take for the refrigerator to cool them? The watermelons can be treated as water whose specific heat is 4.2 kJ/kgK.

Solution: Given data:
   Power input:

$$W = 450\,\text{W} = 0.45\,\text{kW}$$
$$\text{COP} = 2.5$$

also

$$\text{COP} = \frac{\text{Heat removed} : Q_2}{\text{Power input} : W}$$

$$2.5 = \frac{Q_2}{0.45}$$

or $Q_2 = 2.5 \times 0.45 = 1.125$ kW
   Number of watermelons: $n = 5$
   Mass of each watermelon $= 10$ kg
   $\therefore$ Net mass of watermelons: $M = 5 \times 10 = 50$ kg
   Specific heat of the watermelons: $c = 4.2$ kJ/kgK
   Temperature required to be cooled: $T_2 = 8\,°C = (8 + 273)\,K = 281$ K
   Temperature of the watermelons: $T_1 = 30\,°C = (30 + 273)\,K = 303$ K
   Heat removed from watermelons: $Q_2 = mc(T_1 - T_2)$

$$Q_2 = \frac{M}{t} c(T_1 - T_2)$$

$$1.125 = \frac{50}{t} \times 4.2(303 - 281)$$

or

$$t = 4106.66 \, \text{s} = \mathbf{1.14 \, hr}$$

Problem 5.11 A domestic refrigerator maintains a temperature of $-10\,°C$. Every time the door is opened, warm material is placed inside, introducing an average of 410 kJ, but making only a small change in the temperature of the refrigerator. The door is opened 24 times a day, and the refrigerator operates at 25% of the ideal COP. The cost of work is 50 paise per kWh. What is the monthly bill for this refrigerator? The atmospheric temperature is 32 °C.

Solution: Given data:

Low temperature: $T_2 = -10\,°C = (273 - 10)\,K = 263\,K$
Heat removed from refrigerator space per day:

$$Q_2 = 410 \times 24 \, \text{kJ} = 9840 \, \text{kJ}$$

Actual $(COP)_R = 25\%$ of theoretical $(COP)_R$

$$(COP)_R = 0.25 \times \frac{T_2}{T_1 - T_2}$$

$$\text{Cost of work} = 50 \, \text{paise per kWh} = \text{Rs.}\,0.5 \, \text{per kWh}$$

Atmospheric temperature: $T_1 = 32\,°C = (32 + 273)\,K = 305\,K$
Actual $(COP)_{\text{act}} = 0.25 \times \frac{263}{305-265} = 1.64$

also actual

$$(COP)_{\text{act}} = \frac{Q_2}{W}$$

$$1.64 = \frac{9840}{W}$$

or

$$W = \frac{9840}{1.64} = 6000 \, \text{kJ}\,W$$

$$= 6000 \, \frac{\text{kJs}}{\text{s}}$$

$$= \frac{6000 \, \text{kJh}}{3600 \, \text{s}}$$

$$= 1.666 \, \text{kWh}$$

$$\left| \cdot \cdot \, 1\text{s} = \frac{1}{3600} \text{h} \right.$$

Cost of work $= \text{Rs.}\,0.5 \, \text{per kWh}$
Cost of per day $= \text{Rs.}\,0.5 \times 1.666 = \text{Rs.}\,0.833$
Cost of work for one month $= \text{Rs.}\,0.833 \times 30 = \text{Rs.}\,24.99 \approx \mathbf{Rs.\,25}$

Problem 5.12 A Carnot engine operates between two reservoirs at the temperature of T_1 and T_2 K. The work output of the engine is 0.6 times that of heat rejected. Given that the difference of temperature between the source and sink is 200 °C, calculate the source temperature, the sink temperature, and the thermal efficiency.

Solution: Refer Fig. 5.23

$$W = 0.6Q_2$$
$$T_1 - T_2 = 200\,°C$$

Efficiency: $\eta = \frac{T_1 - T_2}{T_1}$

$$\eta = \frac{200}{T_1} \qquad\qquad (5.11)$$

also $\eta = \frac{W}{Q_1}$

$$\eta = \frac{0.6Q_2}{Q_1} = \frac{0.6T_2}{T_1} \qquad\qquad (5.12)$$

Equating Eqs. (5.11) and (5.12), we have $\frac{200}{T_1} = \frac{0.6T_2}{T_1}$

$$\text{or} \quad 200 = 0.6\,T_2$$
$$\text{or} \quad T_2 = \textbf{333.33\,K}$$
$$\text{and} \quad T_1 - T_2 = 200$$
$$T_1 = 200 + T_2 = 200 + 333.33 = \textbf{533.33\,K}$$

From Eq. (5.11),

Fig. 5.23 Schematic for
Problem 5.12

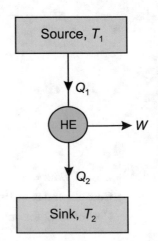

$$\eta = \frac{200}{T_1} = \frac{200}{533.33} = 0.3750 = \mathbf{37.50\%}$$

Problem 5.13 Which is a more effective way to increase the efficiency of the Carnot engine.

(a)   to increase the temperature of the source while the sink temperature is held constant.
(b)   to decrease the temperature of the sink while the source temperature is held constant.

Solution: Let

$T_1 =$ Temperature of the source
$T_2 =$ Temperature of the sink

We know that the efficiency of the Carnot engine (Fig. 5.24).

$$\eta_{Carnot} = 1 - \frac{T_2}{T_1} \tag{5.13}$$

To study the relative influence of the two temperatures, T_1 and T_2 in η_{Carnot}, Eq. (5.13) is differentiated with respect to both the temperatures T_1 and T_2 as follows:

$$\frac{d\eta_{Carnot}}{dT_1} = \frac{d}{dT_1}\left(1 - \frac{T_2}{T_1}\right) \text{ at } T_2 = C$$

$$= 0 - T_2 \frac{d}{dT_1} T_1^{-1}$$

$$= -T_2(-1)T_1^{-2} = \frac{T_2}{T_1^2} \tag{5.14}$$

Fig. 5.24 Schematic for Problem 5.13

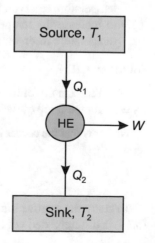

Source, T_1

Q_1

HE $\longrightarrow$ W

Q_2

Sink, T_2

and

$$\frac{d\eta_{\text{Carnot}}}{dT_2} = \frac{d}{dT_2}\left(1 - \frac{T_2}{T_1}\right) \text{ at } T_1 = C$$

$$= 0 - \frac{d}{dT_2}\left(\frac{T_2}{T_1}\right) = -\frac{1}{T_1} \tag{5.15}$$

It is clear from Eq. (5.14) that η_{Carnot} increases with an increase in temperature of the source.

Let $T_1 = 500\,\text{K}, \quad T_2 = 300\,\text{K}$

$\therefore \frac{d\eta_{\text{Carnot}}}{dT_1} = \frac{T_2}{T_1^2} = \frac{300}{(500)^2} = 0.0012$

Let $dT_1 = 10°\text{C}$ or K, increase in temperature of the source.

$\therefore \quad d\eta_{\text{Carnot}} = 0.0012 \times 10$

$$= 0.012 = 1.2\%$$

It is also clear from Eq. (5.15) that η_{Carnot} increases with the decrease in temperature of the sink.

$\therefore \quad \frac{d\eta_{\text{Carnot}}}{dT_2} = -\frac{1}{T_2}$

$$= -\frac{1}{500} = -0.002$$

Let $dT = -10°\text{C}$ or K, decrease in temperature of the sink.

$\therefore d\eta_{\text{Carnot}} = -0.002 \times (-10) = 0.02 = 2\%$

Hence, it is clear from the above analysis that to increase the efficiency of the Carnot engine, decreasing the temperature of the sink is a more effective way. That is, option (b) is the optimum way to increase the efficiency of the Carnot engine.

Problem 5.14 Which is a more effective way to increase the COP of the Carnot refrigerator

(a)   to decrease the temperature of a high-temperature reservoir while the low-temperature reservoir is held constant.

(b)   to increase the temperature of a low-temperature reservoir while the high-temperature reservoir is held constant.

Solution: Let

$T_1 =$ Temperature of the hot space
$T_2 =$ Temperature of the cold space

We know that the coefficient of performance of the Carnot refrigerator, (Fig. 5.25)

$$(\text{COP})_R = \frac{T_2}{T_1 - T_2} \tag{5.16}$$

To study the relative influence of the two temperatures, T_1 and T_2 on COP, Eq. (5.16) is differentiated with respect to both the temperatures T_1 and T_2 as follows:

Fig. 5.25 Schematic for Problem 5.14

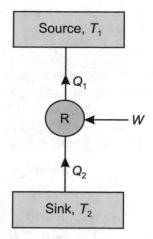

$$\frac{d(COP)_R}{dT_1} = \frac{d}{dT_1}\left[\frac{T_2}{T_1 - T_2}\right] \text{ at } T_2 = C$$

$$= T_2 \frac{d}{dT_1}(T_1 - T_2)^{-1}$$

$$= T_2(-1)(T_1 - T_2)^{-2}$$

$$\frac{d(COP)_R}{dT_1} = -\frac{T_2}{(T_1 - T_2)^2} \qquad (5.17)$$

and

$$\frac{d(COP)_R}{dT_2} = \frac{d}{dT_2}\left[\frac{T_2}{T_1 - T_2}\right]$$

$$= \frac{d}{dT_2}\left[\frac{-T_1 + T_2 + T_1}{T_1 - T_2}\right]$$

$$= \frac{d}{dT_2}\left[\frac{-(T_1 + T_2) + T_1}{T_1 - T_2}\right]$$

$$= \frac{d}{dT_2}\left[-1 + \frac{T_1}{T_1 - T_2}\right]$$

$$= 0 + T_1 \frac{d}{dT_2}(T_1 - T_2)^{-1}$$

$$= T_1(-1)(T_1 - T_2)^{-2}(-1)$$

$$\frac{d(COP)_R}{dT_2} = \frac{T_1}{(T_1 - T_2)^2} \qquad (5.18)$$

It is clear from Eq. (5.17) that $(COP)_R$ increases with a decrease in the temperature of the hot space.

Let $T_1 = 400$ K, $\quad T_2 = 270$ K

$$W_1 = Q_1 - Q_2$$

$$\therefore \quad = 24.21 - 20.21 = 4 \text{ kW}$$

Let $dT_1 = -10°C$ or K, decrease in the temperature of the source.

$$\therefore d(COP)_R = -0.0159 \times (-10) = 0.159 = 15.9\%$$

It is also clear from Eq. (5.18) that $(COP)_R$ increases with the increase in the temperature of the cold space.

$$\frac{d(COP)_R}{dT_2} = \frac{T_1}{(T_1 - T_2)^2} = \frac{400}{(400 - 270)^2} = 0.02366$$

Let $dT_2 = 10\,°C$ or K, increase in the temperature of the cold space.

$$\therefore d(COP)_R = 0.02366 \times 10 = 0.2366 = 23.66\%$$

Hence, it is clear from the above analysis that to increase the COP of the Carnot refrigerator, increasing the temperature of the cold space is a more effective way. That is, option (b) is the optimum way to increase the COP of the Carnot refrigeration.

Problem 5.15 Which is a more effective way to increase the COP of the Carnot heat pump

(a)   to decrease the temperature of a high-temperature reservoir while the low-temperature reservoir is held constant.

(b)   to increase the temperature of a low-temperature reservoir while the high-temperature reservoir is held constant.

Solution: Let

$T_1 = \quad$ Temperature of the hot space, i.e., source

$T_2 = \quad$ Temperature of the cold space, i.e., sink

We know that the coefficient of performance of the Carnot heatpump, (Fig. 5.26).

$$(COP)_{HP} = \frac{T_1}{T_1 - T_2} \tag{5.19}$$

To study the relative influence of the two temperatures, T_1 and T_2 on COP, Eq. (5.19) is differentiated with respect to both the temperatures T_1 and T_2 as follows:

$$\frac{d(COP)_{HP}}{dT_1} = \frac{d}{dT_1}\left[\frac{T_1}{T_1 - T_2}\right] \text{ at } T_2 = C$$

$$= \frac{d}{dT_1}\left[\frac{T_1 - T_2 + T_2}{T_1 - T_2}\right]$$

$$= \frac{d}{dT_1}\left[1 + \frac{T_2}{T_1 - T_2}\right]$$

Fig. 5.26 Schematic for Problem 5.15

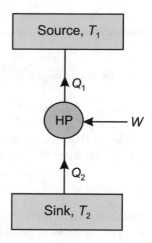

$$= 0 + T_2 \frac{d}{dT_1}(T_1 - T_2)^{-1}$$

$$= T_2(-1)(T_1 - T_2)^{-2}$$

$$\frac{d(COP)_{HP}}{dT_1} = -\frac{T_2}{(T_1 - T_2)^2} \tag{5.20}$$

and

$$\frac{d(COP)_{HP}}{dT_2} = \frac{d}{dT_2}\left(\frac{T_1}{T_1 - T_2}\right) \text{ at } T_1 = C$$

$$= T_1 \frac{d}{dT_2}(T_1 - T_2)^{-1}$$

$$= T_1(-1)(T_2 - T_2)^{-2}(-1)$$

$$\frac{d(COP)_{HP}}{dT_2} = \frac{T_1}{(T_1 - T_2)^2} \tag{5.21}$$

It is clear from Eq. (5.20) that $(COP)_{HP}$ increases with a decrease in the temperature of a high-temperature reservoir.

Let $T_1 = 500$ K, $T_2 = 300$ K

$$\therefore \frac{d(COP)_{HP}}{dT_1} = -\frac{300}{(500-300)^2} = \frac{-300}{(200)^2}$$

$$\frac{d(COP)_{HP}}{dT_1} = -0.0075$$

Let $dT_1 = -10\,°C$ or K, decrease in the temperature of the source

$\therefore d(COP)_{HP} = -0.0075 \times (-10) = 0.075 = 7.5\%$

It is also clear from Eq. (5.21) that $(COP)_{HP}$ increases with the increase in the temperature of a low temperature reservoir.

$$\frac{d}{dT_2}(COP)_{HP} = \frac{500}{(500-300)^2} = \frac{500}{(200)^2} = 0.0125$$

Let $dT_2 = 10\,°C$ or K, increase in the temperature of the sink
$\therefore d(COP)_{HP} = 0.0125 \times 10 = 0.125 = 12.5\%$

Hence, it is clear from the above analysis, to increase the COP of the Carnot heat pump, increasing the temperature of a low-temperature reservoir is a more effective way. That is, option (b) is the optimum way to increase the COP of the Carnot heat pump.

Problem 5.16 A reversible heat engine in a satellite operates between a hot reservoir at T_1 and a radiating panel at $T2$. Radiation from the panel is proportional to its area and T_2^4. For a given work output and value of $T1$ show that the area of the panel will be minimum when $\frac{T_2}{T_1} = 0.75$

Solution: For reversible heat engine: $\eta = \frac{W}{Q_1}$ (Refer Fig. 5.27).

also $\eta = \frac{T_1-T_2}{T_1}$
$\therefore \frac{W}{Q_1} = \frac{T_1-T_2}{T_1}$
or $W = (T_1 - T_2)\frac{Q_1}{T_1}$

$$W = (T_1 - T_2)\frac{Q_2}{T_2}$$

$$\therefore \frac{Q_1}{T_1} = \frac{Q_2}{T_2}$$

Fig. 5.27 Schematic for Problem 5.16

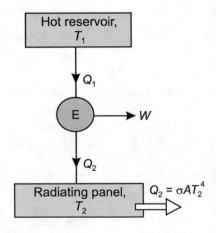

Given condition: $Q_2 \propto A\, T_2^4$ (Stefan Boltzmann's law)

$$Q_2 = \sigma A T_2^4$$

where $\sigma = $ Stefan's constant

$$\therefore W = (T_1 - T_2)\frac{\sigma A T_2^4}{T_2} = (T_1 - T_2)\sigma A T_2^3$$

$$\text{or } A = \frac{W}{\sigma(T_1 - T_2)T_2^3}$$

$$A = \frac{W}{\sigma\left(T_1 T_2^3 - T_2^4\right)}$$

For a minimum area of the radiating panel,

$$\frac{dA}{dT_2} = 0 \quad \text{at } W, \sigma, T_1 \text{ are constant.}$$

$$\frac{d}{dT_2}\left[\frac{W}{\sigma\left(T_1 T_2^3 - T_2^4\right)}\right] = 0$$

$$\frac{W}{\sigma}\frac{d}{dT_2}\left(T_1 T_2^3 - T_2^4\right)^{-1} = 0$$

$$\frac{d}{dT_2}\left(T_1 T_2^3 - T_2^4\right)^{-1} = 0$$

$$(-1)\left(T_1 T_2^3 - T_2^4\right)^{-2} \times \left(3T_1 T_2^2 - 4T_2^3\right) = 0$$

$$-\frac{\left(3T_1 T_2^2 - 4T_2^3\right)}{\left(T_1 T_2^3 - T_2^4\right)^2} = 0$$

$$\left(3T_1 T_2^2 - 4T_2^3\right) = 0$$
$$\text{or } 3T_1 T_2^2 = 4T_2^3$$
$$3T_1 = 4T_2$$
$$\text{or } \frac{T_2}{T_1} = \frac{3}{4} = \mathbf{0.75}$$

Problem 5.17 Block diagrams of two systems are given below:

Giving proper reasons indicate

(i)   Name of the system (*i.e.*, HE, R, or HP).
(ii) Type of cycle is possible or impossible and reversible or irreversible.

Solution: Given data:
  From Fig. 5.28a
  $T_1 = 500$ K
  $Q_1 = 1000$ kJ
  $W = 700$ kJ

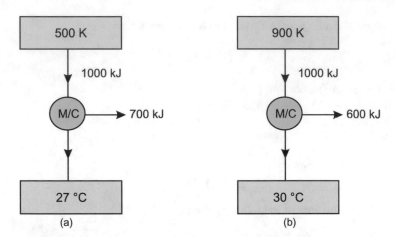

Fig. 5.28 Schematic for Problem 5.17

$$T_2 = 27\,°C = (27 + 273)\,K = 300\,K$$

Carnot efficiency: $\eta_{Camot} = 1 - \frac{T_2}{T_1} = 1 - \frac{300}{500} = 1 - 0.6 = 0.4 = 40\%$

Actual efficiency: $\eta_{act} = \frac{W}{Q_1} = \frac{700}{1000} = 0.7 = 70\%$

(i)   Heat engine: the part of supply heat is converted into useful work and the remaining part is rejected to the sink.

(ii) $\eta_{act} > \eta_{Carnot}$, it is a impossible heat engine.

From Fig. 5.28b
$T_1 = 900\,K$
$Q_1 = 1000\,kJ$
$W = 600\,kJ$

$$T_2 = 30\,°C = (30 + 273)\,K = 303\,K$$

Carnot efficiency: $\eta_{Camot} = 1 - \frac{T_2}{T} = 1 - \frac{303}{900} = 1 - 0.3366 = 0.6634 = 66.34\%$

Actual efficiency: $\eta_{act} = \frac{W}{Q_1} = \frac{600}{1000} = 0.6 = 60\%$

(i)   Heat engine: the part of supply heat is converted into useful work and the remaining part is rejected to the sink.

(ii) $\eta_{act} < \eta_{Carnot}$, it is a possible and irreversible heat engine.

Problem 5.18 A domestic refrigerator provides 80 W of refrigeration at −18 °C for the processes and 40 W of refrigeration at 5 °C for the fresh food compartment. The refrigerator is placed in a room which is maintained at 35 °C. Determine the minimum power input required for the refrigerator.

Solution Given data:

Fig. 5.29 Schematic for
Problem 5.18

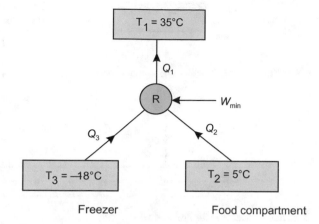

Freezer temperature: $T_3 = -18\ °C = (-18 + 273)\ K = 255\ K$
Freezer capacity: $Q_3 = 80\ W$
Food compartment temperature: $T_2 = 5\ °C = (5 + 273)\ K = 278\ K$
Food compartment capacity: $Q_2 = 40\ W$
Room temperature: $T_1 = 35\ °C = (35 + 273)\ K = 308\ K$
According to the Clausius inequality for reversible cycle, (Fig. 5.29)

$$\oint \frac{\delta Q}{T} = 0$$

$$-\frac{Q_1}{T_1} + \frac{Q_2}{T_2} + \frac{Q_3}{T_3} = 0$$

$$-\frac{Q_1}{308} + \frac{40}{278} + \frac{80}{255} = 0$$

or $\quad -Q_1 + \dfrac{308}{278} \times 40 + \dfrac{308}{255} \times 80 = 0$

$-Q_1 + 44.31 + 96.62 = 0$
or $Q_1 = 140.93\ W$
Applying energy balance equation, we get

$$W_{min} + Q_2 + Q_3 = Q_1$$
$$W_{min} + 40 + 80 = 140.93$$
$$\text{or} \quad W_{min} = 140.93 - 120 = \mathbf{20.93\ W}$$

Problem 5.19 A heat engine is used to drive a heat pump. The heat transfer from the heat engine and from the heat pump is used to heat the circulating water in a plant. The efficiency of the heat engine is 30% whereas the COP of the heat pump is 5.

Determine the ratio of the heat transfer to the circulating water and the heat transfer to the heat engine.

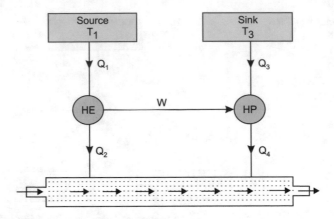

Fig. 5.30 Schematic for Problem 5.19

Solution Given data:

Efficiency of heat engine: $\eta = 30\% = 0.30$

Coefficient of performance of the heat pump: COP = 5 (Fig. 5.30)

$$\frac{Q_2 - Q_4}{Q_1} = ?\tag{5.22}$$

$$\eta = 1 - \frac{Q_2}{Q_1}$$

We know that $\eta = \dfrac{Q_1 - Q_2}{Q_1}$

$$0.30 = \frac{Q_1 - Q_2}{Q_1}$$

or $0.30\,Q_1 = Q_1 - Q_2$

or $Q_2 = Q_1 - 0.30\,Q_2$

$$Q_2 = 0.70\,Q_1\tag{5.23}$$

and COP $= \frac{Q_4}{Q_4 - Q_3}$ $W = Q_1 - Q_2 = Q_4 - Q_3$

$$5 = \frac{Q_4}{Q_1 - Q_2}$$

$$5 = \frac{Q_4}{Q_1 - 0.70Q_1}$$

or

$$Q_4 = 1.5\,Q_1\tag{5.24}$$

Substituting the value of Q_2 from Eq. (5.23) and Q_4 from Eq. (5.24) in Eq. (5.22), we get

$$\frac{Q_2 + Q_4}{Q_1} = \frac{0.70Q_1 + 1.5Q_1}{Q_1} = \frac{2.2Q_1}{Q_1} = 2.2$$

Problem 5.20 A reversible heat engine operating between 875 and 310 K drives a reversible refrigerator operating between 310 and 255 K and also supplies 350 kJ extra work. The engine receives 2000 kJ. Calculate the cooling effect.

Solution Given data (See Fig. 5.31):
$T_1 = 875$ K
$T_2 = 310$ K
$T_3 = 255$ K
Net work input to refrigerator: $W_{net} = W + 350$

$$Q_1 = 200 \text{ kJ}$$

Cooling effect: $Q_3 = ?$
For a reversible heat engine,

$$\eta_{Cannot} = 1 - \frac{T_2}{T_1} = 1 - \frac{310}{875} = 0.6457$$

also $\eta_{Cannot} = \frac{W}{Q_1}$
$\therefore 0.6457 = \frac{W}{Q_1}$
or $W = 0.6457 \times Q_1 = 0.6457 \times 2000 = 1291.4$ kJ

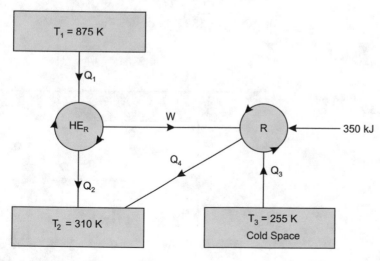

Fig. 5.31 Schematic for Problem 5.20

Net work input to a refrigerator:

$$W_{net} = W + 350 = 1291.4 + 350 = 1641.4 \, kJ$$

Coefficient of performance: $(COP)_R = \frac{T_3}{T_2 - T_3} = \frac{255}{310 - 255} = 4.63$

also
$(COP)_R = \frac{\text{Cooling effect}: Q_3}{\text{Net work input}: W_{net}}$
$(COP)_R = \frac{Q_3}{W_{net}}$

$\therefore 4.63 = \frac{Q_3}{1641.4}$

or $Q_3 = 4.63 \times 1641.4 = \mathbf{7599.68 \, kJ}$

Problem 5.21 A heat pump working on the reversed Carnot cycle takes in heat from a reservoir at 5 °C and delivers heat to a reservoir at 60 °C. The heat pump is driven by a reversible heat engine which takes in heat from a reservoir at 840 °C and rejects heat to a reservoir at 60 °C. The reversible heat engine also drives a machine that absorbs 30 kW. If the heat pump extracts 20.21 kJ/s from 5 °C reservoir, determine

(i)   the rate of heat supply from 840 °C source and
(ii) the rate of heat rejection to the 60 °C sink from the heat engine.

Solution: Given data (Refer Fig. 5.32)

$$T_2 = 5\,°C$$
$$= 5 + 273$$
$$= 278 \, K$$

$$T_1 = 60\,°C$$
$$= 60 + 273$$
$$= 333 \, K$$

Fig. 5.32 Schematic for Problem 5.21

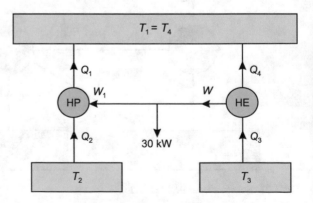

$$T_3 = 840\,°C$$
$$= 840 + 273$$
$$= 1113\,K$$

$$W = W_1 + 30$$
$$Q_2 = 20.21\,kJ/s$$
$$= 20.21\,kW$$

For heat pump: $(COP)_{HP} = \frac{T_1}{T_1-T_2} = \frac{333}{333-278} = 6.05$

also $(COP)_{HP} = \frac{Q_1}{Q_1-Q_2}$

$\therefore 6.05 = \frac{Q_1}{Q_1-20.21}$

or $6.05\,Q_1 - 122.27 = Q_1$

or $5.05\,Q_1 = 122.27$

or $Q_1 = 24.21\,kW$

(a)   Rate of heat supply from 840 °C source: Q_3
      For reversible heat engine:

$$W = W_1 + 30$$
$$= 4 + 30 = 34\,kW\,\eta$$
$$= 1 - \frac{T_4}{T_3} = 1 - \frac{T_4}{T_3} = 1 - \frac{333}{1113}$$
$$= 1 - 0.2991 = 0.7009$$

also

$$\eta = \frac{W}{Q_3}$$

$\therefore$

$$0.7009 = \frac{34}{Q_3}$$

or

$$Q_3 = \frac{34}{0.7009} = \mathbf{48.50\,kW}$$

(b)   Rate of heat rejection to the 60 °C sink from the heat engine: Q_4

$$Q_4 = Q_3 - W = 48.50 - 34 = \mathbf{14.5\,kW}$$

Problem 5.22 A reversible heat engine operates between two reservoirs at temperatures of 600 and 40 °C. The engine drives a reversible refrigerator which operates between reservoirs at temperatures of 40 and −20 °C. The heat transfer to the heat engine is 2000 kJ and the network output of the combined engine-refrigerator plant is 360 kJ.

(i)   Evaluate the heat transfer to the refrigerant and the net heat transfer to the reservoir at 40 °C.
(ii)   Reconsider (a) given that the efficiency of the heat engine and the COP of the refrigerator are each 40% of their maximum possible values.

Solution: Refer Fig. 5.33.
Reversible heat engine operates between two reservoirs at temperatures,

$$T_1 = 600\,°C = 600 + 273 = 873\,K$$
$$\text{and}\quad T_2 = 40\,°C = 40 + 273 = 313\,K$$

Reversible refrigerator operates between two reservoirs at temperatures.

$$T_3 = -20\,°C = -20 + 273 = 253\,K$$
$$\text{and}\quad T_4 = 40\,°C = 40 + 273 = 313\,K$$
$$= T_2$$

Heat supplied to heat engine: $Q_1 = 2000\,kJ$
Net work output of the combined engine-refrigerator plant = 360 kJ
For reversible heat engine, $\eta = 1 - \frac{T_2}{T_1} = 1 - \frac{313}{873} = 0.6414$ or 64.14%.

$$\eta = 1 - \frac{Q_2}{Q_1}$$
also

$$0.6414 = 1 - \frac{Q_2}{2000}$$

or $\frac{Q_2}{2000} = 0.3586$

Fig. 5.33 Schematic for Problem 5.22

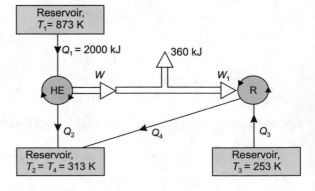

$$Q_2 = 717.2 \, \text{kJ}$$

By the energy balance equation for heat engine:

we get $\quad \dfrac{Q_1 = W + Q_2}{2000 = W + 717.2}$

or $W = 1282.8 \, \text{kJ}$

Work required to drive the refrigerator,

$$W_1 = W - 360 = 1282.8 - 360 = 922.8 \, \text{kJ}$$

For reversible refrigerator, $(COP)_R = \frac{T_3}{T_4 - T_3} = \frac{253}{313 - 253} = 4.21$

also $(COP)_R = \frac{Q_3}{W_1}$

$$4.21 = \frac{Q_3}{922.8}$$

or $Q_3 = 3884.98 \, \text{kJ}$

By the energy balance equation refrigerator:

we get $W_1 + Q_3 = Q_4$

$$922.8 + 3884.98 = Q_4$$

or $Q_4 = 4807.78 \, \text{kJ}$

(i) Net heat transfer to the reservoir at $40°C = Q_2 + Q_4$

$$= 717.2 + 4807.78 = \mathbf{5524.98 \, kJ}$$

(ii) Reconsider: Heat engine and refrigerator are not reversible so the actual efficiency of heat engine is

$$\eta_{\text{act}} = 0.4 \times \eta = 0.4 \times 0.6414 = 0.2565$$

also $\eta_{\text{act}} = \frac{W}{Q_1}$

$$0.2515 = \frac{W}{2000}$$

or $W = 513 \, \text{kJ}$

By the energy balance equation for heat engine:

we get $\quad \dfrac{Q_1 = W + Q_2}{2000 = 513 + Q_2}$

or $Q_2 = 1487 \, \text{kJ}$

Work required to drive the refrigerator,

$$W_1 = W - 360 = 513 - 360 = 153 \, \text{kJ}$$

The actual COP of the refrigerator:

$$(\text{COP})_{\text{act}} = 0.4 \times (\text{COP})_R = 0.4 \times 4.21 = 1.684$$

also

$$(\text{COP})_{\text{act}} = \frac{Q_3}{W_1}$$

$$1.684 = \frac{Q_3}{153}$$

or $Q_3 = 257.65 \, \text{kJ}$

By the energy balance equation for refrigerator
we get $W_1 + Q_3 = Q_4$

$$153 + 257.65 = Q_4$$

or $Q_4 = 410.65 \, \text{kJ}$

Net heat transfer to the reservoir at 40°C $= Q_2 + Q_4 = 1487 + 410.65 =$ **1897.65 kJ**

Problem 5.23 Two Carnot engines A and B are arranged in series between a source at 1500 K and a sink at 300 K. The heat 1000 kJ is supplied to engine A from the source at 1500 K and rejected to the intermediate temperature T_2. Determine

(i)   the work done by the engines A and B when both engines have the same efficiency.

(ii)   the efficiency of the engines A and B when both engines have the same work done.

Solution: Given data (Refer Fig. 5.34):

 Source temperature: $T_1 = 1500 \, \text{K}$
 Sink temperature: $T_3 = 300 \, \text{K}$
 Heat supplied to the engine A: $Q_1 = 1000 \, \text{kJ}$
 Let T_2 be the intermediate temperature

(i) Given condition

$$\eta_A = \eta_B$$

where

$$\eta_A = 1 - \frac{T_2}{T_1}$$

and

$$\eta_B = 1 - \frac{T_3}{T_1}$$

Fig. 5.34 Both engines have the same efficiency

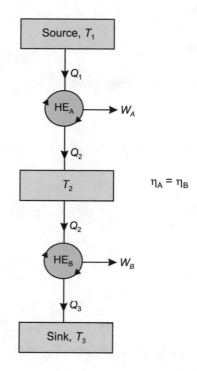

$\eta_A = \eta_B$

$\therefore$

$$1 - \frac{T_2}{T_1} = 1 - \frac{T_3}{T_2}$$

or

$$\frac{T_2}{T_1} = \frac{T_3}{T_2}$$

or

$$T_2^2 = T_1 T_3$$

or

$$T_2 = \sqrt{T_1 T_3}$$

$$T_2 = \sqrt{1500 \times 300}$$
$$= 670.82\,\text{K}$$

∴

$$\eta_A = 1 - \frac{T_2}{T_1}$$
$$= 1 - \frac{670.82}{1500}$$
$$= 1 - 0.4472 = 0.5528$$

also

$$\eta_A = \frac{W_A}{Q_1}$$

∴

$$0.5528 = \frac{W_A}{1000}$$

or

$$W_A = 0.5528 \times 1000 = \mathbf{552.8\,kJ}$$

also

$$W_A = Q_1 - Q_2$$

∴

$$552.8 = 1000 - Q_2$$

or

$$Q_2 = 447.2\,kJ$$

Efficiency of engine B:

$$\eta_B = \frac{W_B}{Q_2}$$
$$0.5528 = \frac{W_B}{447.2} \quad | \because \eta_B = \eta_A$$

or

$$W_B = \mathbf{247.21\,kJ}$$

Fig. 5.35 Both engines have the same work done

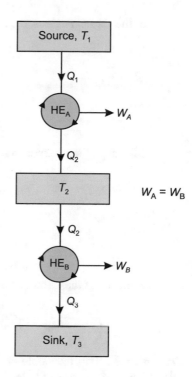

(ii) Given condition $W_A = W_B$ (Refer Fig. 5.35)
where

$$W_A = Q_1 - Q_2$$

and

$$W_B = Q_2 - Q_3$$

$\therefore$

$$Q_1 - Q_2 = Q_2 - Q_3$$

or

$$2Q_2 = Q_1 + Q_3$$

or

$$Q_2 = \frac{Q_1 + Q_3}{2}$$

For the Carnot engines, the above equation is written as

$$T_2 = \frac{T_1 + T_3}{2}$$

$$T_2 = \frac{1500 + 300}{2}$$

$$= 900 \, \text{K}$$

Efficiency of the Carnot engine A,

$$\eta_A = 1 - \frac{T_2}{T_1}$$

$$= 1 - \frac{900}{1500}$$

$$= 1 - 0.6 = 0.4 = \textbf{40\%}$$

Efficiency of the Carnot engine B,

$$\eta_B = 1 - \frac{T_3}{T_2} = 1 - \frac{300}{900} = 1 - 0.3333 = 0.6667 = 66.67\%$$

Problem 5.24 Two reversible engines A and B are arranged in series. Engine A rejects heat directly to engine B. Engine A receives 250 kJ at a temperature of 450 °C from a hot source while engine B is in communication with a cold sink at a temperature of 10 °C. If the work output of engine A is twice that of engine B, find

(a)   Intermediate temperature between A and B.
(b)   Efficiency of each engine.
(c)   Heat rejected to the cold sink.

Solution: Given data (Refer Fig. 5.36):

$$Q_1 = 250 \, \text{kJ}$$
$$T_1 = 450 \,°\text{C} = (450 + 273) \, \text{K} = 723 \, \text{K}$$
$$T_3 = 10 \,°\text{C} = (10 + 273) \, \text{K} = 283 \, \text{K}$$
$$W_A = 2W_B$$

$$(Q_1 - Q_2) = 2(Q_2 - Q_3)$$

$$Q_1 \left(1 - \frac{Q_2}{Q_1} \right) = 2Q_2 \left[1 - \frac{Q_3}{Q_2} \right]$$

$$\left(1 - \frac{Q_2}{Q_1} \right) = 2\frac{Q_2}{Q_1} \left[1 - \frac{Q_3}{Q_2} \right]$$

Fig. 5.36 Two reversible engines in series

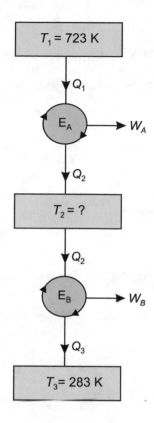

$$\left(1 - \frac{T_2}{T_1}\right) = 2\frac{T_2}{T_1}\left[1 - \frac{T_3}{T_2}\right] \quad \left|\begin{array}{l} \because \frac{Q_2}{Q_1} = \frac{T_2}{T_1} \\ \text{and } \frac{Q_3}{Q_2} = \frac{T_3}{T_2} \end{array}\right.$$

$$T_1\left(1 - \frac{T_2}{T_1}\right) = 2T_2\left[1 - \frac{T_3}{T_2}\right]$$

$$(T_1 - T_2) = 2[T_2 - T_3]$$

$$T_1 - T_2 = 2T_2 - 2T_3$$

or

$$3T_2 = T_1 + 2T_3$$

or

$$T_2 = \frac{T_1 + 2T_3}{3} = \frac{723 + 2 \times 283}{3} = \mathbf{429.67\ K}$$

(a)   Intermediate temperature between A and B: T_2

$$T_2 = \mathbf{429.67\,K}$$

(b)   Efficiency of each engine: η_A, η_B

$$\eta_A = 1 - \frac{T_2}{T_1} = 1 - \frac{429.67}{723} = 1 - 0.5942$$
$$= 0.4058 = \mathbf{40.58\%}$$

and

$$\eta_B = 1 - \frac{I_3}{T_2} = 1 - \frac{283}{429.67} = 1 - 0.6586$$
$$= 0.3414 = \mathbf{34.14\%}$$

$$\eta = 1 - \frac{Q_2}{Q_1}$$

(c)   Heat rejected to cold sink:

$$0.6414 = 1 - \frac{Q_2}{2000}$$

Also efficiency of engine A:

$$\eta_A = \frac{W_A}{Q_1}$$
$$0.4058 = \frac{W_A}{250}$$

or

$$W_A = 101.45\,\text{kJ}$$

also

$$W_A = Q_1 - Q_2$$

$\therefore$

$$101.45 = 250 - Q_2$$

or

$$Q_2 = 148.55\,\text{kJ}$$

we know,

$$W_A = 2\,W_B$$

or

$$W_B = \frac{W_A}{2} = \frac{101.45}{2} = 50.725 \, \text{kJ}$$

also

$$W_B = Q_2 - Q_3$$
$$50.725 = 148.55 - Q_3$$

or

$$Q_3 = \mathbf{92.825 \, kJ}$$

Problem 5.25 A reversible engine operates between reservoirs A, B, and C. The engine receives equal quantities of heat from reservoirs A and B at temperatures T_A and T_B, respectively, and rejects heat to reservoir C at temperature T_C. If the efficiency of the engine is α times the the efficiency of reversible engine which works between two reservoirs A and C only at temperatures T_A and T_C, Prove that:

$$\alpha = \frac{1}{2} \frac{T_A}{T_B} \left[\frac{T_B - T_C}{T_A - T_C} + \frac{T_B}{T_A} \right]$$

Solution: From Fig. 5.37a, we have
  Net heat supplied: $Q_1 = 2Q$
  ∴ Efficiency:

$$\eta_1 = 1 - \frac{Q_2}{Q_1} = 1 - \frac{Q_2}{2Q} \tag{5.25}$$

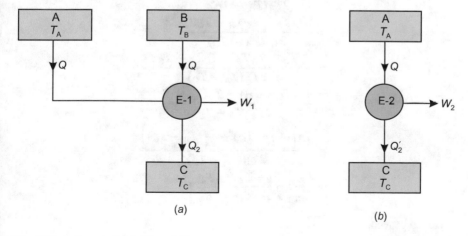

(a)   (b)

Fig. 5.37 Schematic for Problem 5.25

For reversible engine

$$\oint \frac{\delta Q}{T} = 0$$

$$\frac{Q}{T_A} + \frac{Q}{T_B} - \frac{Q_2}{T_C} = 0$$
$$Q\left[\frac{1}{T_A} + \frac{1}{T_B}\right] = \frac{Q_2}{T_C}$$

or

$$\frac{Q_2}{Q} = T_C\left[\frac{1}{T_A} + \frac{1}{T_B}\right] \tag{5.26}$$

Substituting Eq. (5.26) in Eq. (5.25), we have $\eta_1 = 1 - \frac{T_C}{2}\left[\frac{1}{T_A} + \frac{1}{T_B}\right]$
From Fig. 5.37b,

Efficiency: $\eta_2 = 1 - \frac{T_C}{T_A}$

$\eta_1 = \alpha\eta_2$ (given condition)

Substituting the values of η_1 and η_2 in the above equation

$$1 - \frac{T_C}{2}\left[\frac{1}{T_A} + \frac{1}{T_B}\right] = \alpha\left[1 - \frac{T_C}{T_A}\right]$$
$$\frac{2T_A T_B - T_C[T_B + T_A]}{2T_A T_B} = \alpha\frac{[T_A - T_C]}{T_A}$$
$$\frac{2T_A T_B - T_C[T_B + T_A]}{2T_B} = \alpha[T_A - T_C]$$

or

$$\begin{aligned}
\alpha &= \frac{2T_A T_B - T_C[T_B + T_A]}{2T_B(T_A - T_C)} \\
&= \frac{2T_A T_B - T_C T_B - T_C T_A}{2T_B(T_A - T_C)} \\
&= \frac{T_A T_B + T_A T_B - T_C T_B - T_C T_A}{2T_B(T_A - T_C)} \\
&= \frac{T_A(T_B - T_C) + T_B(T_A - T_C)}{2T_B(T_A - T_C)} \\
&= \frac{T_A\left[(T_B - T_C) + \frac{T_B}{T_A}(T_A - T_C)\right]}{2T_B(T_A - T_C)} \\
&= \frac{T_A}{2T_B}\left[\frac{T_B - T_C}{T_A - T_C} - \frac{T_B}{T_A}\right]
\end{aligned}$$

Hence proved.

Problem 5.26 Three Carnot engines between temperatures of 1000 K and 300 K. Determine the intermediate temperature if the work produced by the engines is in the ratio of 4:3:2.

Solution: Given data:
$T_1 = 1000$ K (Refer Fig. 5.38)
$T_4 = 300$ K
$W_1:W_2:W_3$
4:3:2

Fig. 5.38 Schematic for Problem 5.26

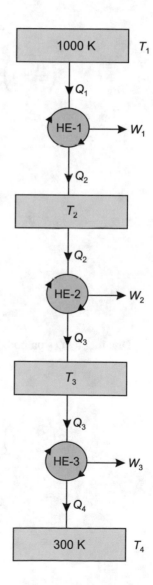

Let T_2 and T_3 be the intermediate temperatures in K.

$$\frac{W_1}{W_2} = \frac{4}{3}$$
$$3W_1 = 4W_2$$
$$3(Q_1 - Q_2) = 4(Q_2 - Q_3)$$

Dividing by Q_2 on both sides, we have

$$3\left(\frac{Q_1}{Q_2} - 1\right) = 4\left(1 - \frac{Q_3}{Q_2}\right)$$

$$3\left(\frac{T_1}{T_2} - 1\right) = 4\left(1 - \frac{T_3}{T_2}\right) \quad \left| \begin{array}{l} \text{For Carnot engines} \\ \frac{Q_1}{Q_2} = \frac{T_1}{T_2}, \frac{Q_3}{Q_2} = \frac{T_3}{T_2} \end{array} \right.$$

$$3(T_1 - T_2) = 4(T_2 - T_3)$$
$$3(1000 - T_2) = 4T_2 - 4T_3$$
$$3000 - 3T_2 = 4T_2 - 4T_3$$

or

$$7 T_2 - 4T_3 = 3000 \tag{5.27}$$

$$\frac{W_2}{W_3} = \frac{3}{2}$$
$$2W_2 = 3W_3$$
$$2(Q_2 - Q_3) = 3(Q_3 - Q_4)$$

Dividing by Q_3 on both sides, we have

$$2\left(\frac{Q_2}{Q_3} - 1\right) = 3\left(1 - \frac{Q_4}{Q_3}\right)$$

$$2\left(\frac{T_2}{T_3} - 1\right) = 3\left(1 - \frac{T_4}{T_3}\right) \quad \left| \begin{array}{l} \text{For Carnot engines} \\ \frac{Q_2}{Q_3} = \frac{T_2}{T_3}, \frac{Q_4}{Q_3} = \frac{T_4}{T_3} \end{array} \right.$$

$$2(T_2 - T_3) = 3(T_3 - T_4)$$
$$2T_2 - 2T_3 = 3T_3 - 3 \times 300$$
$$2T_2 - 2T_3 = 3T_3 - 900$$
$$2T_2 - 5T_3 = -900$$

Multiply by 3.5 on both sides, we have

$$7\,T_2 - 17.5\,T_3 = -3150 \tag{5.28}$$

Eqs. (5.27) and (5.28), we have

$$-4T_3 + 17.5T_3 = 3000 + 3150$$
$$13.5T_3 = 6150$$
$$\text{or} \quad T_3 = \mathbf{455.55\,K}$$

Substituting the value of T_3 in Eq. (5.28),
we have

$$7T_2 - 17.5 \times 455.55 = -3150$$
$$7T_2 - 7972.125 = -3150$$

or

$$7T_2 = 4822.125$$

or

$$T_2 = \mathbf{688.87\,K}$$

Problem 5.27 Find the coefficient of performance and heat transfer rate in the condenser of a refrigerator in kJ/h which has a refrigeration capacity of 12000 kJ/h when power input is 0.75 kW.

Solution: Given data (Refer Fig. 5.39):
  Refrigeration capacity:

$$Q_2 = 12000\,\text{kJ/h}$$

Power input:

$$W = 0.75\,\text{kW}$$
$$= 0.75 \times 3600\,\text{kJ/h}$$
$$= 2700\,\text{kJ/h}$$

Coefficient of performance:

$$COP = \frac{\text{Refrigeration capacity}: Q_2}{\text{Power input}: W}$$

Fig. 5.39 Schematic for
Problem 5.27

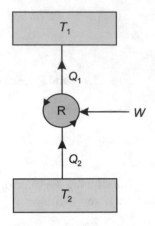

$$= \frac{12000}{2700} = \textbf{4.44}$$

Heat transfer in the condenser: Q_1
By energy balance equation, we get

$$Q_2 + W_2 = Q_1$$
$$12000 + 2700 = Q_1$$

or

$$Q_1 = 14700 \, \textbf{kJ/h}$$

Summary

1. **Thermal Reservoir**: Thermal reservoir is a system having a very large heat capacity that can supply or absorb finite amounts of heat without any appreciable change in temperature. It is also called the thermal energy reservoir or simply reservoir. The reservoir which is at high temperature and supplies energy in the form of heat is called a source or high-temperature reservoir. The reservoir which is at low temperature and absorbs energy in the form of heat is called a sink or low-temperature reservoir.

2. **Limitations of First Law of Thermodynamics**:

   (i)   It is impossible to convert the heat supplied into an equivalent amount of work.

   (ii) It does not specify the direction of heat flow whether the heat flows from low-temperature body to high-temperature body or from high-temperature body to Low-temperature body.

3. **Second Law of Thermodynamics**: There are two statements of the second law of thermodynamics (i) Kelvin-Planck statement, (ii) Clausius statement.

   (i) Kelvin-Planck statement: It is impossible to construct a heat engine that works in a cycle and converts net heat supplied into an equivalent amount of work.

   (ii) Clausius statement: It is impossible to construct a device that operates in a cycle and produces no effect other than the transfer of heat from a low-temperature body to a high-temperature body without external aid of work.

4. **Perpetual-Motion Machine of the Second Kind—PMM II**: Perpetual-motion machine of the second kind is a hypothetical device that takes heat from a source and converts it completely into work. That is, the thermal efficiency of PMM II is 100%. It violates the second law of thermodynamics.

5. **Heat Engine**: It is a mechanical work producing device used for converting a part of heat into work, while the remaining waste heat is rejected to a sink. Thermal efficiency: It is defined as the ratio of the net work output to heat supplied. Mathematically,
   Thermal efficiency:

$$\eta_{th} = \frac{\text{net work output}}{\text{heat supplied}}$$

$$= \frac{W}{Q_1} = \frac{Q_1 - Q_2}{Q} = 1 - \frac{Q_2}{Q_1}$$

6. **Refrigerator**: It is a device that is used to attain and maintain a temperature below that of the surrounding, the aim being to cool some product or space to the required low temperature.
   Coefficient of performance of a refrigerator:

$$(COP)_R = \frac{\text{Desired effect}}{\text{Workinput}} = \frac{\text{Cooling effect}: Q_2}{\text{Work input}: W}$$

$$(COP)_R = \frac{Q_2}{W}$$

7. **Heat Pump**: It is a device that is used to attain and maintain a temperature higher than that **of the surround**ings, the aim being to heat up some space to the required high temperature.
   Coefficient of performance of a heat pump:

$$(COP)_{HP} = \frac{\text{Desired effect}}{\text{Work input}} = \frac{\text{Heating effect}: Q_1}{\text{Work input}: W}$$

$$(COP)_{HP} = \frac{Q_1}{W}$$

$$(\mathbf{COP})_{\mathbf{HP}} = \mathbf{1} + (\mathbf{COP})_{\mathbf{R}}$$

The coefficient of performance of a heat pump is equal to one more than the coefficient of performance of a refrigerator.

8.   **Reversible and Irreversible processes**:
       If the system and its surroundings can be restored to its initial state on reversing
       the process, such a process is called reversible process.
       A process is said to be irreversible if the system and surroundings cannot
       restored to initial state by reversing the process, it is called irreversible.

9.   **Carnot Cycle and Carnot Heat Engine**: The theoretical heat engine that
       operates on the Carnot cycle is called the Carnot heat engine. The Carnot
       cycle is composed of the following four reversible processes:

       (i)   Reversible isothermal compression
       (ii)   Reversible adiabatic compression
       (iii)   Reversible isothermal expansion
       (iv)   Reversible adiabatic expansion.
               Thermal efficiency of the Carnot engine,

$$\eta_{Carnot} = 1 - \frac{T_L}{T_H}$$

       where
       $TL =$ Low temperature of the sink
       $TH =$ High temperature of the source.
       For reversible heat engine,

$$\eta_{Carnot} = \eta_{act}$$

       i.e.,

$$\eta_{Carnot} = 1 - \frac{T_L}{T_H} = 1 - \frac{Q_L}{Q_H}$$

       where

       $QH =$ Heat supplied by the source
       $QL =$ Heat rejection to the sink.

10.   **Carnot Theorem**:
       It states that the efficiency of an irreversible heat engine is always less than the
       efficiency of a reversible heat engine operating between the same temperature
       limits.

11.   **Corollaries of Carnot's Theorem**:
       Corollary-1: *The efficiency of all reversible heat engines operating between
       the same two reservoirs are the same.*
       Corollary-2: *The efficiency of any reversible heat engine does not depend on
       the working fluid in the cycle and depends only on the temperature of the source
       and the sink.*

12.   **Thermodynamic Temperature Scale**:

A temperature scale that is independent of the properties of the substance that are used to measure temperature is called a thermodynamic temperature scale. For reversible heat engine,

$$\frac{Q_1}{Q_2} = \frac{T_1}{T_2}$$

where T_1 and T_2 are temperatures in absolute thermodynamic scale or Kelvin scale. On the Kelvin scale, the temperature ratio depends on the ratio of heat transfer between a reversible heat engine and the reservoirs and is independent of the physical properties of the working substance.

Assignment–1

1. State the limitations of the first law of thermodynamics.
2. What is a thermal energy reservoir? Define in terms of the source and the sink.
3. Discuss the need of the second law of thermodynamics.
4. List the suitable examples of reversible and irreversible processes.
5. Give the following statements of the second law of thermodynamics.

   (i)   Kelvin-Planck statement.
   (ii) Clausius statement.

6. Define the thermal efficiency of a heat engine. Can it be 100%?
7. What are the four processes that make up the Carnot cycle?
8. Explain the difference between a reversible and an irreversible process.
9. What is PMM II? Why is it impossible?
10. State the Kelvin-Planck and Clausius statement of the second law of thermodynamics.
11. Show that the violation of the Kelvin-Planck statement leads to the violation of the Clausius statement and vice versa.
12. Which is a more effective way to increase the efficiency of a Carnot engine: to increase source temperature T_1, keeping sink temperature T_2 constant, or to decrease sink temperature T_2, keeping source temperature T_1 constant?
13. Define heat engine, refrigerator, and heat pump.
14. Show that the efficiency of a reversible heat engine depends only on the temperatures of the source and the sink.
15. What is a heat pump? How does it differ from the refrigerator?
16. Show that the COP of a heat pump is greater than the COP of a refrigerator by unity.
17. Explain the following:

    (i)   PMM II
    (ii) Carnot theorem.
          State and prove Carnot theorem.

18. Is it possible for a heat engine to operate without rejecting any heat to a low-temperature reservoir? Explain.

19. In the absence of any friction and other irreversibilities, can a heat engine have an efficiency of 100%. Explain.

20. An inventor claims to have developed a new reversible heat engine cycle that has the same theoretical efficiency as the Carnot cycle operating between the same temperature limits. Is this a reasonable claim?

Assignment–2

1. A Carnot engine received 200 J of heat from its higher temperature 250 °C reservoir and rejects 120 J to its sink. Determine the temperature of the sink. [**Ans.** 40.8 °C].

2. A Carnot heat engine operates between a source at 800 K and a sink at 315 K. If the heat engine is supplied with the heat at a rate of 600 kJ/min, determine

   (i)   the thermal efficiency and
   (ii) the power output of the heat engine. [**Ans.** (i) 60.2% (ii) 6.06 kW]

3. A heat engine operates between a source at 513 °C and a sink at 35 °C. If heat is supplied to the heat engine at a steady rate of 1150 kJ/min, determine the maximum power output of this heat engine. [**Ans.** 11.65 kW]

4. An innovative way of power generation involves the utilization of geothermal energy, the energy of hot water that exists naturally underground as the heat source. If a supply of hot water at 150 °C is discovered at a location where the environmental temperature is 25 °C, determine the maximum thermal efficiency a geothermal power plant built at that location can have. [**Ans.** 29.55%]

5. A Carnot heat engine receives heat from a reservoir at 900 °C at a rate of 800 kJ/min and rejects the waste heat to the ambient air at 27 °C. The entire work output of the heat engine is used to drive a refrigerator that removes heat from the refrigerated space at −5 °C and transfers it to the same ambient air at 27 °C. Determine (a) the maximum rate of heat removal from the refrigerated space and (b) the total rate of heat rejection to the ambient air. [**Ans.** (a) 4982 kJ/min, (b) 5782 kJ/min

6. Determine the COP of a refrigerator that removes heat from the food compartment at a rate of 8000 kJ/h for each kW of electric power it draws. Also, determine the rate of energy absorption from the outdoor air. [**Ans.** 2.22, 4400 kJ/h]

7. A heat pump working on a reversed Carnot cycle takes in energy from a reservoir maintained at 3 °C and delivers it to another reservoir where the temperature is 77 °C. The heat pump drives power for its operation from a reversible engine operating within the higher and lower temperature limits of 1077 and 77 °C. For 100 kJ/s of energy supplied to the reservoir at 77 °C, estimate the energy taken from the reservoir at 1077 °C.
[**Ans.** 26.71 kJ/s]

8. A reversible engine is used for only driving a reversible refrigerator. Engine is supplied 2000 kJ/s heat from a source at 1500 K and rejects some energy to a low-temperature sink. Refrigerator is desired to maintain the temperature of 15 °C while rejecting heat to the same low-temperature sink. Determine the temperature of sink if total 3000 kJ/s heat is received by the sink. [**Ans.** 351.27 K]

9. A reversible heat engine operates within the higher and lower temperature limits of 1400 K and 400 K, respectively. The entire output from this engine is utilized to operate a heat pump. The pump works on reversed Carnot cycle, extracts heat from a reservoir at 300 K, and delivers it to the reservoir at 400 K. If 100 kJ/s of net heat is supplied to the reservoir at 400 K, calculate the heat supplied to the engine by the reservoir at 1400 K. [**Ans.** 29.17 kW]

10. A reversible heat engine working between two thermal reservoirs at 875 and 315 K drives a reversible refrigerator which operates between the same 315 K reservoir and a reservoir at 260 K. The engine is supplied with 2000 kJ of heat and the net work output from the composite system is 350 kJ. Make calculations for the heat transf**er to** the refrigerator and the net heat interaction with the reservoir at 315 K temperature.
    [**Ans.** 3496 kJ, 6046 kJ]

11. Three Carnot engines *A, B, and C* working between temperatures of 1000 K and 300 K are in a series combination. The work produced by these engines are in the ratio of 5:4:3. Make calculations of temperature for the intermediate reservoirs.
    [**Ans.** 708 K, 475 K]

12. Two Carnot engine works in series between the source and sink temperature of 550 K and 350 K. If both engines develop equal power, determine the intermediate temperature.
    [**Ans.** 450 K]

13. An air-conditioning system is used to maintain a house at a constant temperature of 20 °C. The house is gaining heat from outdoors at a rate of 20,000 kJ/h, and the heat generated in the house from the people, lights, and appliances amounts to 8000 kJ/h. For a COP of 2.5, determine the required power input to this air-conditioning system.
    [**Ans.** 3.11 kW]

14. Two reversible heat engines *A* and *B* are arranged in series. *A* rejects heat directly to *B*. Engine *A* receives 200 kJ at a temperature of 421 °C from the hot source while engine *B* is in communication with a cold sink at a temperature of 5 °C. If the work output of *A* is twice that of *B*, find

    (i)   Intermediate temperature between *A* and *B*.
    (ii) Efficiency of each engine.
    (iii) Heat rejected to the sink. [**Ans.** (i) 416.7 K (ii) 40%, 33.28% (iii) 80 kJ]

15. Two Carnot heat engines operate so that the exhaust of one is the intake of the other; the highest and lowest temperature limits are 1200 and 400 K. Determine

the intermediate temperature if (i) both the engines give the same work output (ii) both the engines have the same thermal efficiency. [**Ans.** 800 K, 692.82 K]

16. A reversible heat engine operating between thermal reservoirs at 900 and 300 K is used to drive a reversible refrigerator for which the temperature limits are 300 and 250 K. The engine absorbs 1800 kJ of energy as heat from the reservoir at 900 K and the net output from the engine-refrigerator system is 360 kJ. Determine the heat extracted from the refrigerator cabinet and the net heat rejected to the reservoir at 300 K. [**Ans.** 5040 kJ, 5640 kJ]

17. A reversible heat engine receives heat from two thermal reservoirs at 870 and 580 K and rejects 50 kW of heat to a sink at 290 K. If the engine output is 85 kW, determine the engine efficiency and heat supplied by each reservoir. [**Ans.** 62.96%, 105 kW, 30 kW]

Chapter 6
Entropy

Nomenclature

The following is a list of the nomenclature introduced in this chapter:

S	kJ/K	Entropy
s	kJ/kgK	Specific entropy
W	N	Weight
W_s	kJ	Shaft work
dS or δS	kJ/K	Entropy change
ds or δs	kJ/kgK	Specific entropy change
δQ	kJ	Small heat transfer
ΔS_{system}	kJ/K	Entropy change of the system
$\Delta S_{surrounding}$	kJ/K	Entropy change of the surroundings
$\Delta S_{universe}$	kJ/K	Entropy change of the universe
$\Delta S_{isolated}$	kJ/K	Entropy change of the isolated system
δQ	kJ	Small work transfer
$Q_{surrounding}$	kJ	Heat transfer between the system and surroundings
m	kg	Mass of the system
η	no unit	Actual efficiency of the heat engine
η_{carnot}	no unit	Efficiency of the Carnot heat engine
S_{gen}	kJ/K	Entropy generation
dU	kJ	Change in internal energy
c_v	kJ/kgK	Specific heat at constant volume
c_p	kJ/kgK	Specific heat at constant pressure
R	kJ/kgK	Gas constant

(continued)

© The Author(s) 2022
S. Kumar, *Thermal Engineering Volume 1*,
https://doi.org/10.1007/978-3-030-67274-4_6

(continued)

dV	m^3	Change in volume
p	kPa	Pressure
V	m^3	Volume
γ	no unit	Adiabatic index
n	no unit	Polytropic index
s_f	kJ/kgK	Specific entropy of saturated liquid
s_g	kJ/kgK	Specific entropy of saturated vapour
x	no unit	Dryness fraction
L	kJ/kg	Latent heat of fusion
h_{fg}	kJ/kg	Latent heat of evaporation

6.1 Introduction

We know that the first law of thermodynamics introduced a very useful property called **internal energy**. The second law of thermodynamics introduces a new property called **entropy**. Even the entropy is difficult to give a physical description of it without considering the microscopic state of the system but the change in entropy is a very powerful tool; by use of this, the engineers describe.

(i)   The feasibility of the process.
(ii)   The level of irreversibility associated with any process.
(iii) A definite limit on the maximum amount of work that can be obtained from the system during a given change of state.
(iv)   The nature of the process; either reversible or irreversible.
(v)   The possibility of a cycle.

6.2 Entropy

Entropy is defined as a measure of molecular disorder or molecular randomness (i.e., level of irreversibility) in the system. The entropy is denoted by S. It is difficult to calculate the entropy of the system at a particular state because it requires the physical description of the microscopic state of the system, i.e., the entropy of the system requires how many molecules are in disorder. The calculation of the number of molecules in disorder is impossible. But the entropy change dS of the system is calculated in this chapter with the help of mathematical expressions. For the reversible process, the small increase of entropy, dS of the system is defined as the ratio of small addition of heat, δQ to the absolute temperature, T of the system at which the heat is supplied. The entropy of the system increases with the addition of heat and decreases with the removal of heat (Fig. 6.1).

Mathematically,

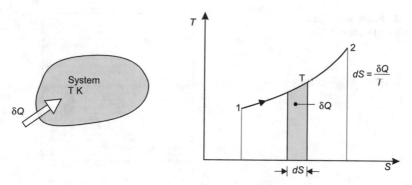

Fig. 6.1 Entropy

$$dS = \int_1^2 \left(\frac{\delta Q}{T}\right)_{nw} \quad kJ/K \tag{6.1}$$

Entropy is an extensive property of a system. Entropy per unit mass is called the specific entropy, denoted by s, which is an intensive property and has the unit kJ/kgK.

The entropy change of a system during a process can be determined by integrating Eq. (6.1) between the initial state 1 and final state 2:

$$\int_1^2 dS = \int_1^2 \left(\frac{\delta Q}{T}\right)_{rev}$$

$$S_2 - S_1 = \int_1^2 \left(\frac{\delta Q}{T}\right)_{rev}$$

Unlike energy, entropy is a nonconserved property. An interesting fact is that the entropy can be created but cannot be destroyed. A more interesting fact is that the entropy of the universe continuously increases.

As the system becomes more disordered, the position of the molecules become less predictable and the entropy increases. Thus, the entropy of a substance is lowest in the solid phase and the highest in the gas phase. The gas molecules possess a considerable amount of kinetic energy. But we know that the gas molecules cannot rotate a paddle inserted into a gas tank and produce work, as shown in Fig. 6.2; however, large may be their kinetic energy. This is due to the fact that the energy of the gas molecules is disorganized. Some molecule trying to rotate the wheel and some molecule trying to prevent the wheel from rotating match all the time, this not enabling the wheel to rotate. This shows that we cannot extract any useful work directly from disorganized energy.

As another example, we consider a rotating shaft as shown in Fig. 6.3. The energy of the molecules is completely organized since the molecules of the shaft are rotating together in the same direction. The organized energy can be used to perform useful tasks such as lifting a weight and generating electricity. The work, being an organized form of energy, is free of molecular disorder and thus free of entropy. There is no entropy transfer associated with energy transfer as work. Assuming that there is no

Fig. 6.2 Paddle wheel in a
gas tank; no work output

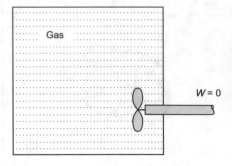

Fig. 6.3 A rotating shaft
used to lifting a weight

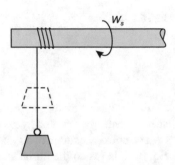

friction present in the rotating shaft process, we can say that there is no entropy associated with this process. Hence, this process can be reversed and used for lowering
the weight, since any process that does produce a net entropy is reversible. Therefore,
during this process, the energy is not degraded and the potential to do work is not
lost.

6.3   Entropy of the Universe

A system and its surroundings can be viewed as an isolated system or universe.

The entropy change of the universe is equal to the sum of the entropy change of
the system and the entropy change of the surroundings. The entropy change of the
universe is also called the entropy change of the isolated system (or total entropy).

Mathematically, $\Delta S_{\text{universe}} = \Delta S_{\text{isolated}} = \Delta S_{\text{system}} + \Delta S_{\text{surrounding}}$

where

$$\Delta S_{system} = S_2 - S_1$$
$$= m(s_2 - s_1),$$

entropy change of the system. It is calculated by mathematical relations. We will discuss the mathematical relations function of the entropy change of the system in this chapter.

$$\Delta S_{\text{surrounding}} = \frac{Q_{\text{surrounding}}}{T_0},$$

entropy change of the surroundings, it is $-ve$ value when heat is supplied to the system (*i.e.*, heat loss by the surroundings) and $+ve$ value when heat is lost by the system or heat is gained by the surroundings.

From Fig. 6.4,

$$\Delta S_{\text{surrounding}} = -\frac{Q_{\text{surrounding}}}{T_0}$$

$$\therefore \Delta S_{\text{universe}} = m(s_2 - s_1) - \frac{Q_{\text{surrounding}}}{T_0}$$

From Fig. 6.5,

$$\Delta S_{surrounding} = -\frac{Q_{surrounding}}{T_0}$$

$$\therefore \Delta S_{universe} = m(s_2 - s_1) - \frac{Q_{surrounding}}{T_0}$$

Thus for a reversible process, the entropy changes of the system and its surrounding are equal with opposite sign due to heat transfer (i.e., the universe entropy is zero).

Fig. 6.4 Entropy decreases of the surroundings

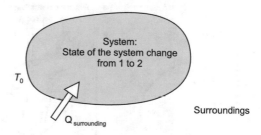

System:
State of the system change
from 1 to 2

T_0

$Q_{\text{surrounding}}$

Surroundings

Fig. 6.5 Entropy increases
of the surroundings

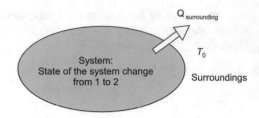

6.4 Clausius Theorem

The Clausius theorem states that the cycle integral of $\delta Q/T$ for a reversible cycle is
equal to zero.

Mathematically,

$$\oint_R \frac{\delta Q}{T} = 0$$

The symbol $\oint$ (integral symbol with a circle in the middle) is used to indicate that
the integration is to be performed over the entire cycle. The letter R stands for a
reversible cycle.

Proof: Let a smooth closed curve represent a reversible cycle as shown in Fig. 6.6a.
Let the closed cycle be divided into a large number of strips by means of reversible
adiabatics. Each strip is closed at the top and bottom by means of reversible isothermals as shown in Fig. 6.6b. The original closed cycle is thus replaced by a zig-zag
closed path consisting of alternate adiabatic and isothermal processes, such that the
heat transferred during all the isothermal processes is equal to the heat transferred in
the original cycle.

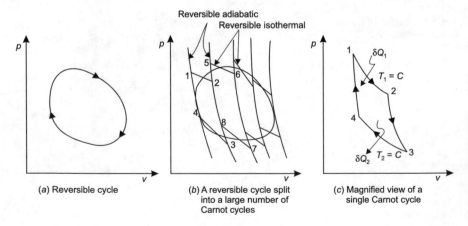

Fig. 6.6 Demonstration of Clausius's theorem

Now a magnified view of one Carnot cycle 1–2–3–4 is shown in Fig. 6.6c. Let δQ_1 be the heat supplied at temperature T_1 and δQ_2 be the heat rejected at temperature T_2.

Then

$$\frac{\delta Q_1}{T_1} = \frac{\delta Q_2}{T_2}$$

If heat supplied is taken as $+ve$ sign and heat rejected as $-ve$ sign, then

$$\frac{\delta Q_1}{T_1} = -\frac{\delta Q_2}{T_2}$$

or $\frac{\delta Q_1}{T_1} + \frac{\delta Q_2}{T_2} = 0$.

Similarly, for a Carnot cycle 5–6–7–8,

$$\frac{\delta Q_3}{T_3} = \frac{\delta Q_4}{T_4}$$

Applying the sign convection, heat supplying the sign convection, heat supplied is taken as $+ve$ sign and heat rejected as $-ve$ sign, then

$$\frac{\delta Q_3}{T_3} = -\frac{\delta Q_4}{T_4}$$

or $\frac{\delta Q_3}{T_3} + \frac{\delta Q_4}{T_4} = 0$.

If similar equations are written for all the Carnot cycles in Fig. 6.6b, then for the original cycle in Fig. 6.6a,

$$\frac{\delta Q_1}{T_1} + \frac{\delta Q_2}{T_2} + \frac{\delta Q_3}{T_3} + \frac{\delta Q_4}{T_4} + \cdots = 0$$

or $\oint_R \frac{\delta Q}{T} = 0$.

6.5 Clausius Inequality

The Clausius inequality states that in a cyclic process the algebraic sum of the ratio of heat interaction to the absolute temperature at which heat interaction occurs over the complete cycle is less than or equal to zero.

Mathematically,

Fig. 6.7 Reversible heat engine

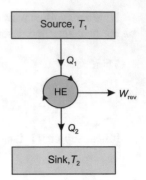

$$\oint \frac{\delta Q}{T} \leq 0$$

This inequality is valid for all cycles, reversible or irreversible, including the refrigeration and heat pump cycles. It provides the criterion to check the types of the cycle, i.e., the cycle is reversible, irreversible, or impossible.

$$\oint \frac{\delta Q}{T} \begin{cases} < 0, \text{ the cycle is irreversible and possible} \\ = 0, \text{ the cycle is reversible} \\ > 0, \text{ the cycle is impossible and it violates the} \\ \quad\text{2nd law of thermodynamics} \end{cases}$$

For Reversible Heat Engine.

We know that the Carnot efficiency of the reversible heat engine is equal to the actual efficiency of the reversible heat engine (Fig. 6.7).

Mathematically,

$$\eta_{Cannot} = \eta$$

$$1 - \frac{T_2}{T_1} = 1 - \frac{Q_2}{Q_1}$$

or $\frac{T_2}{T_1} = \frac{Q_2}{Q_1}$.

or $\frac{Q_1}{T_1} = \frac{Q_2}{T_2}$.

If heat supplied is taken as $+ve$ sign and heat rejected as $-ve$ sign, then

$$\frac{Q_1}{T_1} = -\frac{Q_2}{T_2}$$

or $\frac{Q_1}{T_1} + \frac{Q_2}{T_2} = 0.$

For small amount of heat δQ,

$$\frac{\delta Q_1}{T_1} + \frac{\delta Q_2}{T_2} = 0$$

or $\oint \frac{\delta Q}{T} = 0$.
for a reversible heat engine.

For Irreversible Heat Engine

According to the Carnot theorem, we know that the Carnot efficiency is more than the actual efficiency of the irreversible heat engine (Fig. 6.8).

Mathematically, $\eta_{\text{Carnot}} > \eta$ for an irreversible heat engine.

$$1 - \frac{T_2}{T_1} > 1 - \frac{Q_2}{Q_1}$$

or $\frac{T_2}{T_1} < \frac{Q_2}{Q_1}$.

or $\frac{Q_1}{T_1} < \frac{Q_2}{T_2}$.

or $\frac{Q_1}{T_1} - \frac{Q_2}{T_2} < 0$.

If heat supplied is taken as $+ve$ and heat rejected as $-ve$, then

$$\frac{Q_1}{T_1} - \left(\frac{-Q_2}{T_2}\right) < 0$$

or $\frac{Q_1}{T_1} + \frac{Q_2}{T_2} < 0$.

For a small amount of heat δQ,

$$\frac{\delta Q_1}{T_1} + \frac{\delta Q_2}{T_2} < 0$$

Fig. 6.8 Irreversible heat engine

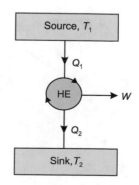

or $\oint \frac{\delta Q}{T} < 0$ for an irreversible heat engine.

For reversible and irreversible heat engines, we get from the above two equations $\oint \frac{\delta Q}{T} \leq 0$ for any heat engine. This is called the Clausius inequality. According to the Clausius inequality,

(a) $\oint \frac{\delta Q}{T} < 0$ for an irreversible and possible cycle.

(b) $\oint \frac{\delta Q}{T} = 0$ for a reversible cycle.

(c) $\oint \frac{\delta Q}{T} > 0$ for an impossible cycle, this shows that the efficiency of the heat engine is more than that of Carnot efficiency or efficiency of the reversible heat engine. Thus, it violates the second law of thermodynamics.

Note:

(i)   All temperatures must be in kelvin.

(ii)   Heat supplied to the heat engine is taken as + ve.

(iii) Heat rejected by the heat engine is taken as −ve.

6.6   Entropy—A Property of the System

Entropy is a property of the system, it means that the change of entropy does not depend upon the path but it depends upon the end states.

Consider a system that undergoes a state change from state 1 to state 2 by following the reversible path A and B. The system is returned back from state 2 to state 1 by following the reversible path C as shown in Fig. 6.9.

For reversible cycle 1–2–1 via A, applying the Clausius theorem,

$$\oint \frac{\delta Q}{T} = 0$$

or $\int_{1A}^{2} \frac{\delta Q}{T} + \int_{2C}^{1} \frac{\delta Q}{T} = 0$

$$(S_2 - S_1)_A + (S_1 - S_2)_C = 0$$

Fig. 6.9 Reversible cycle

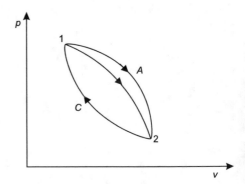

$$\text{or} \quad (S_1 - S_2)_C = -(S_2 - S_1)_A \tag{6.2}$$

For reversible cycle 1–2–1 via B, applying the Clausius theorem,

$$\oint \frac{\delta Q}{T} = 0$$

$$\text{or} \quad \int_{1A}^{2} \frac{\delta Q}{T} + \int_{2C}^{1} \frac{\delta Q}{T} = 0$$

$$(S_2 - S_1)_A + (S_1 - S_2)_C = 0$$

$$\text{or} \quad (S_1 - S_2)_C = -(S_2 - S_1)_A \tag{6.3}$$

Equating Eqs. (6.2) and (6.3), we get

$$-(S_2 - S_1)_A = -(S_2 - S_1)_B$$

$$\text{or} \quad (S_2 - S_1)_A = (S_2 - S_1)_B \tag{6.4}$$

Equation (6.4) shows that the change in entropy between two states of the system is the same, whether the system follows the path A or path B, i.e., entropy is independent of the path followed by the system. So, entropy is a point function and a property of the system.

6.7 Entropy Change of an Irreversible Process

We know that the entropy change of a reversible process is given by

$$dS = \frac{\delta Q}{T} \tag{6.5}$$

For reversible process 2–1,

$$\int_{2}^{1} dS = \int_{2}^{1} \left(\frac{\delta Q}{T}\right)_R \tag{6.6}$$

Now to find the energy change of an irreversible process, consider a cycle composed of a reversible and an irreversible process as shown in Fig. 6.10.
Applying the Clausius inequality for an irreversible cycle,

$$\int \frac{\delta Q}{T} < 0$$

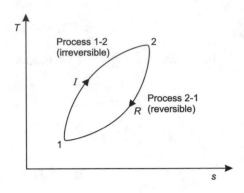

Fig. 6.10 Irreversible cycle: A cycle composed of a reversible and an irreversible process

For cycle 1–2–1,

$$\int_1^2 \left(\frac{\delta Q}{T}\right)_I + \int_2^1 \left(\frac{\delta Q}{T}\right)_R < 0 \tag{6.7}$$

From Eq. (6.6),

$$\int_2^1 \left(\frac{\delta Q}{T}\right)_R = \int_2^1 dS$$

Therefore, Eq. (6.7) becomes

$$\int_1^2 \left(\frac{\delta Q}{T}\right)_I + \int_2^1 dS < 0$$

or $\int_1^2 \left(\frac{\delta Q}{T}\right)_I + \int_2^1 dS < 0$.
or $\int_1^2 \left(\frac{\delta Q}{T}\right)_I < \int_1^2 dS$

$$\text{or} \quad \int_1^2 dS > \int_1^2 \left(\frac{\delta Q}{T}\right)_I \tag{6.8}$$

Equation (6.8) can be written in differential form as

$$dS > \frac{\delta Q}{T} \quad \text{for an irreversible process}$$

and

$$dS = \frac{\delta Q}{T} \quad \text{for an reversible process}$$

In the general case, we can write

$$dS \geq \frac{\delta Q}{T} \quad \text{for a reversible and an irreversible process}$$

where the equality sign is for a reversible process and the inequality sign for an irreversible process.

Due to irreversibility, entropy of the system always increases.

For process 1–2,

$$S_2 - S_1 \geq \int_1^2 \frac{\delta Q}{T} \tag{6.9}$$

We may conclude from the above equation that the entropy change of a closed system during an irreversible process is greater than $\int_1^2 \frac{\delta Q}{T}$, which represents the entropy transfer with heat. That is, some entropy is generated or created during an irreversible process, and this generation is due to the presence of irreversibilities. The entropy generated during a process is called entropy generation and is denoted by S_{gen}.

Equation (6.9) can be rewritten as an equality as

$$\Delta S_{system} = S_2 - S_1 = \int_1^2 \frac{\delta Q}{T} + S_{gen}$$

where S_{gen} is the entropy generation and its value always a $+ ve$ or zero, i.e., $+ ve$ value for an irreversible process and zero for a reversible process.

The value of S_{gen} depends on the process, and this is not a property of the system (i.e., entropy generation is a path function). Also, in the absence of any entropy transfer, the entropy change of the system is equal to the entropy generation.

6.8 Increase of Entropy Principle

We know that the entropy change of a reversible and an irreversible process is given by

$$dS \geq \frac{\delta Q}{T}$$

For an isolated system,

$$\delta Q = 0$$

$$\therefore \quad dS_{iso} \geq 0 \tag{6.10}$$

$$\left. \begin{array}{l} dS_{\mathrm{iso}} = 0 \\ S = C \end{array} \right\} \quad \text{for a reversible process}$$

and $dS_{\mathrm{iso}} > 0$ for an irreversible process

Equation (6.10) can be expressed as the entropy of an isolated system; during an irreversible process, it always increases and remains constant only when the process is reversible. In other words, the entropy of an isolated system can never decrease. This is known as the increase of entropy principle, or simply the entropy principle. The system and surroundings may be treated as a universe or an isolated system. From the entropy principle, we have

$$dS_{\mathrm{uni}} \geq 0 \tag{6.11}$$

which means that the entropy of the universe increases continuously during an irreversible process and remains constant during a reversible process,

$$i.e., \ dS_{\mathrm{uni}} > 0 \text{ for an irreversible process}$$

$$i.e., \ dS_{\mathrm{uni}} > 0 \text{ for an reversible process}$$

Equation (6.11) is rewritten as

$$dS_{\mathrm{sys}} + dS_{\mathrm{surr}} \geq 0$$

where.
$dS_{\mathrm{sys}} =$ entropy change in system and.
$dS_{\mathrm{surr}} =$ entropy change in surroundings.
Note: The entropy change in an isolated system is also called the entropy of the universe or total entropy,

$$i.e., \quad dS_{\mathrm{iso}} = dS_{\mathrm{uni}} = dS_{\mathrm{total}}$$

6.9   Third Law of Thermodynamics

The third law of thermodynamics states that the **entropy of a pure perfect crystal substance is zero at absolute zero temperature.** That is,

$$S = 0 \quad \text{at} \quad T = 0\,\mathrm{K} = -273.15°\mathrm{C}$$

A pure perfect crystal is the one in which every molecule is identical, and the molecular alignment is perfectly even throughout the substance. For non-pure crystals, or those with less than perfect alignment, there will be some energy associated with the imperfections, so the entropy cannot become zero.

The third law of thermodynamics refers to a state known as 'absolute zero'. This is the bottom point on the Kelvin temperature scale. The Kelvin scale is absolute, meaning 0 K is mathematically the lowest possible temperature in the universe.

The third law of thermodynamics can be visualized by thinking about water. Water in vapour form has molecules that can move around very freely. Water vapour has very high entropy (randomness). As the vapour cools, it becomes liquid. The liquid water molecules can still move around, but not as freely. They have lost some entropy. When the water cools further, it becomes solid ice. The solid water molecules can no longer move freely, but can only vibrate with the ice crystals. The entropy is now very low. As the ice is cooled more and closer to absolute zero temperature, the vibration of the molecules diminishes. If the solid water (ice) reached absolute zero, all molecular motions would stop completely. At this point, the water would have no entropy (randomness) at all.

The third law provides an absolute reference point (0 K) for the determination of entropy. The entropy determined relative to this point is called the absolute entropy.

6.10 General Equations for the Entropy Change of an Ideal Gas

We know that the first law for non-flow process

$$\delta Q = dU + \delta W$$

where $\delta Q = TdS$ for reversible process,

$$\delta W = pdV \quad \text{for non} - \text{flow reversible process}$$

and $dU = mc_v dT$

$$\therefore \quad TdS = mc_v dT + pdV$$

or $$dS = mc_v \frac{dT}{T} + \frac{pdV}{T}$$
$$= mc_v \frac{dT}{T} + \frac{mRdV}{V}$$

$$\begin{vmatrix} pv = mRT \\ or \ \frac{p}{T} = \frac{mR}{V} \end{vmatrix}$$

Integrating between states 1 and 2 gives

$$\int_1^2 dS = mc_V \int_1^2 \frac{dT}{T} + mR \int_1^2 \frac{dV}{V}$$

$$S_2 - S_1 = mc_v \log_e \frac{T_2}{T_1} + mR \log_e \frac{V_2}{V_1}$$

For unit mass,

$$s_2 - s_1 = c_v \log_e \frac{T_2}{T_1} + R \log_e \frac{V_2}{V_1} \tag{6.12}$$

Equation (6.12) gives entropy change for unit mass, in terms of temperature and volume ratios. From the equation of state for process 1–2 gives

$$\frac{p_1 V_1}{T_1} = \frac{p_2 V_2}{T_2}$$

$$\text{or} \quad \frac{T_2}{T_1} = \frac{p_2 V_2}{p_1 V_1} \tag{6.13}$$

Substituting Eqs. (6.13) in (6.12), we get

$$s_2 - s_1 = c_v \log_e \frac{p_2 V_2}{p_1 V_1} + R \log_e \frac{V_2}{V_1}$$

$$= c_v \log_e \frac{p_2}{p_1} + c_v \log_e \frac{V_2}{V_1} + R \log_e \frac{V_2}{V_1}$$

$$= c_v \log_e \frac{p_2}{p_1} + (c_v + R) \log_e \frac{V_2}{V_1}$$

$$= c_v \log_e \frac{p_2}{p_1} + c_p \log_e \frac{V_2}{V_1}$$

$$s_2 - s_1 = c_v \log_e \frac{p_2}{p_1} + c_p \log_e \frac{V_2}{V_1} \tag{6.14}$$

$$\begin{vmatrix} \because \ c_p - c_V = R \\ or \ c_p = c_V + R \end{vmatrix}$$

Equation (6.14) gives entropy change for unit mass, in terms of pressure and volume ratios.

To eliminate the volume ratio by using the equation of state for process 1–2,

$$\frac{p_1 V_1}{T_1} = \frac{p_2 V_2}{T_2}$$

$$\text{or} \quad \frac{V_2}{V_1} = \frac{T_2}{T_1} \frac{p_1}{p_2} \tag{6.15}$$

Substituting Eqs. (6.15) in (6.12), we get

$$s_2 - s_1 = c_v \log_e \frac{T_2}{T_1} + R \log_e \frac{T_2}{T_1} \frac{p_1}{p_2}$$

$$= c_v \log_e \frac{T_2}{T_1} + R \log_e \frac{T_2}{T_1} + R \log_e \frac{p_1}{p_2}$$

$$= (c_v + R) \log_e \frac{T_2}{T_1} + R \log_e \frac{p_1}{p_2}$$

$$= c_p \log_e \frac{T_2}{T_1} + R \log_e \frac{p_1}{p_2}$$

$$s_2 - s_1 = c_p \log_e \frac{T_2}{T_1} - R \log_e \frac{p_2}{p_1} \tag{6.16}$$

$$\left[\because c_p = c_v + R \right.$$

Equation (6.16) gives entropy change for unit mass, in terms of temperature and pressure ratios.

6.11   Entropy Change During Various Process

In Sect. 6.10, the general equations of entropy change have been derived as follows:

$$s_2 - s_1 = c_v \log_e \frac{T_2}{T_1} + R \log_e \frac{V_2}{V_1} \tag{6.17}$$

$$s_2 - s_1 = c_v \log_e \frac{p_2}{p_1} + c_p \log_e \frac{V_2}{V_1} \tag{6.18}$$

$$s_2 - s_1 = c_p \log_e \frac{T_2}{T_1} - R \log_e \frac{p_2}{p_1} \tag{6.19}$$

(i)   Constant Volume Process (Isometric or Isochoric)

This process takes place at constant volume,

i.e., $V = C$

For process $1 - 2$, $V_1 = V_2$

∴ Entropy change for this process is obtained from Eq. (6.17)

$$s_2 - s_1 = c_v \log_e \frac{T_2}{T_1} \tag{6.20}$$

Equation (6.18) becomes, for this process,

$$s_2 - s_1 = c_v \log_e \frac{p_2}{p_1} \tag{6.21}$$

Equation (6.19) remains the same because in the constant volume process, both the temperature and pressure are variable.

$$\therefore \quad s_2 - s_1 = c_p \log_e \frac{T_2}{T_1} - R \log_e \frac{p_2}{p_1} \tag{6.22}$$

Equations (6.20), (6.21), and (6.22) are used to determine the entropy change for the constant volume process. The entropy change must be the same either using Eqs. (6.20) or (6.21) or (6.22). The choice of the equation is a matter of convenience.

Calling Eq. (6.20)

$$s_2 - s_1 = c_v \log_e \frac{T_2}{T_1}$$

For constant volume process 1–2,

$$\frac{T_2}{T_1} = \frac{p_2}{p_1}$$

$$\therefore \quad s_2 - s_1 = c_v \log_e \frac{p_2}{p_1}$$

We know that

$$R = c_p - c_v$$

$$\text{or} \quad c_v = c_p - R$$

$$\therefore \quad s_2 - s_1 = \left(c_p - R\right) \log_e \frac{p_2}{p_1}$$

$$= c_p \log_e \frac{p_2}{p_1} - R \log_e \frac{p_2}{p_1} s_2 - s_1$$

$$= c_p \log_e \frac{T_2}{T_1} - R \log_e \frac{p_2}{p_1}$$

(ii) Constant Pressure Process (*Isobaric*)

This process takes place at constant pressure,

$$i.e., \quad p = C$$
$$\text{For process } 1 - 2, \quad p_1 = p_2$$

∴ Entropy change for this process is obtained from Eq. (6.18)

$$s_2 - s_1 = c_p \log_e \frac{V_2}{V_1} \tag{6.23}$$

Equation (6.19) becomes, for this process,

$$s_2 - s_1 = c_p \log_e \frac{T_2}{T_1} \tag{6.24}$$

Equation (6.17) remains the same because in constant pressure process, both the temperature and volume are variables.

$$\therefore \quad s_2 - s_1 = c_v \log_e \frac{T_2}{T_1} + R \log_e \frac{V_2}{V_1} \tag{6.25}$$

Equations (6.23), (6.24), and (6.25) are used to determine the entropy change for the constant pressure process.

(iii) Isothermal Process

This process takes place at a constant temperature,

$$i.e. \quad T = C$$
$$\text{For process } 1 - 2, \quad T_1 = T_2$$

∴ Entropy change for this process is obtained from Eq (6.17)

$$s_2 - s_1 = R \log_e \frac{V_2}{V_1} \tag{6.26}$$

Equation (6.18) remains the same because in the constant pressure process, both the pressure and volume are variables.

$$s_2 - s_1 = c_v \log_e \frac{p_2}{p_1} + c_p \log_e \frac{V_2}{V_1} \tag{6.27}$$

Equation (6.19), for this process, becomes

$$s_2 - s_1 = -R \log_e \frac{p_2}{p_1}$$

$$\text{or} \quad s_2 - s_1 = R \log_e \frac{p_1}{p_2} \tag{6.28}$$

Equations (6.26), (6.27), and (6.28) are used to determine the entropy change for the isothermal process. This is the only single reversible process in which the entropy increases with the addition of heat.

(iv)   Reversible Adiabatic Process

This process follows the law of

$$pV^\gamma = C$$

For adiabatic process,

$$\frac{T_2}{T_1} = \left(\frac{p_2}{p_1}\right)^{\frac{\gamma-1}{\gamma}} = \left(\frac{V_2}{V_1}\right)^{1-\gamma}$$

$$\text{or} \quad \frac{T_2}{T_1} = \left(\frac{V_2}{V_1}\right)^{1-\gamma}$$

$$\text{or} \quad \frac{V_2}{V_1} = \left(\frac{T_2}{T_1}\right)^{\frac{1}{1-\gamma}}$$

Substituting the value of in Eq. (6.17), we get

$$s_2 - s_1 = c_v \log_e \frac{T_2}{T_1} + R \log_e \left(\frac{T_2}{T_1}\right)^{\frac{1}{1-\gamma}}$$

$$= c_v \log_e \frac{T_2}{T_1} + \frac{R}{1-\gamma} \log_e \frac{T_2}{T_1}$$

$$= c_v \log_e \frac{T_2}{T_1} - \frac{R}{\gamma-1} \log_e \frac{T_2}{T_1}$$

$$= c_v \log_e \frac{T_2}{T_1} - c_v \log_e \frac{T_2}{T_1}$$

$$\left|\because c_V = \frac{R}{\gamma-1}\right.$$

$$s_2 - s_1 = 0$$

$$\text{or} \quad s_2 = s_1$$

$$\text{or} \quad s = \textbf{constant}$$

Thus, a reversible adiabatic process is a frictionless adiabatic process in which the entropy remains constant. This process is also called as an adiabatic isentropic process or an simply isentropic process.

(v) Polytropic Process

This process follows the law of

$$pV^n = C$$

For polytropic process,

$$\frac{T_2}{T_1} = \left(\frac{p_2}{p_1}\right)^{\frac{n-1}{n}} = \left(\frac{V_2}{V_1}\right)^{1-n}$$

$$\text{or} \quad \frac{T_2}{T_1} = \left(\frac{V_2}{V_1}\right)^{1-n}$$

$$\text{or} \quad \frac{V_2}{V_1} = \left(\frac{T_2}{T_1}\right)^{\frac{1}{1-n}}$$

Substituting the value of in Eq. (xi), we get

$$s_2 - s_1 = c_v \log_e \frac{T_2}{T_1} + R \log_e \left(\frac{T_2}{T_1}\right)^{\frac{1}{1-n}}$$

$$= c_v \log_e \frac{T_2}{T_1} + \frac{R}{1-n} \log_\varepsilon \frac{T_2}{T_1}$$

$$= \left[1 + \frac{R}{c_v(1-n)}\right] c_v \log_e \frac{T_2}{T_1}$$

$$= \left[1 + \frac{(\gamma-1)}{(1-n)}\right] c_v \log_e \frac{T_2}{T_1}$$

$$= \frac{[1-n+\gamma-1]}{1-n} c_v \log_e \frac{T_2}{T_1}$$

$$s_2 - s_1 = \left(\frac{\gamma-n}{1-n}\right) c_v \log_e \frac{T_2}{T_1} \tag{6.29}$$

$$\left| \begin{array}{l} c_v = \dfrac{R}{\gamma-1} \\[2mm] \text{or} \ \dfrac{R}{c_v} = \gamma - 1 \end{array} \right.$$

Equation (6.29) is the entropy change for unit mass in polytropic process in terms of temperature ratio.

Substituting the value of $\frac{T_2}{T_1} = \left(\frac{V_2}{V_1}\right)^{1-n}$ in Eq. (6.29), we get

$$s_2 - s_1 = \left(\frac{\gamma - n}{1-n}\right) c_v \log_e \left(\frac{V_2}{V_1}\right)^{1-n}$$
$$s_2 - s_1 = \frac{(\gamma - n)}{1-n} \times (1-n)\left(c_v \log_e \frac{V_2}{V_1}\right) \qquad (6.30)$$
$$s_2 - s_1 = (\gamma - n) c_v \log_e \frac{V_2}{V_1}$$

Equation (6.30) is the entropy change for unit mass in the polytropic process in terms of volume ratio.

Substituting the value of $\frac{T_2}{T_P} = \left(\frac{p_2}{p_1}\right)^{\frac{n-1}{n}}$ in Eq. (xiii), we get

$$s_2 - s_1 = \left(\frac{\gamma - n}{1 - n}\right) c_v \log_e \left(\frac{p_2}{p_1}\right)^{\frac{n-1}{n}}$$

$$s_2 - s_1 = \left(\frac{\gamma - n}{1 - n}\right) c_v \frac{(n-1)}{n} \log_e \frac{p_2}{p_1}$$

$$= -\frac{(\gamma - n)}{n} c_v \log_e \frac{p_2}{p_1}$$

$$s_2 - s_1 = \left(\frac{n - \gamma}{n}\right) c_v \log_e \frac{p_2}{p_1} \qquad (6.31)$$

Equation (6.31) is the entropy change for unit mass in the polytropic process in terms of pressure ratio.

6.12 Combined Statement of the First and Second Laws of Thermodynamics

For an incremental change in the state of a closed system, the first law of thermodynamics gives

$$\delta Q = dU + \delta W$$

and, if the process occurs reversibly, the second law of thermodynamics gives

$$dS = \frac{\delta Q}{T}$$

$$\text{or} \quad \delta Q = T dS$$
$$\text{or} \quad \delta W = p dV$$

Combination of the two laws gives the equation

$$T dS = dU + p dV \qquad (6.32)$$

Equation (6.32) is known as the first *TdS* or **Gibbs equation**. This equation is applicable for a closed system when work due to change in volume is the only form of work performed by the system. There is no path function term in Eq. (6.32), hence applicable for any process reversible or irreversible.

By definition of the enthalpy,

$$H = U + pV$$

On differentiating, we get

$$dH = dU + pdV + Vdp$$

From Eq. (i),

$$dU + pdV = TdS$$

$$\therefore dH = TdS + Vdp$$
$$\text{or } TdS = dH - Vdp \tag{6.33}$$

Equation (6.32) is known as the second *TdS* or **Gibbs equation**. This equation applicable for any process reversible or irreversible, because there is no path function term in the equation and closed system.

Some other equations are obtained from the first law and second law.

$$\delta Q = dE + \delta W \tag{6.33}$$

Equation (6.33) applicable for any process and any system.

$$\delta Q = dU + \delta W \tag{6.34}$$

Equation (6.34) applicable for any process undergone by a closed stationary system.

$$\delta Q = dU + pdV \tag{6.35}$$

Equation (6.35) applicable for a reversible process undergone by a closed system.

6.13   Entropy Change for Substance Such as Steam, Liquids, and Solids

At any state between saturated points f and g, shown in Fig. 6.11, liquid and vapour mixture is called wet steam.

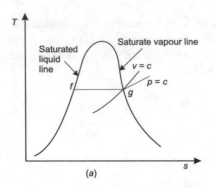

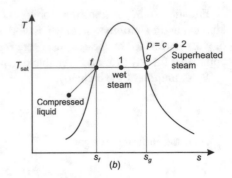

Fig. 6.11 Schematic *T-s* diagrams for water

Let x_1 be the dryness fraction of wet steam at state 1. It is defined as the ratio of the mass of vapour to the total mass of the wet steam.

Entropy at state 1,

$$s_1 = s_f + x_1\left(s_g - s_f\right)$$

$$s_1 = s_f + x_1 s_{fg} \tag{6.36}$$

where.

s_f = specific entropy of saturated liquid.

s_g = specific entropy of saturated vapour.

$$s_{fb} = s_g - s_f$$

x_1= dryness fraction.

Equation (6.36) is used to determine the entropy of the wet steam.

During evaporation at constant saturation temperature between saturated states *f* and *g*, the entropy change is given by

$$ds_{fg} = s_g - s_f = \frac{\text{latent heat}}{\text{saturatedtemperature}}$$
$$ds_{fg} = \frac{h_{fg}}{T_{\text{sat}}}$$

If the steam is heated above the saturated temperature, it is called superheated steam.

Let steam be heated from saturated temperature to superheated temperature T_1 as shown in Fig. 6.11b,

Entropy change:

$$ds_{\text{sup}} = c_{pv} \log_e \frac{T_1}{T_{\text{sat}}}$$

For a solid or a liquid, the entropy change can be found quite easily if we can assume the specific heat to be constant. Calling Eq. (*i*) of Sect. 6.12,

$$T dS = dU + p dV$$

assuming the solid or liquid to be incompressible so that $dV = 0$

$$\therefore \quad T dS = dU$$
$$T dS = mc dT$$
$$dS = mc \frac{dT}{T}$$

For unit mass,

$$ds = c \frac{dT}{T}$$

where we have dropped the subscript on the specific heat since for solids and liquids $c_p \approx c_v$. For solids and liquid, there is only a single specific heat c.

Entropy change:

$$ds = \int_1^2 c \frac{dT}{T}$$
$$ds = c \log_e \frac{T_2}{T_1}$$

6.14 Isentropic Process for Liquids and Solids

We know that the entropy change for solids and liquids

$$ds = c \log_e \frac{T_2}{T_1}$$

For isentropic process, the entropy change is zero,

$$i.e., \quad ds = 0$$
$$\therefore \quad 0 = c \log_e \frac{T_2}{T_1}$$
$$\text{or} \quad \log_e \frac{T_2}{T_1} = 0$$

$$\text{or } \frac{T_2}{T_1} = 1$$

$$T_2 = T_1$$

$$\text{or } T = \textbf{constant}$$

Thus, the isentropic process of an incompressible substance (like solid and liquid) is also called the isothermal process.

Problem 6.1: What is the entropy change when a gas expands isothermally and reversibly at 300 K? The work done by the system is 12.54 kJ.

Solution: Given data:
Temperature: $T = 300\,\text{K}$.
Work done: $W = 12.54\,\text{kJ}$.
For isothermal process, heat transfer = work done

$$Q = W$$

$$\text{or } Q = 12.54\,\text{kJ}$$

The entropy change of the system:

$$\Delta S = \frac{Q}{T} = \frac{12.54}{300}$$
$$= 0.0418\,\text{kJ/K}$$
$$= \textbf{41.8 J/K}$$

Problem 6.2: Using a Carnot cycle, show that the integral of $\frac{\delta Q}{T}$ around a cycle is zero (Fig. 6.12).

Solution: For 1–2 isothermal process,
Heat supplied:

Fig. 6.12 Carnot cycle

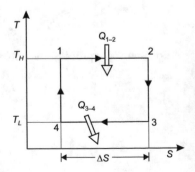

$$Q_{1-2} = \text{area under process } 1 - 2$$
$$= T_H \Delta S$$

For 2–3 adiabatic process,
Heat transfer: $Q_{2-3} = 0$.
For 3–4 isothermal process,
Heat rejected:

$$Q_{3-4} = \text{area under process } 3 - 4$$
$$= T_L \Delta S$$

For 4–1 adiabatic process,
Heat transfer:

$$Q_{4-1} = 0$$
$$\oint \frac{\delta Q}{T} = \frac{Q_{1-2}}{T_H} + 0 - \frac{Q_{3-4}}{T_L} + 0$$
$$= \frac{T_H \Delta S}{T_H} - \frac{T_L \Delta S}{T_L}$$
$$= \Delta S - \Delta S = 0$$

Problem 6.3: When air is throttled, its entropy increases by 0.87 kJ/kmol K. Determine the pressure ratio.

Solution: Given data:

$$\text{Entropy increase} : \Delta s = 0.87 \, \text{kJ/kmol K}$$
$$\text{Entropy increase} : \Delta S = -mR \log_e \frac{p_2}{p_1}$$

where

$$\text{Gas constant} : R = \frac{\text{Universal gas constant} : \bar{R}}{\text{Molecular weight} : M}$$

$$\therefore \quad \Delta S = -\frac{mR}{M} \log_e \frac{p_2}{p_1}$$
$$\Delta S = -n\bar{R} \log_e \frac{p_2}{p_1}$$

where

$$n = \frac{m}{M}, \text{number of moles}$$
$$\bar{R} = 8.314 \, \text{kJ/kmol K}$$

Entropy increase per unit mole:

$$\Delta s = -\,\bar{R}\log_e \frac{p_2}{p_1}$$

$$0.87 = -\,8.314\log_e \frac{p_2}{p_1}$$

or $\log_e \dfrac{p_2}{p_1} = -0.1046$

or $\dfrac{p_2}{p_1} = 0.9$

Problem 6.4: A nitrogen gas ($R = 297$ J/kgK) at 1 bar and 27 °C is compressed adiabatically up to 10 bar and then expanded isothermally up to initial specific volume and then cooled at constant volume to initial condition, find work, heat, change in internal energy, and change in entropy for each process and for entire processes. Take γ as 1.4.

Solution: Given data:

$$R = 297\,\text{J/kgK} = 0.297\,\text{kJ/kgK}$$
$$p_1 = 1\,\text{bar}$$
$$T_1 = 27°\text{C} = 27 + 273 = 300\,\text{K}$$
$$p_2 = 10\,\text{bar}$$

Process 1–2 adiabatic

$$\frac{T_2}{T_1} = \left(\frac{p_2}{p_1}\right)^{\frac{\gamma-1}{\gamma}}$$

$$\frac{T_2}{300} = \left(\frac{10}{1}\right)^{1.4-1}$$

Fig. 6.13 *p-v* diagram for Problem 6.4

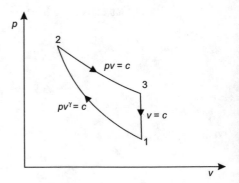

$$\frac{T_2}{300} = (10)^{0.285}$$

$$\text{or} \quad T_2 = 578.25 \, \text{K}$$
$$p_1 v_1 = RT_1$$

where p_1 in kPa, R in kJ/kgK, and T in K.

$$100 \times v_1 = 0.297 \times 300$$

$$\text{or} \quad v_1 = 0.891 \, \text{m}^3/\text{kg}$$

$$\frac{T_2}{T_1} = \left(\frac{v_1}{v_2}\right)^{\gamma-1}$$

$$\frac{578.25}{300} = \left(\frac{0.891}{v_2}\right)^{1.4-1}$$

$$\text{or} \quad v_2 = 0.1727 \, \text{m}^3/\text{kg}$$

Work:

$$w_{1-2} = \frac{R(T_2 - T_1)}{1 - \gamma}$$
$$= \frac{0.297(578.25 - 300)}{1 - 1.4}$$
$$= -206.60 \, \text{kJ/kg}$$

Heat transfer: $q_{1-2} = 0$.
Change in internal energy:

$$u_2 - u_1 = -w_{1-2}$$
$$= -(-206.60) \, \text{kJ/kg} = \mathbf{206.60 \, kJ/kg}$$

Change in entropy:

$$s_2 - s_1 = c_v \log_e \frac{T_2}{T_1} + R \log_e \frac{v_2}{v_1}$$

$$\frac{c_p}{c_v} = \gamma$$
$$c_p = \gamma c_v = 1.4 c_v$$
$$\text{and} \quad R = c_p - c_v$$

$$0.297 = 1.4c_v - c_v$$
$$0.297 = 0.4c_v$$
$$\text{or} \quad c_v = 0.7425 \, \text{kJ/kgK}$$
$$\text{and} \quad c_p = 1.4 \times 0.7415$$
$$= 1.0395 \, \text{kJ/kgK}$$

$$s_2 - s_1 = 0.7425 \log_e\left(\frac{578.25}{300}\right) + 0.297 \log_e\left(\frac{0.1727}{0.891}\right)$$
$$= 0.487 - 0.487 = 0$$

Process 2–3 isothermal: $p_2 v_2 = p_3 v_3$

$$10 \times 0.1727 = p_3 \times 0.891$$

$$\because v_3 = v_1$$

or $\quad p_3 = 1.938 \, \text{bar}$

Work: $\quad w_{2-3} = RT_2 \log_e \frac{p_2}{p_1} = 0.297 \times 578.25 \log_e\left(\frac{0.891}{0.1723}\right) = \mathbf{282.18 \, kJ/kg}$

Heat transfer: $q_{2-3} = w_{2-3} = 282.18 \, \textbf{kJ/kg}$.
Change in internal energy: $du = 0$

$$u_3 - u_2 = 0$$

Change in entropy:

$$s_3 - s_2 = R \log_e \frac{v_3}{v_0} = 0.297 \log_e\left(\frac{0.891}{0.1723}\right)$$
$$= 0.972 \times 1.64 = \mathbf{0.4870 \, kJ/kgK}$$

Process 3–1 isochoric:
Work: $w_{3-1} = 0$.

Heat transfer: $\quad q_{3-1} = c_v(T_1 - T_3) = 0.7425 \times (300 - 578.28)$
$$= -\mathbf{206.62 \, kJ/kg}$$

Change in internal energy: $(u_1 - u_3) = q_{3-1} = -\mathbf{206.62 \, kJ/kg}$.
Change in entropy:

$$s_1 - s_3 = c_v \log_e \frac{T_1}{T_3}$$

$$=0.7425 \times \log_e \left(\frac{300}{578.28} \right)$$

$$=0.7425 \times (-0.6560) = -\mathbf{0.4870\,kJ/kgK}$$

Net work:

$$w_{net} = w_{1-2} + w_{2-3} + w_{3-1}$$
$$= -206.60 + 282.18 + 0 = 75.58\,kJ/kg$$

Net heat:

$$q_{net} = q_{1-2} + q_{2-3} + q_{3-1}$$
$$= 0 + 282.18 - 206.62 = \mathbf{75.56\,kJ/kg}$$

Also we know that

$$\oint \delta w = \oint \delta q$$

Net change in internal energy $=(u_2 - u_1) + (u_3 - u_2) + (u_1 - u_3)$
$$=206.6 + 0 - 206.62 = 0$$

Also we know that

$$\oint du = 0$$

Net change in entropy $=(s_2 - s_1) + (s_3 - s_2) + (s_1 - s_3)$
$$=0 + 0.4870 - 0.4870 = 0$$

Also we know that

$$\oint ds = 0$$

for reversible cycle.

Problem 6.5: One kg of water at 0 °C is brought into contact with a heat reservoir at 100 °C. When the water has reached 100 °C, find the entropy change of the universe. If water is heated from 0 to 100 °C by first bringing it in contact with a reservoir at 50 °C and then with a reservoir at 100 °C, what will be the entropy change of the universe? Explain how water might be heated from 0 to 100 °C with no change in entropy of the universe.

Solution: Given data:

Mass of water: $m = 1$ kg.

Initial temperature of water: $T_1 = 0°C = (0 + 273)K = 273$ K.

Temperature of a heat reservoir: $T_R = 100°C = (100 + 273)K = 373$ K.

Final temperature of water: $T_2 =$ temperature of a heat reservoir $= 373$ K.

Entropy change of the water:

$$(\Delta S)_{\text{water}} = mc_p \log_e \frac{T_2}{T_1} = 1 \times 4.18 \log_e \frac{373}{273}$$

$$= 1.3046 \, \text{kJ/K}$$

Heat transferred between the water and a heat reservoir:

$$Q = mc_p(T_R - T_1) = 1 \times 4.18 \times (373 - 273) = 418 \, \text{kJ}$$

Entropy change of the reservoir:

$$(\Delta S)_{\text{reservoir}} = -\frac{Q}{T_s} = -\frac{418}{373} = -1.1206 \, \text{kJ/K}$$

Entropy change of the universe:

$$(\Delta S)_{\text{universe}} = (\Delta S)_{\text{water}} + (\Delta S)_{\text{reservoir}} = 1.3046 - 1.1208$$
$$= \mathbf{0.1838 \, kJ/K}$$

When water is heated in two stages by receiving heat from reservoir at 50 and 100 °C,

Temperature of the first reservoir:

$$T_{R1} = 50°C = (50 + 273)K = 323 \, \text{K}$$

Temperature of the second reservoir:

$$T_{R2} = 100°C = (100 + 273)K = 373 \, \text{K}$$

Entropy change of the water:

$$(\Delta S)_{\text{water}} = mc_p \log_e \frac{T_{R1}}{T_1} + mc_p \log_e \frac{T_{R1}}{T_1}$$

$$= mc_p \left[\log_e \frac{T_{R1}}{T_1} + \log_e \frac{T_{R2}}{T_{R1}} \right]$$

$$= mc_p \log_e \left(\frac{T_{R1}}{T_1} \times \frac{T_{R2}}{T_{R1}} \right)$$

$$= mc_p \log_e \frac{T_{R2}}{T_1}$$

$$= 1 \times 4.18 \times \log_e \frac{373}{273} = 1.3046 \, \text{kJ/K}$$

Heat transferred between the water and first reservoir:

$$Q_1 = mc_p(T_{R1} - T_1)$$
$$= 1 \times 4.18(323 - 273) = 209 \, \text{kJ}$$

Heat transferred between the water at 323K and second reservoir:

$$Q_2 = mc_p(T_{R2} - T_{R1}) = 1 \times 4.18 \times (373 - 323) = 209 \, \text{kJ}$$

Entropy change of the reservoirs:

$$(\Delta S)_{\text{reservoirs}} = -\frac{Q_1}{T_{R1}} - \frac{Q_2}{T_{R2}}$$
$$= -\frac{209}{323} - \frac{209}{373} = -0.6470 - 0.5603 = -1.2073 \, \text{kJ/K}$$

Entropy change of the universe:

$$(\Delta S)_{universe} = (\Delta S)_{water} + (\Delta S)_{reservoirs}$$
$$= 1.3046 - 1.2073 = 0.0973 \, kJ/K$$

The entropy change of the universe would be less and less if the water is heated more and more in stages, by bringing the water in contact successively with more and more heat reservoirs, each succeeding reservoir being at a higher temperature than the preceding one.

When the water is heated in infinite steps, by bringing in contact with an infinite number of reservoirs in succession, so that at any instant the temperature difference between the water and the reservoir in contact is infinitesimally small, then the net entropy change of the universe would be zero.

Problem 6.6: A reversible heat engine interacts with three thermal reservoirs at 500 K, 400 K, and 300 K, respectively. The engine does 300 kJ of net work and absorbs 900 kJ of energy as heat from the reservoir at 500 K. Determine the magnitude and direction of heat interactions of the engine with the other two reservoirs (Fig. 6.14).

Solution: Given data:
Temperatures of three reservoirs:

$$T_1 = 500 \, \text{K}$$
$$T_2 = 400 \, \text{K}$$
$$T_3 = 300 \, \text{K}$$

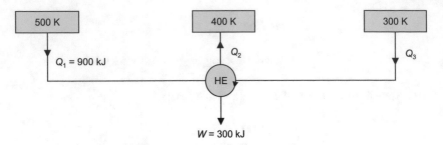

Fig. 6.14 Schematic diagram for Problem 6.6

Net work output:

$$W = 300\,\text{kJ}$$

Heat absorbs from the reservoir at 500 K: Q_1.
Let

$Q_1 = 900\,\text{kJ}$
$Q_2 = $ heat rejects from the heat engine to the reservoir at 400 K
$Q_3 = $ heat absorbs from the heat engine from the reservoir at 300 K

By energy balance equation, we have

$$Q_1 + Q_3 = Q_2 + W$$
$$900 + Q_3 = Q_2 + 300$$
$$\text{or} \quad Q_3 = Q_2 - 600 \tag{6.37}$$

By Clausius's inequality, we have

$$\oint \frac{\delta Q}{T} \leq 0$$
$$\oint \frac{\delta Q}{T} = 0$$

for reversible heat engine.

$$\frac{Q_1}{T_1} - \frac{Q_2}{T_2} + \frac{Q_3}{T_3} = 0$$
$$\frac{900}{500} - \frac{Q_2}{400} + \frac{Q_3}{300} = 0 \tag{6.38}$$

Substituting the value of Q_3 from Eq. (6.37) in the above Eq. (6.38), we have

$$\frac{900}{500} - \frac{Q_2}{400} + \frac{(Q_2 - 600)}{300} = 0$$
$$\frac{9}{5} - \frac{Q_2}{400} + \frac{Q_2}{300} - 2 = 0$$

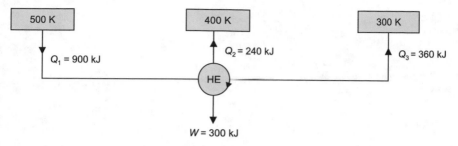

Fig. 6.15 Schematic diagram for Problem 6.6

$$1.8 - \frac{Q_2}{400} + \frac{Q_2}{300} - 2 = 0$$

$$-\frac{Q_2}{400} + \frac{Q_2}{300} = 0.2$$

$$\frac{-300Q_2 + 400Q_2}{400 \times 300} = 0.2$$

$$100Q_2 = 0.2 \times 400 \times 300$$

$$\text{or} \quad Q_2 = \mathbf{240\,kJ}$$

The $+ve$ sign with Q_2 shows that the assumed direction of heat Q_2 is correct and heat Q_2 will flow from the heat engine to the reservoir at 400 K.

Substituting the value of $Q_2 = 204$ kJ in Eq. (6.37), we have

$$Q_3 = 240 - 600 = -360\,kJ$$

The $-ve$ sign with Q_3 shows that the assumed direction of heat Q_3 is not correct and actually heat Q_3 will flow from the heat engine to the reservoir.

The actual sign of heat transfers and magnitude are shown in Fig. 6.15.

$$Q_2 = \mathbf{240\,kJ} \text{ from the heat engine.}$$
$$Q_3 = \mathbf{360\,kJ} \text{ from the heat engine.}$$

Problem 6.7: A piston-cylinder device contains 1.2 kg of nitrogen gas at 120 kPa and 27 °C. The gas is now compressed slowly in a polytropic process during which $pV^{1.3}$ = constant. The process ends when the volume is reduced by one-half. Determine the entropy change of nitrogen during this process. Take $\gamma = 1.4, R = 0.296$ kJ/kgK, $c_p = 1.039$ kJ/kgK.

Solution: Given data:

$$m = 1.2\,kg$$
$$p_1 = 120\,kPa$$
$$T_1 = 27°C = 27 + 273 = 300\,K$$

Let

$$V_1 = V$$

and

$$V_2 = \frac{1}{2}V$$
$$p_1 V_1^{1.3} = p_2 V_2^{1.3}$$

or

$$p_2 = p_1 \left(\frac{V_1}{V_2}\right)^{1.3} = 120(2)^{1.3} = 295.47 \, \text{kPa}$$

$$\frac{T_2}{T_1} = \left(\frac{p_2}{p_1}\right)^{\frac{n-1}{n}}$$

$$T_2 = T_1 \left(\frac{p_2}{p_1}\right)^{\frac{n-1}{n}}$$

$$= 300 \left(\frac{295.47}{120}\right)^{\frac{1.3-1}{1.3}} = 300(2.462)^{0.23}$$

$$= 300 \times 1.23 = 369 \, \text{K}$$

The entropy change during the process:

$$\Delta S = mc_v \log_e \frac{T_2}{T_1} + mR \log_e \frac{V_2}{V_1}$$

$$= 1.2 \times 0.742 \log_e \left(\frac{369}{300}\right) + 1.2 \times 0.296 \log_e \left(\frac{1}{2}\right)$$

$$\left| \because \gamma = \frac{c_p}{c_v}, c_v = \frac{c_p}{\gamma} = \frac{1.039}{1.4} = 0.742 \, kJ/kg\,K \right.$$

$$= 0.1843 - 0.2462 = -0.0619 \, kJ/K$$

or

$$\Delta S = mc_p \log_e \frac{T_2}{T_1} - mR \log_e \frac{p_2}{p_1}$$

$$= 1.2 \times 1.039 \log_e \left(\frac{369}{300}\right) - 1.2 \times 0.296 \log_e \left(\frac{295.47}{120}\right)$$

$$= 0.2581 - 0.3200 = -0.0619 \, \textbf{kJ/K}$$

or

$$\Delta S = \left(\frac{n-\gamma}{n-1}\right) mc_v \log_e \frac{T_2}{T_1}$$
$$= \left(\frac{1.3-1.4}{1.3-1}\right) \times 1.2 \times 0.742 \log_e \frac{369}{300}$$
$$= -0.0619 \, \text{kJ/K}$$

or

$$\Delta S = \left(\frac{n-\gamma}{n}\right) mc_v \log_e \frac{p_2}{p_1}$$
$$= \left(\frac{1.3-1.4}{1.3}\right) \times 1.2 \times 0.742 \log_e \left(\frac{295.47}{120}\right)$$
$$= -0.0617 \, kJ/K$$

or

$$\Delta S = (\gamma - n) mc_v \log_e \frac{V_2}{V_1}$$
$$= (1.4 - 1.3) \times 1.2 \times 0.742 \log_e \left(\frac{1}{2}\right)$$
$$= -0.0617 \, kJ/K$$

Problem 6.8: A 1.5 m³ insulated rigid tank contains 2.7 kg of carbon dioxide ($R = 0.1889$ kJ/kgK, $c_v = 0.657$ kJ/kgK) at 100 kPa. Now paddle wheel work is done on the system until the pressure in the tank rises to 150 kPa. Determine the entropy change of carbon dioxide during this process (Fig. 6.16).

Solution: Given data:

$$V = 1.5 \, \text{m}^3$$
$$m = 2.7 \, \text{kg}$$

Fig. 6.16 Schematic for
Problem 6.8

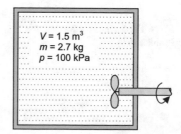

$$V = 1.5 \, m^3$$
$$m = 2.7 \, kg$$
$$p = 100 \, kPa$$

$$p_1 = 100\,\text{kPa}$$
$$p_2 = 150\,\text{kPa}$$

Change in entropy during the process:

$$\Delta S = mc_v \log_e \frac{p_2}{p_1} + mc_p \log_e \frac{V_2}{V_1}$$

$$= mc_v \log_e \frac{p_2}{p_1} \quad \because V_1 = V_2 = V$$

$$= 2.7 \times 0.657 \log_e \frac{150}{100} = 2.7 \times 0.657 \times 0.4054 = 0.7191\,\textbf{kJ/K}$$

Problem 6.9: An insulated rigid tank is divided into two equal parts by a partition. Initially, one part contains 5 kmol of an ideal gas at 250 kPa and 40 °C, and the other side is evacuated. The partition is now removed, and the gas fills the entire tank. Determine the total entropy change during this process (Fig. 6.17).

Solution: Given data:
Let V = volume of each part,
i.e., $V_1 = V$.
and $V_2 = 2\,V$.

Number of moles:

$$n = 5\,\text{kmol}$$
$$p_1 = 250\,\text{kPa}$$
$$T_1 = 40°C$$

When the partition is removed, the gas fills the entire tank,
i.e., $T_2 = T_1 = 40°C$.
Change in entropy during the process:

$$\Delta S = mc_v \log_e \frac{T_2}{T_l} + mR \log_e \frac{V_2}{V_1}$$

$$= mR \log_e \frac{V_2}{V_1}$$

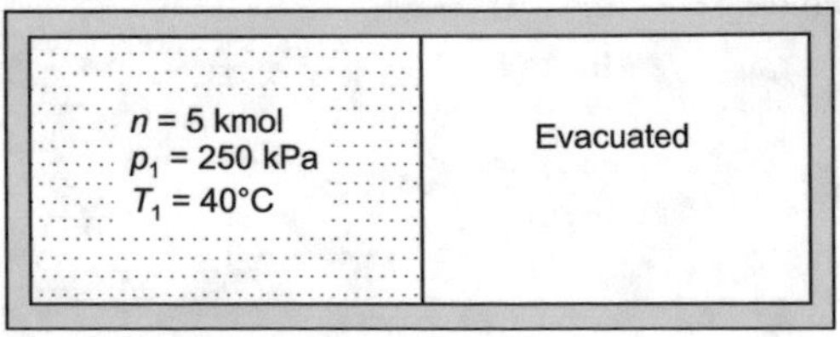

Fig. 6.17 Schematic for Problem 6.9

$$= n\bar{R}\log_e \frac{V_2}{V_1}$$

$$\because T_1 = T_2$$

where

$$\bar{R} = 8.314\,\text{kJ/kmolK, universal gas constant.}$$
$$\Delta S = 5 \times 8.314 \times \log_0 2 = \mathbf{28.81\,kJ/K}$$

Problem 6.10: An ideal gas undergoes a reversible adiabatic compression from 5 bar, 0.2 to 0.05 m^3 according to the law $pV^{1.3} = $ constant. Determine the change in enthalpy, internal energy, and entropy and the heat and work transfer during the process (Fig. 6.18).

Solution: Given data for adiabatic process:

$$pV^{1.3} = C$$

At initial state, $\begin{aligned}p_1 &= 5\,\text{bar}\\ V_1 &= 0.2\,\text{m}^3\end{aligned}$.

At final state, $V_2 = 0.05\,\text{m}^3$.

Adiabatic index: $\gamma = 1.3$

For adiabatic process 1–2, $\begin{aligned}p_1 V_1^{\gamma} &= p_2 V_2^{\gamma}\\ 5 \times (0.2)^{1.3} &= p_2 \times (0.05)^{1.3}\end{aligned}$

or $p_2 = 5 \times \left(\frac{0.2}{0.05}\right)^{1.3} = 30.31\,\text{bar.}$

(i)   Change in enthalpy: $H_2 - H_1 = mc_p(T_2 - T_1)$

where

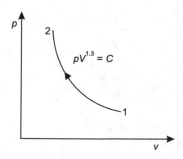

Fig. 6.18 p-v diagram for Problem 6.10

$$c_p = \frac{\gamma R}{\gamma - 1}$$

$$\therefore \quad H_2 - H_1 = m\frac{\gamma R}{\gamma - 1}(T_2 - T_1) = \frac{\gamma R}{\gamma - 1}(mRT_2 - mRT_1)$$

$$= \frac{\gamma}{\gamma - 1}(p_2V_2 - p_1V_1)$$

where

$$p_1 = 5 \times 10^2 \text{ kPa, and}$$
$$p_2 = 30.31 \times 10^2 \text{ kPa}$$

$$\therefore \quad H_2 - H_1 = \frac{1.3}{1.3 - 1}\left(30.31 \times 10^2 \times 0.05 - 5 \times 10^2 \times 0.2\right)$$
$$= \mathbf{223.38\,kJ}$$

(ii)   Change in internal energy:

$$U_2 - U_1 = mc_v(T_2 - T_1)$$

where

$$c_v = \frac{R}{\gamma - 1}$$

$$\therefore \quad U_2 - U_1 = \frac{R}{\gamma - 1}(T_2 - T_1)$$
$$= \frac{mR}{\gamma - 1}(mRT_2 - mRT_1)$$
$$= \frac{1}{\gamma - 1}(p_2V_2 - p_1V_1)$$
$$= \frac{\gamma}{\gamma(\gamma - 1)}(p_2V_2 - p_1V_1)$$
$$= \frac{H_2 - H_1}{\gamma} = \frac{223.38}{1.3} = 171.83\,kJ$$

(iii)   Change in entropy: $S_2 - S_1 = 0$
(iv)   Heat transfer: $Q_{1-2} = 0$
(v)   Work transfer: $W_{1-2} = \frac{p_1V_1 - p_2V_2}{\gamma - 1} = -(U_2 - U_1) = -171.83 \text{ kJ}$

The $-ve$ sign indicated that work was done on the system.

Problem 6.11: $0.34 \, \text{m}^3$ of a perfect gas is heated from 100 to 300 °C at a constant pressure of 2.8 bar and is then cooled at constant volume to its initial temperature. Calculate the change in entropy. Given $c_p = 1.05 \, \text{kJ/kgK}$ and $c_v = 0.75 \, \text{kJ/kgK}$. Represent the processes on p–v and T-s diagrams (Fig. 6.19).

Solution: Given data:

$$V_1 = 0.34 \, \text{m}^3$$
$$T_1 = 100°\text{C} = 100 + 273 = 373 \, \text{K}$$
$$T_2 = 300°\text{C} = 300 + 273 = 573 \, \text{K}$$

$$p_1 = p_2 = 2.8 \, \text{bar} = 280 \, \text{kPa}$$
$$V_2 = V_3$$
$$c_p = 1.05 \, \text{kJ/kgK}$$
$$c_v = 0.75 \, \text{kJ/kgK}$$
$$\therefore \quad = c_p - c_v = 1.05 - 0.75 = 0.3 \, \text{kJ/kgK}$$
$$p_1 V_1 = mRT_1$$
$$280 \times 0.34 = m \times 0.3 \times 373$$
$$\text{or} \quad m = 0.85 \, \text{kg}$$

Process 1–2 isobaric: $\frac{V_1}{T_1} = \frac{V_2}{T_2}$

$$\frac{0.34}{373} = \frac{V_2}{573}$$

or $V_2 = 0.522 \, \text{m}^3$.
Change in entropy:

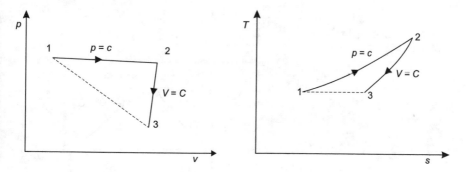

Fig. 6.19 p-v and T-s diagrams for Problem 6.11

$$S_2 - S_1 = mc_v \log_e \frac{T_2}{T_1} + mR \log_e \frac{V_2}{V_1}$$

$$= 0.85 \times 0.75 \log_e \frac{573}{373} + 0.85 \times 0.3 \log_e \frac{0.522}{0.34}$$

$$= 0.273 + 0.109 = \mathbf{0.382\,kJ/K}$$

or

$$S_2 - S_1 = mc_p \log_e \frac{T_2}{T_1} = 0.85 \times 1.05 \log_e \frac{573}{373}$$

$$= 0.85 \times 1.05 \times 0.429 = \mathbf{0.382\,kJ/K}$$

Process 2–3 isochoric:
Change in entropy:

$$S_3 - S_2 = mc_v \log_e \frac{T_3}{T_2} = 0.85 \times 0.75 \log_e \frac{373}{573}$$

$$= -0.273\,\mathbf{kJ/K}$$

$$\text{Net change in entropy} = (S_2 - S_1) + (S_3 - S_2) = 0.382 - 0.273$$

$$= \mathbf{0.109\,kJ/K}$$

Problem 6.12: An ideal gas is heated from temperature T_1 to T_2 by keeping its volume constant. The gas is expanded back to its initial temperature according to the law $pv^n =$ constant. If the entropy changes in the two processes are equal, find the value of n in terms of the adiabatic index γ (Fig. 6.20).

Solution: Given processes:
    1–2, heat addition at constant volume.
    2–3, the gas expands polytropically.

For process 1–2,

Fig. 6.20 *T-s* diagram for
Problem 6.12

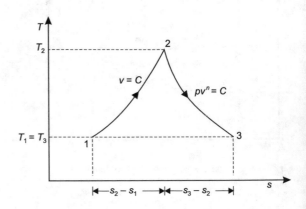

Entropy change per unit mass:

$$s_2 - s_1 = c_v \log_e \frac{T_2}{T_1} \quad \text{for unit mass}$$

For process 2–3,
Entropy change per unit mass:

$$s_3 - s_2 = \left(\frac{\gamma - n}{1 - n}\right) c_v \log_e \frac{T_3}{T_2}$$

$$s_3 - s_2 = \left(\frac{\gamma - n}{1 - n}\right) c_v \log_e \frac{T_1}{T_2} \quad \because T_3 = T_1$$

Given condition: the entropy changes in the two processes are equal,
i.e., $s_2 - s_1 = s_3 - s_2$.

or

$$c_v \log_e \frac{T_2}{T_1} = \left(\frac{\gamma - n}{1 - n}\right) c_v \log_e \frac{T_3}{T_2}$$

$$- c_v \log_e \frac{T_2}{T_1} = \left(\frac{\gamma - n}{1 - n}\right) c_v \log_e \frac{T_1}{T_2}$$

or

$$-1 = \frac{\gamma - n}{1 - n}$$

or

$$- 1(1 - n) = \gamma - n$$
$$- 1 + n = \gamma - n$$

or

$$2n = \gamma + 1$$
$$= \frac{\gamma + 1}{2}$$

Problem 6.13: Consider two finite bodies of same mass m and specific heat c_p used as source and sink for a heat engine. The first body is initially at an absolute temperature T_1 while the second body is at a lower absolute temperature T_2. Heat is transferred from the first body to the heat engine, which rejects the waste heat to the second body. The process continues until the temperature of the two bodies T_f becomes equal. Show that $T_f = \sqrt{T_1 T_2}$ when the heat engine produces the maximum work. Also show that the maximum work output, $V_{\text{Max}} = mc_p \left(\sqrt{T_1} - \sqrt{T_2}\right)^2$ (Fig. 6.21).

Fig. 6.21 Schematic for
Problem 6.13

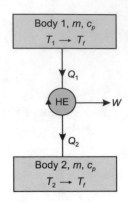

Solution: Let.

T_1 = absolute temperature of body 1,

T_2 = absolute temperature of body 2,

m = mass of each body,

c_p = specific heat of each body,

Q_1 = heat supplied to the heat engine from body 1,

Q_2 = heat rejected by the heat engine to body 2, and.

T_f = final equilibrium temperature attained by bodies 1 and 2.

When the two bodies achieve the equilibrium temperature condition, then work output is stopped.

Heat supplied to the heat engine from body 1:

$$Q_1 = mc_p(T_1 - T_f)$$

Heat rejected from the heat engine to body 2:

$$Q_2 = mc_p(T_f - T_2)$$

∴ Work delivered by the heat engine:

$$\begin{aligned} W &= Q_1 - Q_2 \\ &= mc_p(T_1 - T_f) + mc_p(T_f - T_2) \\ W &= mc_p[T_1 + T_2 - 2T_f] \end{aligned} \tag{i}$$

The entropy change of body 1:

$$\Delta S_1 = mc_p \log_e \frac{T_f}{T_1}$$

The entropy change of body 2:

$$\Delta S_2 = mc_p \log_e \frac{T_f}{T_2}$$

$\therefore$ Entropy change of the universe:

$$\Delta S_{\text{universe}} = \Delta S_1 + \Delta S_2$$

$$= mc_p \log_e \frac{T_f}{T_1} + mc_p \log_e \frac{T_f}{T_2}$$

For maximum work output, the heat engine must be reversible.
For reversible heat engine,

$$\Delta S_{\text{universe}} = 0$$

$$mc_p \log_e \frac{T_f}{T_1} + mc_p \log_e \frac{T_f}{T_2} = 0$$

or

$$\log_e \frac{T_f}{T_1} + \log_e \frac{T_f}{T_2} = 0$$

or

$$\log_e \frac{T_f^2}{T_1 T_2} = 0$$

or

$$\frac{T_f^2}{T_1 T_2} = e^0$$

$$\frac{T_f^2}{T_1 T_2} = 1$$

$$T_f = \sqrt{T_1 T_2}$$

Substituting $T_f = \sqrt{T_1 T_2}$ in Eq. (i), we get maximum work output:

$$W_{\text{Max}} = mc_p \left[T_1 + T_2 - 2\sqrt{T_1 T_2} \right]$$
$$W_{\text{Max}} = mc_p \left[\sqrt{T_1} - \sqrt{T_2} \right]^2$$

Problem 6.14: A 1 kg of ice at –5 °C is exposed to the atmosphere which is at 20 °C. The ice melts and comes into thermal equilibrium with the atmosphere (Figs. 6.22 and 6.23).

Fig. 6.22 Schematic for
Problem 6.14

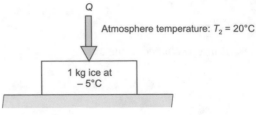

Fig. 6.23 Schematic for
Problem 6.14

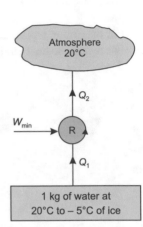

(i)   Determine the entropy change of the universe.
(ii) What is the minimum amount of work necessary to convert the water back into
      ice at –5 °C?

Take the specific heat of ice as 2.093 kJ/kgK and latent heat of fusion of ice as
333.3 kJ/kg.

Solution: Given data:
Mass of ice: $m = 1$ kg.
Temperature: $T_1 = -5\,°C = (-5 + 273)\,K = 268\,K$.
Temperature of the atmosphere: $T_2 = 20\,°C = (20 + 273)\,K = 293\,K$.
Specific heat of ice: $c_{ice} = 2.093$ kJ/kgK.
Latent heat of ice: $L = 333.3$ kJ/kg.

Let Q = net heat absorbed by ice from the atmosphere air to come into thermal
equilibrium with the atmosphere and.
∴ Q = sensible heat absorbed by ice from –5 °C to 0 °C ice + latent heat absorbed
by melt of ice from 0 °C to 0 °C water + sensible heat absorbed by water from 0 °C
to 20 °C.

$$= mc_{ice}[0 - (-5)] + mL + mc_w(20 - 0)$$

where

$$c_w = 4.187 \, \text{kJ/kgK, specific heat of water}$$
$$\therefore \quad Q = 1 \times 2.093 \times (0+5) + 1 \times 333.3 + 1 \times 4.187 \times 20$$
$$= 10.465 + 333.3 + 83.74 = 427.505 \, \text{kJ}$$

The entropy change of ice from –5 °C to 20 °C of water:

$$(\Delta S)_{\text{ice}} = m c_{\text{ice}} \log_e \frac{T_0}{T_2} - \frac{mL}{T_0} + m c_w \log_e \frac{T_1}{T_0}$$

where

$$T_0 = 0°C = 273 \, \text{K}$$
$$T_1 = 268 \, \text{K, and}$$
$$T_2 = 293 \, \text{K}$$

$$\therefore \quad (\Delta S)_{\text{ice}} = 1 \times 2.093 \log_e \frac{273}{268} + \frac{1 \times 333.3}{273} + 1 \times 4.187 \log_e \frac{293}{273}$$
$$= 0.0386 + 1.2208 + 0.2960 = 1.5554 \, \text{kJ/K}$$

The entropy change of the atmosphere:

$$(\Delta S)_{\text{atm}} = -\frac{Q}{T_2} = -\frac{427.505}{293} = -1.4590 \, \text{kJ/K}$$

(i)   The entropy change of the universe:

$$(\Delta S)_{\text{universe}} = (\Delta S)_{\text{ice}} + (\Delta S)_{\text{atm}}$$
$$= 1.5554 - 1.4590 = 0.0964 \, \text{kJ/K}$$
$$= \mathbf{96.4 \, J/K}$$

(ii)   The minimum work required to drive the reversible refrigerator which converts 1 kg of water from 20 °C to –5 °C of ice: $W_{\min}$

The amount of heat is removed to produce 1 kg of ice at – 5 °C from 20 °C of water,

$$Q_1 = m c_w (20 - 0) + mL + m c_{\text{ice}}[0 - (-5)]$$
$$= 1 \times 4.187 \times 20 + 1 \times 333.3 + 1 \times 2.093 \times 5$$
$$= 83.74 + 333.3 + 10.465 = 427505 \, \text{kJ}$$

The entropy change of water from 20 °C to –5 °C of ice:

$$(\Delta S)_{\text{water}} = mc_w \log_e \frac{T_0}{T_2} - \frac{mL}{T_0} + mc_w \log_e \frac{T_1}{T_0}$$

$$(\Delta S)_{\text{water}} = 1 \times 4.187 \log_e \frac{273}{293} - 1 \times \frac{333.3}{273} + 1 \times 2.093 \log_e \frac{268}{273}$$

$$= -0.2960 - 1.2208 - 0.0386 = -1.5554 \,\text{kJ/K}$$

The entropy change of the atmosphere:

$$(\Delta S)_{\text{atm}} = \frac{Q_2}{T_2} = \frac{Q_2}{293}$$

$\therefore$ The entropy change of the universe:

$$(\Delta S)_{\text{universe}} = (\Delta S)_{\text{water}} + (\Delta S)_{\text{atm}} = -1.5554 + \frac{Q_2}{293}$$

According to the increase of entropy principle for reversible process,

$$(\Delta S)_{\text{universe}} = 0$$

$$\therefore \quad -1.5554 + \frac{Q_2}{293} = 0$$

or

$$\frac{Q_2}{293} = 1.5554$$

or

$$Q_2 = 455.732 \,\text{kJ}$$

So, the minimum amount of work that is required to drive the refrigerator,

$$W_{\text{min}} = Q_2 - Q_1$$
$$= 455.732 - 427.505$$
$$= \mathbf{28.227 \,kJ}$$

Problem 6.15: A metal block of 6 kg and at 250 °C is cooled in a surrounding of air which is at 25 °C. If the specific heat of metal is 0.42 kJ/kgK, determine the following (Fig. 6.24):

(i)   entropy change of metal block and

Fig. 6.24 Schematic for
Problem 6.15

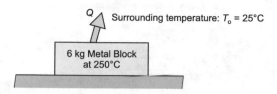

(ii)   entropy change of the universe.

Solution: Given data:

  Mass of metal block: $m = 6$ kg.
  Temperature: $T = 250\,°C = (250 + 273)\,K = 523\,K$.

  Temperature of a surrounding air: $T_0 = 25\,°C = (25 + 273)\,K = 298\,K$.
  Specific heat of metal: $c = 0.42$ kJ/kgK.
  Heat gained by the surrounding air: $Q =$ heat lost by the metal block

$$Q = mc(T - T_0) = 6 \times 0.42(523 - 298) = 567\,kJ$$

(i)   Entropy change of metal block: $(\Delta S)_{\text{metal}}$

$$(\Delta S)_{\text{metal}} = mc\log_e \frac{T_0}{T} = 6 \times 0.42 \log_e \frac{298}{523}$$
$$= -1.417\,kJ/K$$

(ii)   Entropy change of universe: $(\Delta S)_{\text{uni}}$

$$(\Delta S)_{\text{uni}} = (\Delta S)_{\text{metal}} + (\Delta S)_{\text{surr}}$$
$$= mc\log_e \frac{T_0}{T} + \frac{Q}{T_o} = -1.417 + \frac{567}{298} = -1.417 + 1.902$$
$$= 0.485\,kJ/K$$

Problem 6.16: One kg of ice at –5 °C is mixed with four kg of water at 40 °C in
an insulated chamber at atmospheric pressure. The ice melts and comes into thermal
equilibrium with the water. Assuming ice-water system as isolated, determine the
entropy increase of the universe. The latent heat of fusion of ice is 335 kJ/kg, the
specific heat of ice is 2.09 kJ/kgK, and the specific heat of water is 4.18 kJ/kgK
(Fig. 6.25).

Solution: Given data:

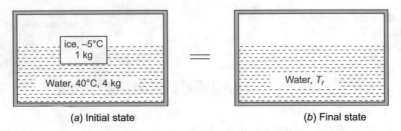

(a) Initial state   (b) Final state

Fig. 6.25   Schematic for Problem 6.16

For ice	For water
$m_{\text{ice}} = 1\,\text{kg}$	$m_w = 4\text{kg}$
$T_1 = -5°\text{C} = 268\,\text{K}$	$T_2 = 40°\text{C}$
Latent heat : $L = 335\,\text{kJ/kg}$	Specific heat : $c_w = 4.18\,\text{kJ/kgK}$
Specific heat : $c_{\text{ice}} = 2.09\,\text{kJ/kgK}$	

Let T_f = final temperature of water after mixing of ice-water.
Applying the energy balance equation, heat gained by ice = heat lost by water.

$$m_{\text{ice}}\,c_{\text{ice}}[0 - (-5)] + m_{\text{ice}}L + m_{\text{ice}}c_w(T_f - 0) = m_w c_w(40 - T_f)$$
$$1 \times 2.09 \times 5 + 1 \times 335 + 1 \times 4.18 \times T_f = 4 \times 4.18(40 - T_f)$$
$$10.45 + 335 + 4.18T_f = 668.8 - 16.72T_f$$

or $20.9T_f = 323.35$.

or $T_f = 15.47°\text{C}$.
Entropy change of ice:

$$\Delta S_{\text{ice}} = m_{\text{ice}}c_{\text{ice}}\log_e \frac{T_0}{T_1} + \frac{m_{ice}L}{T_0} + m_{ice}c_w \log_e \frac{T_f}{T_0}$$

where

$$T_1 = -5°\text{C} = 268\,\text{K}$$
$$T_0 = 0°\text{C} = 273\,\text{K, and}$$
$$T_f = 15.47°\text{C} = 288.47\,\text{K}$$

$$\therefore \quad \Delta S_{\text{ice}}\,1 \times 2.09 \times \log_e \frac{273}{268} + \frac{1 \times 335}{273} + 1 \times 4.18\log_e \frac{288.47}{273}$$
$$= 0.0386 + 1.2271 + 0.230 = 1.496\,\text{kJ/K}$$

Entropy change of water:

$$\Delta S_{\text{water}} = m_w c_w \log_e \frac{T_f}{T_2}$$

where

$$T_f = 288.47 \text{ K, and}$$
$$T_2 = 40°C = 313 \text{ K}$$

$$\Delta S_{\text{water}} = 4 \times 4.18 \log_e \frac{288.47}{313} = -1.364 \text{ kJ/K}$$

Entropy change of the universe:

$$\Delta S_{\text{uni}} = \Delta S_{\text{ice}} + \Delta S_{\text{water}}$$
$$= 1.496 - 1.364 = 0.132 \text{ kJ/K} = \mathbf{132 \, J/K}$$

Problem 6.17: A 50 kg block of iron having a temperature of 150 °C is immersed in 100 kg of water at a temperature of 25 °C. What will be the entropy change of the combined system of iron and water? The specific heat of iron and water is 0.44 kJ/kgK and 4.18 kJ/kgK, respectively (Fig. 6.26).

Solution: Given data:

For iron   **For water**

Mass : $m_1 = 50 \text{ kg}$ Mass : $m_2 = 100 \text{ kg}$

Initial temperature : $T_1 = 150°C$ Initial temperature : $T_2 = 25°C$

Specific heat : $c_1 = 0.44 \text{ kJ/kgK}$ Specific heat : $c_2 = 4.18 \text{ kJ/kgK}$

Let $T_f =$ final equilibrium temperature of iron and water. Applying the energy balance equation for final state of the system, heat lost by a block of iron = heat gained by the water.

$$m_1 c_1 (T_1 - T_f) = m_2 c_2 (T_f - T_2)$$

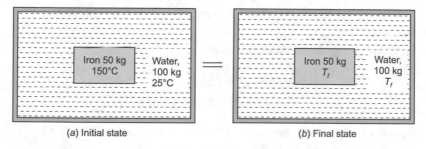

(a) Initial state   (b) Final state

Fig. 6.26   Schematic for Problem 6.17

$$50 \times 0.44 \times \left(150 - T_f\right) = 100 \times 4.18\left(T_f - 25\right)$$
$$3300 - 22T_f = 418T_e - 10450$$

or
$$3300 + 10450 = 418T_f + 22T_f$$
$$13750 = 440T_f$$

or $T_f = 31.25°C$.

Entropy change of iron:

$$(\Delta S)_{\text{iron}} = m_1 c_1 \log_e \frac{T_f}{T_2}$$

where

$$T_f = 31.25°C = (31.25 + 273)K = 304.25 \text{ K}$$

and

$$T_1 = (150 + 273)K = 423 \text{ K}$$

$$\therefore \quad (\Delta S)_{\text{iron}} = 50 \times 0.44 \times \log_e \frac{304.25}{298} = -7.249 \text{ kJ/K}$$

Entropy change of water:

$$(\Delta S)_{\text{water}} = m_2 c_2 \log_e \frac{T_1}{T_2}$$

where

$$T_f = 304.25 \text{ K and } T_2 = (25 + 273)K = 298 \text{ K}$$

$$\therefore \quad (\Delta S)_{\text{water}} = 100 \times 4.18 \log_e \frac{304.25}{298} = 8.676 \text{ kJ/K}$$

$\therefore$ Entropy change of the combined system of iron and water:

$$\Delta S = (\Delta S)_{\text{iron}} + (\Delta S)_{\text{water}}$$
$$= -7.249 + 8.676 = 1.427 \, \textbf{kJ/K}$$

Problem 6.18: One kg ice block at $-10 °C$ is brought into contact with 5 kg copper block at $30 °C$ in an insulated container. Determine the entropy change of (*i*) ice block, (*ii*) copper block, and (*iii*) the universe. Specific heats of ice and copper are

Fig. 6.27 Initial state of the
system

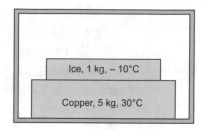

Ice, 1 kg, – 10°C

Copper, 5 kg, 30°C

2.057 kJ/kgK and 0.385 kJ/kgK, respectively. The latent heat of ice is 333.32 kJ/kg
(Fig. 6.27).

Solution: Given data:

	For ice	**For copper**
	$m_1 = 1\,\text{kg}$	$m_2 = 5\,\text{kg}$
	$T_1 = -10°C$	$T_2 = 30°C$
	$c_1 = 2.057\,\text{kJ/kgK}$	$c_2 = 0.385\,\text{kJ/kgK}$

Latent heat: $L = 333.32$ kJ/kg.

Now to check the final equilibrium state of the system.

The amount of heat required to change 1 kg of ice at –10 °C to 1 kg of water at
0 °C,

$$= m_1 c_1 [0 - (-10)] + m_1 L$$
$$= 1 \times 2.057 \times 10 + 1 \times 333.32$$
$$= 20.57 + 333.32 = 353.89\,\text{kJ}$$

and the amount of heat loss to change 5 kg of copper at 30 to 0 °C,

$$= m_2 c_2 [30 - 0] = 5 \times 0.385 \times 30 = 57.75\,\text{kJ}$$

Thus, the amount of heat required to decrease the temperature of 5 kg copper from
30 to 0 °C is less than the amount of heat required to change 1 kg of ice at –10 °C to
1 kg of water at 0 °C.

But the heat gain to change 1 kg ice at –10 °C to ice at 0 °C is less than the heat
loss by 5 kg copper from 30 to 0 °C. It means that some amount of ice is melted to
gain the net heat loss of 5 kg copper by changing the temperature from 30 to 0 °C.
It is clear from the above analysis that the final condition of the system is water–ice
and copper mixture at 0 °C temperature.

Let $m =$ mass of ice is melted at 0 °C.

Applying the energy balance equation, heat gain by ice when m kg of ice is melted
at 0 °C = heat loss of 5 kg copper by changing the temperature from 30 to 0 °C.

$$m_1 c_1 [0 - (-10)] + mL = m_2 c_2 (T_2 - 0)$$

$$1 \times 2.057 \times 10 + m \times 333.32 = 5 \times 0.385 \times 30$$

$$20.57 + 333.32 \, m = 57.75$$

or $333.32 \, m = 37.18$

or $m = 0.1115 \, kg$

Mass of ice at 0°C after mixing $= 1 - 0.1125 = 0.8875 \, kg$

(i)   The entropy change of ice block:

$$(\Delta S)_{\text{ice}} = m_1 c_1 \log_e \frac{T_f}{T_1} + \frac{mL}{T_f}$$

where

$$T_f = 0°C = 273 \, K, \text{ and}$$
$$T_1 = -10°C = 263 \, K$$

$$\therefore \quad (\Delta S)_{\text{ice}} = 1 \times 2.057 \log_e \frac{273}{263} + 0.115 \times \frac{333.32}{273}$$
$$= 0.076 + 0.136 = \mathbf{0.212 \, kJ/K}$$

(ii)   The entropy change of copper block:

$$(\Delta S)_{\text{copper}} = m_2 c_2 \log_e \frac{T_2}{T_1}$$

where

$$T_f = 273 \, K, \text{ and}$$
$$T_2 = 30°C = 303 \, K$$

Fig. 6.28 Final state of the system

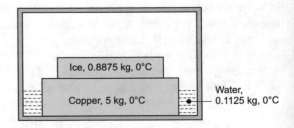

Ice, 0.8875 kg, 0°C

Copper, 5 kg, 0°C

Water, 0.1125 kg, 0°C

$$\therefore \quad (\Delta S)_{\text{copper}} = 5 \times 0.385 \log_e \frac{273}{283} = -0.200 \, \text{kJ/K}$$

(iii)   The entropy change of the universe:

$$\Delta S_{\text{universe}} = (\Delta S)_{\text{ice}} + (\Delta S)_{\text{copper}}$$
$$= 0.212 - 0.200 = 0.012 \, \text{kJ/K} = 12 \, \text{J/K}$$

Problem 6.19: 10 kg of pure ice at –10 °C is separated from 6 kg of pure water at 10 °C in an adiabatic chamber using a thin adiabatic membrane. Upon rupture of the membrane, ice and water mix uniformly at constant pressure. At this pressure, the melting temperature of ice is 0 °C and the latent heat of melting is 335 kJ/kg. The mean specific heat at constant pressure for ice and water are, respectively, 2.1 and 4.2 kJ/kgK (Fig. 6.29).

(i)   Sketch the systems before and after mixing.
(ii)   What is the final equilibrium temperature of the system after the completion of the mixing process?
(iii) Estimate the change of entropy of the universe due to the mixing.
(iv)   Is the final phase of the system solid ice, liquid water, or ice-water mixture?

Solution: Given data:

For ice	For water
$m_{\text{ice}} = 10 \, \text{kg}$	$m_w = 6 \, \text{kg}$
$T_1 = -10°C$	$T_2 = 10°C$

Latent heat : $L = 335 \, \text{kJ/kg}$ $c_w = 4.2 \, \text{kJ/kgK}$
$c_{\text{ice}} = 2.1 \, \text{kJ/kgK}$

The amount of heat required to change 10 kg ice at –10 °C to 10 kg of water at 0 °C,

$$= m_{\text{ice}} c_{\text{ice}} [0 - (-10)] + m_{\text{ice}} L$$
$$= 10 \times 2.1 \times 10 + 10 \times 335 = 210 + 3350 = 3560 \, \text{kJ}$$

and the amount of heat loss to change 6 kg of water at 10 to 0 °C of water,

Fig. 6.29 System before mixing

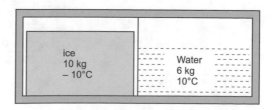

ice
10 kg
– 10°C

Water
6 kg
10°C

$$= m_w c_w (10 - 0) = 6 \times 4.2 \times 10 = 252 \, kJ$$

Thus, the amount of heat required to change 10 kg of ice at –10 °C to 10 kg of water at 0 °C is more than the amount of heat loss to change 6 kg of water at 10 °C to water at 0 °C. But the heat loss to change 6 kg of water at 10 to 0 °C of water is more than the heat gain to change 10 kg of ice at –10 °C to 10 kg of ice at 0 °C. It means that some amount of ice is melted to gain the net heat loss to change 6 kg of water at 10 to 0 °C of water. It is clear from the above analysis that, upon rupture of the membrane, the final condition of the system is the ice-water mixture at 0 °C temperature.

Let $m =$ mass of ice melted at 0 °C.

Applying the energy balance equation, heat gain by ice when m kg of ice is melted at 0 °C = heat loss when 6 kg of water at 10 °C is cooled to 0 °C of water (Fig. 6.30).

$$m_{ice} c_{ice} [0 - (-10)] + mL = m_w c_w (T_2 - 0)$$
$$10 \times 2.1 \times 10 + m \times 335 = 6 \times 4.2 \times 10$$
$$210 + 335m = 252$$

or $335m = 42$

or $m = 0.125 \, kg$

Mass of ice at 0°C after mixing $= 10 - 0.125 = 9.875 \, kg$

(ii) Final temperature of the system after the completion of the mixing process is 0 °C, *i.e.*,

$$T_f = 0°C = 273 \, K$$

(iii) Change of entropy of the universe due to the mixing: $\Delta S_{universe}$

Entropy change of the ice,

$$\Delta S_{ice} = m_{ice} c_{ice} \log_e \frac{T_f}{T_1} + \frac{mL}{T_f}$$

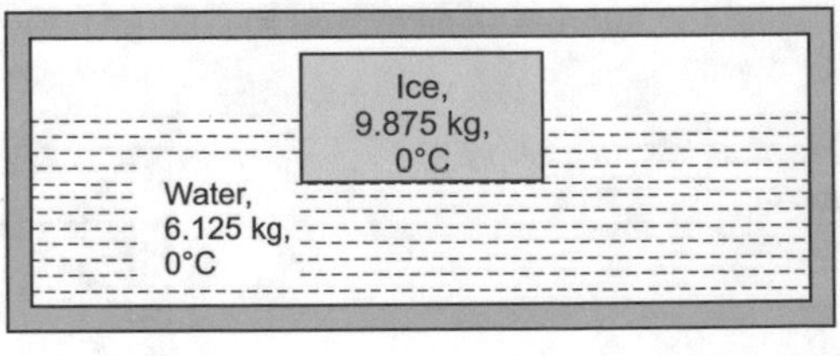

Fig. 6.30 Condition of the system after mixing

Ice,
9.875 kg,
0°C

Water,
6.125 kg,
0°C

where

$$T_1 = -10°C = 263 \text{ K, and}$$
$$T_f = 273 \text{ K}$$

$$\therefore \quad \Delta S_{ice} = 10 \times 2.1 \log_e \frac{273}{263} + \frac{0.125 \times 335}{273}$$
$$= 0.7836 + 0.1533 = 0.9369 \text{ kJ/K}$$

Entropy change of the water,

$$\Delta S_{water} = m_w c_w \log_e \frac{T_f}{T_2}$$

where

$$T_f = 273 \text{ K, and } T_2 = 10°C = 283 \text{ K}$$

$$\therefore \quad \Delta S_{water} = 6 \times 4.2 \log_e \frac{273}{283} = -0.9065 \text{ kJ/K}$$

Change of entropy of the universe:

$$\Delta S_{universe} = \Delta S_{ice} + \Delta S_{water}$$
$$= 0.9369 - 0.9065 = 0.0304 \text{ kJ/K} = 30 \text{ J/K}$$

(iv)   Final condition of the system is **ice-water mixture.**

Summary

1.   The entropy is a property of a system that is a measure of the microscopic disorder within the system. It is defined for a reversible process as

$$dS = \frac{\delta Q}{T}$$

For process 1–2,

$$S_2 - S_1 = \int_1^2 \frac{\delta Q}{T}$$

The SI units of entropy are kJ/K or J/K.

2.   The entropy of a system increases when the system undergoes an irreversible process.
3.   The entropy is not created when a system undergoes a reversible process; entropy is simply transferred from one part of the system to another part.
4.   The entropy of a system in an adiabatic enclosure can never decrease; it increases during an irreversible process and remains constant during a reversible process.
5.   The entropy is a state function, then in undergoing any specific change of state from 1 to 2, the entropy change of the system is the same whether the process is reversible or irreversible.
6.   The entropy change of the universe is equal to the sum of the entropy change of the system and the entropy change of the surroundings.

Mathematically,

$$\Delta S_{\text{universe}} = \Delta S_{\text{system}} + \Delta S_{\text{surrounding}}$$

7.   **Clausius Theorem:** It states that the cycle integral of $\delta Q/T$ for a reversible cycle is equal to zero.

Mathematically,

$$\oint_R \frac{\delta Q}{T} = 0$$

8.   **Clausius Inequality:** According to the Clausius inequality, the cycle integral of $\delta Q/T$ is always less than and equal to zero.

Mathematically,

$$\oint \frac{\delta Q}{T} \leq 0$$

$$\oint \frac{\delta Q}{T} \begin{cases} < 0, & \text{the cycle is irreversible and possible.} \\ = 0, & \text{the cycle is reversible.} \\ > 0, & \text{the cycle is imposible and it violates, the 2nd law of thermodynamics.} \end{cases}$$

9.   **Increase of Entropy Principle.** The entropy of an isolated system can never decrease. This is known as the increase of entropy principle, or simply the entropy principle.
10. **Third Law of Thermodynamics.** It states that the entropy of a pure perfect crystal substance is zero at absolute zero temperature.

Mathematically,

$$S = 0 \text{ at } T = 0\,\text{K} = -273.15°\text{C}$$

11. General equation for the entropy change of an ideal gas.

For unit mass,

$$s_2 - s_1 = c_v \log_e \frac{T_2}{T_1} + R \log_e \frac{V_2}{V_1}$$

$$s_2 - s_1 = c_v \log_e \frac{p_2}{p_1} + c_p \log_e \frac{V_2}{V_1}$$

and

$$s_2 - s_1 = c_p \log_e \frac{T_2}{T_1} - R \log_e \frac{p_2}{p_1}$$

12. Entropy Change during Various Processes:
(i)   Constant volume process (Isometric or Isochoric):

$$s_2 - s_1 = c_v \log_e \frac{T_2}{T_1} - c_v \log_e \frac{p_2}{p_1}$$

(ii) Constant pressure process (Isobaric):

$$s_2 - s_1 = c_p \log_e \frac{V_2}{V_1} = c_p \log_e \frac{T_2}{T_1}$$

(iii) Constant temperature process (Isothermal):

$$s_2 - s_1 = R \log_e \frac{V_2}{V_1} = R \log_e \frac{p_1}{p_2}$$

$$= c_v \log_e \frac{p_2}{p_1} + c_p \log_e \frac{V_2}{V_1}$$

(v)   Reversible adiabatic process:

$$s_2 - s_1 = 0$$
$$s_1 = s_2 \quad i.e., s = C$$

or.

(vi) Polytropic process:

$$s_2 - s_1 = \left(\frac{\gamma - n}{1 - n}\right) c_v \log_e \frac{T_2}{T_1}$$

$$= (\gamma - n) c_v \log_e \frac{V_2}{V_1} = \left(\frac{n - \gamma}{n}\right) c_v \log_e \frac{p_2}{p_1}$$

13.   Combined statement of the first and second laws of thermodynamics.

$TdS = dU + pdV$ for closed system and any process reversible or irreversible

$TdS = dH - Vdp$ for closed system and any process

$\delta Q = dE + \delta W$ for any process and any system

$\delta Q = dU + \delta W$ for any process and closed stationary system

$\delta Q = dU + pdV$ for a reversible process and closed system.

14.   Entropy change for substances such as steam, solids, and liquids.

For wet steam,

$$s_1 = s_f + x_1 s_{fk}$$

where.
s_f = specific entropy of saturated liquid.
s_g = specific entropy of saturated vapour.

$$s_{fg} = s_g - s_f$$

x_1 = dryness fraction.
The entropy change when phase change occurs:

$$ds_{fg} = \frac{\text{latent heat}}{\text{saturatedtemperature}} = \frac{h_{fg}}{T_{sat}}$$

The entropy change of the steam heated from saturated temperature T_{sat} to superheated temperature T_1 per unit mass:

$$ds_{mp} = c_{pv} \log_t \frac{T_1}{T_{sa}}$$

The entropy change of the solid body is heated from temperature T_1 to temperature T_2 per unit mass:

$$ds = c \log_e \frac{T_2}{T_1}$$

where.
c = specific heat of the solid body.

Assignment 1

1. What do you understand by the term 'Entropy'?
2. Define the entropy change of the system, surroundings, and universe.
3. Explain the importance of entropy.
4. Give the criteria of reversibility, irreversibility, and impossibility of a thermo-dynamic cycle.
5. Under what condition $dS = \frac{\delta Q}{T}$? How do you calculate the entropy change if the condition stated above is not satisfied?
6. State the Clausius inequality. What is its practical importance?
7. State and prove the Clausius theorem.
8. Write down the name of a reversible process in which the entropy of an ideal gas will increase during the heat addition.
9. Is it possible to create entropy? Is it possible to destroy it?
10. The entropy of a hot coffee decreases as it cools. Is this a violation of the increase of entropy principle? Explain.
11. Is it possible for an entropy change of a closed system to be zero during an irreversible process? Explain.
12. Is a process that is internally reversible and adiabatic necessarily isentropic? Explain.
13. Show that entropy is a property of a system.
14. Show that when a perfect gas changes from a state p_1, V_1, T_1 to another state p_2, V_2, T_2, the entropy change per unit mass is given by

$$s_2 - s_1 = c_v \log_e \frac{p_2}{p_1} + c_p \log_e \frac{V_2}{V_1}$$

15. Derive an expression for entropy change of the universe.
16. Derive an expression to determine the entropy change during a process in terms of pressure and temperature ratio.
17. $Tds = du + pdv$ is true for which type of process and system.
18. Match **list-I** (Equation) with **list-II** (used for).

LIST-I	LIST-II
(Equation)	(Used for)
1. $\delta Q = dE + \delta W$	(a) Any process, closed system
2. $\delta Q = dU + \delta W$	(b) Only for reversible process, closed system
3. $\delta Q = dU + pdV$	(c) Only for reversible process
4. $\delta Q = TdS$	(d) Any process, any system
5. s $TdS = dU + pdV$	(e) Any process undergone by a closed stationary system

19. Explain the 'increase of entropy principle'.

20. What do you understand by the term entropy transfer? Why is entropy transfer associated with heat transfer and not with work transfer?

21. Prove that the entropy change for unit mass in polytropic process

$$s_2 - s_1 = \frac{\gamma - n}{1 - n} c_v \log_e \frac{T_2}{T_1}$$

22. Derive an expression of the entropy change for unit mass in the polytropic process in terms of pressure ratio.

Assignment-2

1. A 0.25 m^3 of air at a pressure of 10 bar and temperature 300 °C expands to six times its original volume and the final temperature after expansion is 30 °C. Determine the entropy change of air during the process. Take gas constant $R =$ 0.287 kJ/kgK.
   [**Ans.** 0.0863 kJ/K].

2. During a free expansion of 1.5 kg air expands from 2 to 4 m^3 volume in an insulated vessel. Determine the entropy change of (a) the air, (b) the surroundings, and (c) the universe.
   [**Ans.** (a) 0.298 kJ/K (b) 0 (c) 0.298 kJ/K].

3. Determine the entropy change of air, if it is throttled from 1.1 MPa, 30 °C to 2.2 bar.
   [**Ans.** 0.4619 kJ/kgK].

4. A 50 kg block of iron casting at 227 °C is thrown into a large lake that is at a temperature of 12 °C. The iron block eventually reaches thermal equilibrium with lake water. Assuming an average specific heat of 0.45 kJ/kgK for iron, determine

   (a) the entropy change of the iron block,
   (b) the entropy change of the lake water, and
   (c) the entropy change of the universe.
       [**Ans.**(a) −12.647 kJ/K (b) 16.973 kJ/K (c) 4.326 kJ/K].

5. A 20 kg copper block initially at 200 °C is brought into constant with a 20 kg block of iron at 100 °C in an insulated system. Determine the final equilibrium temperature and the total entropy change of this process. Take average specific heats of copper and iron as 0.36 kJ/kg K and 0.44 kJ/kgK, respectively. [**Ans.** 145 °C, 112.34 J/K]

6. A 3 kg of water at 80 °C is mixed adiabatically with 6 kg of water at 25 °C at a constant pressure. Determine the final equilibrium temperature and increase in the entropy of the mixing process. Take specific heat of water as 4.18 kJ/kgK.
   [**Ans.** 43.33 °C, 122 J/K]

7. An insulated vessel of capacity 0.056 m^3 is divided into two compartments A and B by a conducting diaphragm. Each compartment has a capacity of 0.028

m^3. The compartment A contains air at a pressure of 1.5 bar and 25 °C and the compartment B contains air at a pressure of 4.2 bar and 175 °C. Determine

(a) The final equilibrium temperature,
(b) The final pressure on each side of the diaphragm, and
(c) The change in entropy of the system.
   [**Ans.** (*a*) 122.5 °C (*b*) 1.99 bar, 3.70 bar (*c*) 1.81 J/K].

8. An adiabatic vessel contains 2.5 kg of water at 30 °C. By paddle wheel work transfer, the temperature of the water is increased to 55 °C. Determine the entropy change of the universe. Take specific heat of water as 4.2 kJ/kgK. [**Ans.** 0.8324 kJ/K]

9. Air is flowing steadily in an insulated pipe. The pressure and temperature measurements of the air at two stations A and B are given below:

	Station A	Station B
Pressure	130 kPa	100 kPa
Temperature	50°C	13°C

Determine the direction of the flow of air in the pipe. Assume that for air, specific heat $c_p = 1.005$ kJ/kgK and gas constant $R = 0.287$ kJ/kgK.
   [**Ans.** Flow takes place from B to A].

10. A reversible heat engine has heat interaction from three reservoirs at 327, 427, and 527 °C. The heat engine rejects 10 kW to the sink at 47 °C after doing 20 kW of work. The heat supplied by the reservoir at 527 °C is 70% of the heat supplied by the reservoir at 427 °C, determine the exact amount of heat interaction with each high-temperature reservoir.

> **Ans.** Heat supplied to reservoir at 327°C = 120.42 kW
> Heat supplied by reservoir at 427°C = 88.48 kW
> Heat supplied by reservoir at 527°C = 61.48 kW

11. A 1 kg of air at 310 K is heated at constant pressure by bringing it in contact with a hot reservoir at 1150 K. Determine the entropy change of air, hot reservoir, and the universe.
   If the air is heated from 310 to 1150 K by first bringing it in contact with a reservoir at 730 K and then with a reservoir at 1150 K, what will be the entropy change of the universe? [**Ans.** 1.317 kJ/K, – 0.734 kJ/K, 0.583 kJ/K, 0.3718 kJ/K].

12. A certain quantity of an ideal gas is heated in a reversible isothermal process from 1 bar and 40 °C to 10 bar. Find the work done per kg of gas and the entropy change per kg of gas.
   Take $R = 0.287$ kJ/kgK. [**Ans.** –206.84 kJ/kg, – 0.6608 kJ/kgK].

13. A fluid undergoes a reversible adiabatic compression from 4 bar, 0.3 m³ to 0.08 m³ according to the law $pV^{1.25}$ = constant. Determine

    (i)   Change in enthalpy.
    (ii) Change in internal energy.
    (iii) Entropy change.
    (iv) Heat transfer.
    (v)   Work transfer. [**Ans.** (i) 234.80 kJ (ii) 187.84 kJ (iii) 0 (iv) 0 (v) – 187.84 kJ]

14. A 3 kg of water at 80 °C is mixed with 4 kg of water at 15 °C in an insulated system. Determine the entropy change due to the mixing process. [**Ans.** 149.5 J/K]

15. During the isothermal heat addition process of a Carnot cycle, 750 kJ of heat is added to the working fluid from a source at 475 °C. Determine

    (i)   the entropy change of the working fluid,
    (ii) the entropy change of the source, and
    (iii) the total entropy change for the process.
    [**Ans.** (i) 1.0026 kJ/K (ii) –1.0026 kJ/K (iii) 0].

17. Determine the entropy change of the universe owing to each of the following processes:

    (i)   A 0.5 kg copper block at 100 °C is placed in a lake of water at 10 °C.
    (ii) The same block at 10 °C is dropped from a height of 100 m into the lake.
    (iii) Two such blocks at 100 °C and 0 °C, respectively, are joined together.
    [**Ans.** (i) 8.3 J/K (ii) 1.73 J/K (iii) 4.7 J/K].

18. A rigid tank contains a certain amount of nitrogen gas, initially at 120 kPa and 30 °C. Two kilograms of nitrogen gas is then added to the tank. If the final pressure and the final temperature of the gas in the tank are 240 kPa and 30 °C, respectively, what is the volume of the tank?
    [**Ans.** 1.5 m³].

19. A quantity of air has a pressure 7 bar and it occupies a volume of 0.014 m³ at a temperature of 150 °C. The gas is then expanded isothermally to a volume of 0.084 m³. Determine the entropy change of the air.
    [**Ans.** 513.4 J/kgK].

20. An iron block of mass 30 kg and a copper block of mass 40 kg, both initially at 110 °C, are dropped into a lake at 20 °C. As a result of heat transfer, thermal equilibrium is established in a short period of time. Determine the total entropy change of this process, taking the average specific heat of iron and copper to be 0.45 kJ/kgK and 0.386 kJ/kgK, respectively.
    [**Ans.** 1.137 kJ/K].

Fig. 6.31 Schematic for
Q21

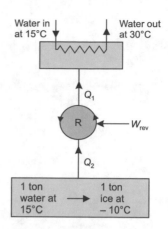

21.   Ice is to be made from water supplied at 15 °C by the process shown in
      Fig. 6.31. The final temperature of the ice is –10 °C, and the final temperature
      of the water that is used as cooling water in the condenser is 30 °C. Determine
      the minimum work required to produce 1 ton of ice. Take the specific heat for
      water as 4.18 kJ/kgK, the specific heat for ice as 2.09 kJ/kgK, and the latent
      of ice as 334.4 kJ/kg.
      [**Ans.** 33,096 kJ].

Chapter 7
Availability and Irreversibility (Exergy and Anergy)

Nomenclature

The following variables are introduced in this chapter:

AE	kJ	Available energy
UE	kJ	Unavailable energy
T_1	K	Temperature of source
T_2	K	Temperature of sink
W_{rev}	kJ	Reversible work
Q_1	kJ	Heat supplied by the source
Q_2	kJ	Heat rejected to the sink
η_{Carnot}	–	Carnot efficiency
η	–	Actual efficiency
T_0	K	Surroundings temperature
p_0	kPa	Surroundings pressure
A	kJ	Availability
a	kJ/kg	Availability per unit mass.
W_{max}	kJ	Maximum/reversible work
$\eta_{rev} = \eta_{Carnot}$	–	Efficiency of reversible heat engine
KE	kJ	Kinetic energy
PE	kJ	Potential energy
W	kJ	Actual work
W_u	kJ	Actual useful work
$W_{u,max}$	kJ	Maximum useful work
w_{rev}	kJ/kg	Reversible work per unit mass
w_u	kJ/kg	Useful work per unit mass.

(continued)

S. Kumar, *Thermal Engineering Volume 1*,
https://doi.org/10.1007/978-3-030-67274-4_7

(continued)

$w_{u,max}$	kJ/kg	Maximum useful work per unit mass
ψ	kJ	Availability function for steady flow system (i.e., open system)
ϕ	kJ	Availability function for closed system (i.e., non-flow system)
U	kJ	Internal energy
u	kJ/kg	Specific internal energy (i.e., internal energy per unit mass)
H	kJ	Enthalpy
h	kJ/kg	Specific enthalpy (i.e., enthalpy per unit mass)
V	m/s and m^3	Velocity and volume
v	m^3/kg	Specific volume
S	kJ/K	Entropy
s	kJ/kgK	Specific entropy
I	kJ	Irreversibility
ΔS_{sys}	kJ/K	Entropy change of the system
ΔS_{surr}	kJ/K	Entropy change of the surroundings
ΔS_{uni}	kJ/K	Universe entropy
η_I	–	First-law efficiency
η_{II}	–	Second-law efficiency
η_{th}	–	Thermal efficiency
COP	–	Coefficient of performance
$(COP)_R$	–	Actual coefficient of performance of the refrigerator
$(COP_{rev})_R$	–	Maximum coefficient of performance of the refrigerator
$(COP)_{HP}$	–	Actual coefficient of performance of the heat pump
$(COP_{rev})_{HP}$	–	Maximum coefficient of performance of the heat pump
F	kJ	Helmholtz function
G	kJ	Gibbs function

7.1   Introduction

Suppose some amount of heat energy (low-grade energy) is in your hand. According to the second law of thermodynamics, the heat energy is not completely converted into useful work (high-grade energy). In this situation, the concept of availability is too much important, and realize the situation that how much of the heat energy (low-grade energy) is converted into useful work (high-grade energy) and how much energy will get lost? How we get maximum work from the system? The part of supplied heat energy converted into useful work is called **available energy**. The remaining part of heat energy that goes waste is termed as **unavailable energy**. The maximum amount of work obtained from the system when it approaches atmospheric condition (p_0, T_0) is called **availability**.

The concept of availability is applied in power plant components like a compressor, a heat exchanger, and a turbine. In case of compressor, we determine the actual available energy required to drive the compressor, the minimum available energy required to drive the compressor in case of ideal condition, and the irreversibility of compressor in actual condition. In case of heat exchanger, we determine irreversibility and second-law efficiency of heat exchanger. In case of turbine, we determine the maximum available energy output in ideal condition, second-law efficiency, the actual available energy output, and irreversibility.

7.2 High-Grade and Low-Grade Energies

The energy can be classified into two categories according to conversion of energy into useful work (*i.e.*, mechanical work).

(i)   High-Grade Energy.
(ii)   Low-Grade Energy.

(iii)   **High-Grade Energy:** The energy which can be converted entirely into useful work is called **high-grade energy**, for examples, Mechanical work, electrical energy, hydraulic energy, tidal energy, etc. The electrical energy is converted into mechanical energy to a maximum 95 to 98% and hydraulic energy [(i) pressure energy (ii) kinetic energy (iii) potential energy] is converted into mechanical work to a maximum 85 to 92%. So a maximum part (more than 85%) of energy converted into mechanical work is considered as high-grade energy.

(iv)   **Low-Grade Energy:** If a small part of energy is converted into mechanical work, it is called **low-grade energy**, for example, Heat energy (sources of heat energy are coal, oil, gas, nuclear, energy, and solar energy.). The heat energy can be converted in mechanical work to a maximum of 30 to 40% with the help of IC engine and gas turbine. So a small part of energy converted into mechanical work is considered as low-grade energy.

High-Grade Energy	Low-Grade Energy
1. Mechanical work,	1. Heat energy (sources of heat energy are coal, oil, gas, nuclear energy, and solar
2. Electrical energy,	
3. Hydraulic energy, and energy)	

(continued)

(continued)

High-Grade Energy	Low-Grade Energy
(i) Pressure energy,	
(ii) Kinetic energy,	
(iii) Potential energy,	
4. Tidal energy	

7.3 Available and Unavailable Energy

According to the second law of thermodynamics, the heat energy is not completely converted into useful work. **The part of heat energy converted into useful work is called available energy. The remaining part of heat energy that goes to the sink (i.e., waste) is called unavailable energy.**

Let us consider two heat engines, one is reversible and other is irreversible as shown in Fig. 7.1.

For reversible heat engine, shown in Fig. 7.1a,

Available energy:

$$AE = W_{\text{rev}} \tag{7.1}$$

Unavailable energy:

$$UE = Q_2 \tag{7.2}$$

Adding Eqs. (7.1) and (7.2), we get

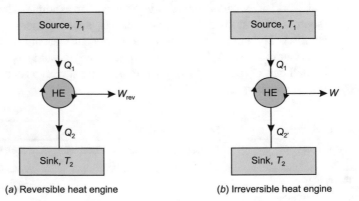

(a) Reversible heat engine   (b) Irreversible heat engine

Fig. 7.1 Available and unavailable energy

$$AE + UE = W_{rev} + Q_2$$
$$AE + UE = Q_1$$

Available energy + unavailable energy = heat supplied.

Carnot efficiency:

$$\eta_{Carnot} = 1 - \frac{T}{T_1}.$$

Actual efficiency:

$$\eta = \frac{W_{rev}}{Q_1}.$$

For reversible heat engine,

$$\eta = \eta_{Carnot}.$$

$$\frac{W_{rev}}{Q_1} = 1 - \frac{T_2}{T_1}$$

or

$$W_{rev} = Q_1\left(1 - \frac{T_2}{T_1}\right).$$

or Available energy:

$$AE = Q_1\left(1 - \frac{T_2}{T_1}\right).$$

For irreversible heat engine, shown in Fig. 7.1b
Available energy:

$$AE = W \tag{7.3}$$

Unavailable energy:

$$UE = Q_{2'} \tag{7.4}$$

Adding Eqs. (7.3) and (7.4), we get

$$AE + UE = W_{rev} + Q_2$$

or

$$AE + UE = Q_1$$

Available energy + unavailable energy = heat supplied.

The available energy in case of reversible heat engine is more than that of irreversible heat engine and unavailable energy is less than that of irreversible heat engine at same heat supply.

7.4 Loss of Available Energy Due to Heat Transfer Through a Finite Temperature Difference

Let heat Q_1 be transferred through a finite temperature difference from the source-I at temperature T_1 and from the source-I' at temperature T'_1 lower than T_1 to the reversible heat engine as shown in Fig. 7.2a and b.

$W =$ Work output when H.E. received Q_1 from source-I,
$W' =$ Work output H.E. received Q_1 from source-I',
$Q_2 =$ Heat rejected to the sink when H.E. connected to source-I, and.
$Q_2' =$ Heat rejected to the sink when H.E. connected to source-I'.
Cycle 1–2–3–4–1 for H.E. when connected to source-I, and.
Cycle 1'–2'–3'–4–1' for H.E. when connected to source-I'.
Heat supplied:

$$Q_1 = T_1 \Delta S = T'_1 \Delta S'.$$

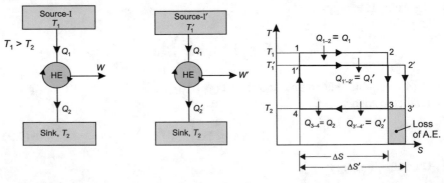

(a) Heat Q_1 is supplied   (b) Heat Q_1 is supplied   (c) T-s diagram for heat transfer through
   to the H.E. from source-I   to the H.E. from Source-I'   a finite temperature difference

Fig. 7.2 Loss of available energy due to heat transfer through a finite temperature difference

Since

$$T_1 > T_1'.$$

∴

$$\Delta S' > \Delta S.$$

and

$$Q_2 = T_2 \Delta S \text{ and } Q_2' = T_2 \Delta S'.$$

∴

$$Q_2' > Q_2$$

$$W = Q_1 - Q_2 = Q_1 - T_2 \Delta S$$

and

$$W' = Q_1 - Q_2' = Q_1 - T_2 \Delta S'.$$

Since $\Delta S' > \Delta S$ at constant Q_1 and T_2.

∴

$$W = W'.$$

Loss of available energy due to irreversible heat transfer through a finite temperature difference between the source and the working fluid during the heat addition process is equal to area under 3'–3.

$$
\begin{aligned}
\text{Loss of available energy} &= \text{Area under } 3' - 3 \\
&= T_2 (\Delta S' - \Delta S) \\
&= T_2 \Delta S' - T_2 \Delta S \\
&= Q_2' - Q_2 = -Q_2 + Q_2' \\
&= Q_1 - Q_2 - (Q_1 - Q_2') = W - W'
\end{aligned}
$$

7.5   Dead State

The dead state is that state at which a system is in thermodynamic equilibrium with its surroundings. At the dead state, the system is at the same temperature and pressure as that of its surroundings, it has no kinetic or potential energy relative to the surroundings, and it does not react with the surroundings. The properties of a system in the dead state are denoted by subscript '0', for example, p_0, T_0, $H0$, S_0, U_0, etc. If the dead state is not specified for numerical problems, we take the values of temperature and pressure as $T_0 = 25$ °C and $p_0 = 101.325$ kPa.

7.6   Availability

The availability (also called exergy) of the system is defined as the maximum useful work obtainable from the system when the system approaches dead state (i.e., atmospheric temperature and pressure) from its initial state. It is denoted by A. The availability is function of both the system and the surroundings. It is important to note that the availability only represents the upper limit on the amount of work a device can deliver without violating the thermodynamics laws, and it does not represent the amount of work that a work-producing device will actually deliver.

The following observations can be made about availability:

(1)   Availability of a system is zero at its dead state since both $T = T_0$ and $p = p_0$, it is not possible for the system to interact with the reference atmosphere and produce useful work. [i.e., after the system and surroundings reach equilibrium, the exergy is zero]

(2)   Availability is never negative. As long as a system is not at its dead state (i.e., at T_0 and p_0), it will always be possible to produce some useful work by interacting with the reference atmosphere.

(3)   Availability is not conserved like mass and energy, and it is destroyed continuously in the actual system.

7.7   Availability of Various Systems

We will discuss the availability in the following cases.

7.7.1   Availability for Heat Engine Cycle

Let Q_1 = heat supplied to the system from heat source at temperature T_1.

Q_2 = heat rejected to the sink a t atmospheric temperature T_0.

Fig. 7.3 Reversible heat engine for Sect. 7.7.1

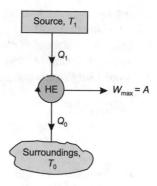

Efficiency of reversible heat engine,

$$\eta_{rev} = \frac{W_{\max}}{Q_1}$$

also

$$\eta_{rev} = 1 - \frac{T_0}{T_1}$$

$$W_{\max} = Q_1 \left(1 - \frac{T_0}{T_1} \right)$$

or

$$\frac{W_{\max}}{Q_1} = 1 - \frac{T_0}{T_1}.$$

or

$$A = W_{\max} = Q_1 \left(1 - \frac{T_0}{T_1} \right)$$

7.7.2 Availability of a Work Reservoir

By definition of a work reservoir, it is of infinite capacity. As such, no final state can be specified for a work reservoir because it never reaches equilibrium with the surroundings (i.e., it never reaches the dead state). Hence, the availability of reservoir is defined as the maximum useful work obtained in relation to a given quantity of energy withdrawn from the reservoir. For example, if we withdraw energy δW from a work reservoir, we talk in terms of the availability of δW rather than the availability of the whole reservoir.

Suppose an amount of mechanical energy δW is withdrawn from a work reservoir. Now, by a reversible process, all this energy can be converted into useful work. Hence, the availability of this energy is 100%. Thus, for energy δW withdrawn from a work reservoir,

Availability:

$$A = \delta W.$$

It is interesting to note that the availability of either kinetic or potential energy is equal to itself, since each is fully available to work. That is,

$$A_{KE} = KE = \frac{1}{2}mV^2$$

and

$$A_{PE} = PE = \text{mgz}.$$

7.7.3 Availability of Heat

Suppose a certain amount of heat δQ_1 is withdrawn from a heat reservoir. What is the availability of this heat?

To answer the above question we should find out the maximum work that can be obtained when this heat δQ_1 is transferred to a cycle. Now, the work of cycle will be maximum when.

(i)   the cycle is reversible, and
(ii)   the heat rejection process of the cycle occurs at temperature T_0 of the atmosphere.

Two distinct cases may arise as follows:

(1) **Heat δQ_1 is withdrawn at a constant temperature T**

This would be the case if the heat is withdrawn from a heat reservoir because the temperature of the reservoir would not change as a result of removing heat δQ_1.

In this case, the reversible cycle will operate between temperatures T_1 and T_0 as shown in Fig. 7.4.

From the first law of thermodynamics,

$$W_{rev} = heat\ added\ to\ cycle - heat\ rejected\ by\ cycle$$

Fig. 7.4 Availability of heat, $T = C$

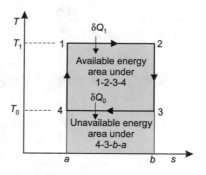

$$= \delta Q_1 - \delta Q_0$$
$$= \int_1^2 T_1 dS + \int_3^4 T_0 dS$$
$$= T_1(S_2 - S_1) + T_0(S_4 - S_3)$$
$$= T_1(S_2 - S_1) - T_0(S_3 - S_4)$$
$$= T_1(S_2 - S_1) - T_0(S_2 - S_1)$$
$$= (T_1 - T_0)(S_2 - S_1) \quad | \because S_3 - S_4 = S_2 - S_1 \qquad (7.5)$$

For a reversible cycle, $\frac{\delta Q_1}{T_1} = \frac{\delta Q_0}{T_0}$.

or

$$\delta Q_0 = T_0 \frac{\delta Q_1}{T_1}.$$

Substituting the value of δQ_0 in Eq. (7.5), we get

$$W_{rev} = \delta Q_1 - T_0 \frac{\delta Q_1}{T_1} = \delta Q_1 \left(1 - \frac{T_0}{T_1}\right)$$

Hence, availability of heat δQ_1,

$$A = W_{rev} = \delta Q_1 \left(1 - \frac{T_0}{T_1}\right)$$

(2) **Heat δQ_1 is withdrawn not at constant temperature**

This would be the case when heat is withdrawn the temperature of the source will decrease T_2 to T_1 and that of the cycle will increase T_1 to T_2.

However, at any instant during the heat transfer from the source to cycle fluid, the temperature of the two must be equal. Otherwise the heat transfer process will not be reversible.

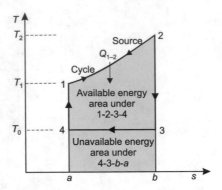

Fig. 7.5 Availability of heat, $T \neq C$

To obtain the maximum work from the heat Q_{1-2}, a reversible cycle should be devised so that the heat rejection takes place at atmospheric temperature T_0. This can be done by using two reversible adiabatic processes 2–3 and 4–1 to complete the cycle as shown in Fig. 7.5. The work of this cycle equals the availability of heat Q_{1-2} (Fig. 7.6).

From the first law of thermodynamics,

$$
\begin{aligned}
W_{\text{rev}} &= \text{heat added to cycle} - \text{heat rejected by cycle} \\
&= Q_{1-2} - Q_{3-4} \\
&= Q_{1-2} - T_0(S_3 - S_4) \\
&= Q_{1-2} - T_0(S_2 - S_1)
\end{aligned}
$$

Hence, availability of heat Q_{1-2},

$$
A = W_{rev} = Q_{1-2} - T_0(S_2 - S_1)
$$

where $(S_2 - S_1)$ represents the increase in entropy of the cycle fluid during the heat addition process.

Unavailable energy:

$$
\begin{aligned}
UE &= Q_{1-2} - A \\
&= Q_{1-2} - Q_{1-2} + T_0(S_2 - S_1) \\
&= T_0(S_2 - S_1)
\end{aligned}
$$

Fig. 7.6   Available energy

7.8   Useful Work

The work done by a system is not always entirely in a usable form. For example, when gas in a piston-cylinder device is expanding, part of the work done by the gas is used to push the atmospheric air out of the way of the piston. This work cannot be recovered and utilized for any useful purpose. This is shown in Fig. 7.8. The pressure inside the system p is resisted by a force F and the atmospheric pressure p_0.

For equilibrium condition, $pA = F + p_0 A$.

where A is the cross-sectional area of the piston.

If the piston moves a distance l, the work done by the various components shown in Fig. 7.8 is

Fig. 7.7 Available energy crisis

Fig. 7.8 Forces acting on a piston

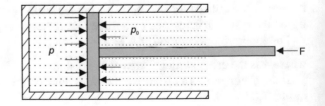

$$pAl = Fl + p_0Al$$
$$pdV = Fl + p_0dV$$

where

$$dV = Al, \text{ changeinvolume,}$$
$$pdV = W, \text{ workdonebythefluidinthesystem,}$$
$$Fl = W_u, \text{ work done against the resisting force, i.e,useful work,}$$

and

$$p_0dV = W_{surr}, \text{ workagainstthesurroundings.}$$

$\therefore$

$$W = W_u + W_{surr}.$$

or

$$W_u = W - W_{surr}.$$

The difference between the actual work W and the surroundings work W_{surr} is called the useful work or actual useful work W_u.

For reversible system,
Maximum useful work:

$$\boldsymbol{W_{u,\max} = W_{rev} - W_{surr}}.$$

For a steady flow process,
Surroundings work:

$$W_{surr} = 0.$$

$\therefore$ Useful work:

$$W_u = W.$$

and maximum useful work:

$$\boldsymbol{W_{u,\max} = W_{rev}}.$$

7.9   Reversible Work

Let

$$m_i, h_i, V_i, Z_i = \text{mass flow rate, specific enthalpy, velocity, datum}$$
$$\text{head at inlet of the system as shown in Fig. 7.9.}$$

and

$$m_e, h_e, V_e, Z_e = \text{mass flow rate, specific enthalpy, velocity, datum}$$
$$\text{head at inlet of the system as shown in Fig.7.9.}$$

$$W = \text{workdonebythesystem}$$
$$\delta Q = \text{heattransferbythesystemandsupply}$$
$$\text{tothereversibleheatengine,}$$
$$W_e = \text{workproducedbytheheatengine,}$$
$$\delta Q_0 = \text{heatrejectedbytheheatenginetothesurrounding,}$$
$$T_0, p_0 = \text{temperatureandpressureofthesurroundings,}$$
$$d\left(U + \frac{mV^2}{2} + mgz\right)_{sys} = \text{energychangeofthesystem.}$$

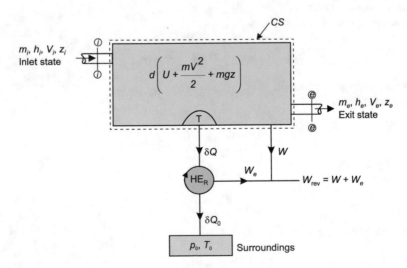

Fig. 7.9 Reversible work of an open system which transfer heat with the surroundings via a small reversible heat engine

By application of the first law of thermodynamics, we get

$$m_i\left(h_i + \frac{V_i^2}{2} + gz_i\right) - m_e\left(h_e + \frac{V_e^2}{2} + gz_e\right) - \delta Q - W = d\left[U + \frac{mV^2}{2} + mgz\right]_{sys}$$

or

$$W = m_i\left(h_i + \frac{V_i^2}{2} + gz_i\right) - m_e\left(h_e + \frac{V_e^2}{2} + gz_e\right) - \delta Q - d\left[U + \frac{mV^2}{2} + mgz\right]_{sys}$$

(7.6)

For reversible heat engine, $\eta_{rev} = 1 - \frac{T_0}{T}$.
also

$$\eta_{rev} = \frac{W_e}{\delta Q}.$$

$\therefore$

$$\frac{W_e}{\delta Q} = 1 - \frac{T_0}{T}.$$

or

$$W_e = \left(1 - \frac{T_0}{T}\right)\delta Q$$

$$W_e = \delta Q - T_0\frac{\delta Q}{T}$$

(7.7)

Now, the process is reversible, entropy change of the system will be equal to the net entropy transfer.

$\therefore$

$$dS_{sys} = m_i s_i - m_e s_e - \frac{\delta Q}{T}.$$

or

$$\frac{\delta Q}{T} = m_i s_i - m_e s_e - dS_{sys}.$$

Substituting the above value of $\frac{\delta Q}{T}$ in Eq. (7.7), we get

$$W_e = \delta Q - T_0(m_i s_i - m_e s_e - dS_{sys})$$

(7.8)

The reversible work is the sum of the work done by the system W and the work done by reversible heat engine W_e.

$\therefore$

$$W_{rev} = W + W_e \tag{7.9}$$

Substituting the values of W from Eq. (7.6) and W_e from Eq. (7.8) in above Eq. (7.9), we get

$$W_{rev} = m_i \left(h_i + \frac{V_i^2}{2} + g z_i \right) - m_e \left(h_e + \frac{V_e^2}{2} + g z_e \right) - \delta Q - d \left[U + \frac{m V^2}{2} + mgz \right]_{sys}$$

$$+ \delta Q - T_0 (m_i s_i - m_e s_e - dS_{sys})$$

$$W_{rev} = m_i \left(h_i + \frac{V_i^2}{2} + g z_i \right) - m_e \left(h_e + \frac{V_e^2}{2} + g z_e \right) - d \left[U + \frac{m V^2}{2} + mgz \right]_{sys}$$

$$- T_0 (m_i s_i - m_e s_e - dS_{sys})$$

$$W_{rev} = m_i \left(h_i - T_0 s_i + \frac{V_i^2}{2} + g z_i \right) - m_e \left(h_e - T_0 s_e + \frac{V_e^2}{2} + g z_e \right) - d \left[U - T_0 S + \frac{m V^2}{2} + mgz \right]_{sys} \tag{7.10}$$

Equation (7.10) is general expression for the reversible work of an open system which transfers heat with the surroundings via a small reversible heat engine.

7.9.1 Reversible Work in a Steady Flow Process

For a steady flow process,

$$m_i = m_e = m$$

and

$$d \left[U - T_0 S + \frac{m V^2}{2} + mgz \right]_{sys} = 0.$$

Equation (7.10) becomes to $W_{rev} = m \left(h_i - T_0 s_i + \frac{V_i^2}{2} + g z_i \right) - m \left(h_e - T_0 s_e + \frac{V_e^2}{2} + g z_e \right)$.

For unit mass,

$$w_{rev} = \left(h_i - T_0 s_i + \frac{V_i^2}{2} + g z_i \right) - \left(h_e - T_0 s_e + \frac{V_e^2}{2} + g z_e \right)$$

$$w_{rev} = (h_i - h_e) - T_0 (s_i - s_e) + \frac{(V_i^2 - V_e^2)}{2} + g(z_i - z_e) \tag{7.12}$$

If changes in kinetic energy and potential energy are neglected, Eq (7.11) becomes

$$w_{rev} = (h_i - h_e) - T_0 (s_i - s_e)$$

Let subscripts $i = 1$ and $e = 2$.

$\therefore$

$$w_{rev} = (h_1 - h_2) - T_0(s_1 - s_2).$$

Maximum useful work:

$$w_{u,\max} = w_{rev} = (h_1 - h_2) - T_0(s_1 - s_2).$$

where surroundings work:

$$w_{\text{surr}} = 0 \text{ for a steady flow process.}$$

7.9.2 Availability Function for a Steady Flow Process:

The availability function for a steady flow process is given by

$$\Psi = H - T_0 S + \frac{mV^2}{2} + mgz$$

If kinetic and potential energies are neglected, then

$$\Psi = H - T_0 S$$

For unit mass $\Psi = h - T_0 s$.

The availability function at inlet state 1 and exit state 2 given by

$$\Psi_1 = h_1 - T_0 s_1 \text{ and } \Psi_2 = h_2 - T_0 s_2$$

$\therefore$ Reversible/maximum work: $w_{rev} = \Psi_1 - \Psi_2$, change in availability.

$$w_{rev} = (h_1 - T_0 s_1) - (h_2 - T_0 s_2)$$
$$w_{rev} = (h_1 - h_2) - T_0(s_1 - s_2)$$

7.9.3 Reversible Work in a Closed System

We know that there is no mass transfer across the boundary of a closed system.

$\therefore m_i = m_e = 0$.

Equation (v) becomes $W_{rev} = -d\left[U - T_0S + \frac{mV^2}{2} + mgz\right]_{sys}$.

For a change of state of the system from the initial state 1 to the final state 2,

$$W_{rev} = -\left[(U_2 - U_1) - T_0(S_2 - S_1) + \frac{m}{2}(V_2^2 - V_1^2) + mg(z_2 - z_1)\right]$$

$$or \quad W_{rev} = (U_1 - U_2) - T_0(S_1 - S_2) + \frac{m}{2}(V_1^2 - V_2^2) + mg(z_1 - z_2)$$

For unit mass $w_{rev} = (u_1 - u_2) - T_0(s_1 - s_2) + \frac{(V_1^2 - V_2^2)}{2} + g(z_1 - z_2)$ (7.13)

If changes in kinetic and potential energies are neglected, Eq. (7.13) becomes

$$w_{rev} = (u_1 - u_2) - T_0(s_1 - s_2)$$

Maximum useful work: $w_{u,max} = w_{rev} - p_0(v_2 - v_1)$

$$= w_{rev} + p_0(v_1 - v_2)$$

where p_0 = atmosphere pressure.

v_2 = final specific volume of the system.

v_1 = initial specific volume of the system

$$w_{u,max} = (u_1 - u_2) - T_0(s_1 - s_2) - p_0(v_2 - v_1)$$

$$= (u_1 - u_2) - T_0(s_1 - s_2) + p_0(v_1 - v_2)$$

7.9.4 Availability Function for a Closed System: ϕ

The availability function for a closed system is given by

$$\phi = U + p_0V - T_0S$$

Availability function at initial state 1 and final state 2 is given by

$$\phi_1 = U_1 + p_0V_1 - T_0S_1 \text{ and } \phi_2 = U_2 + p_0V_2 - T_0S_2$$

Maximum useful work:

$$W_{u,max} = \phi_1 - \phi_2$$

$$= (U_1 + p_0V_1 - T_0S_1) - (U_2 + p_0V_2 - T_0S_2)W_{u,max}$$

$$= (U_1 - U_2) + p_0(V_1 - V_2) - T_0(S_1 - S_2)$$

where $V_1 =$ initial volume of the system,
   $V_2 =$ final volume of the system.
   For unit mass

$$w_{u,\max} = (u_1 - u_2) + p_0(v_1 - v_2) - T_0(s_1 - s_2)$$
$$= w_{rev} + p_0(v_1 - v_2)$$

7.10   Heat Transfer with Other System

Let us consider heat transfer of a system with other bodies at different temperature as well as with its surrounding as shown in Fig. 7.10. The kinds of process can be handled easily by treating the main system and the individual bodies as separate systems and then superimposing the result.
   That is,

$$\psi_{Total} = \phi_{sys} + \phi_{body}$$

and

$$W_{\max,Total} = W_{rev,sys} + W_{rev,body}$$

Consider a closed system receiving heat Q_R from a reservoir at T_R. Let the system be located in an environment at temperature T_0 and let it changes its state from state 1 to state 2 owing to the heat transfer process, the reversible work can be expressed as

$$W_{rev} = (U_1 - U_2) - T_0(S_1 - S_2) - Q_R\left(1 - \frac{T_0}{T_R}\right)$$
$$= mc_v(T_1 - T_2) - T_0(S_1 - S_2) - Q_R\left(1 - \frac{T_0}{T_R}\right)$$

Fig. 7.10 Heat transfer of a system with the other bodies as well as with the surroundings

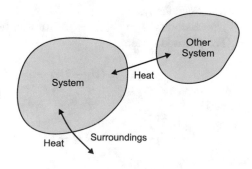

The irreversibility relation remains the same. But the ΔS_{uni} becomes

$$\Delta S_{uni} = (S_2 - S_1)_{sys} + \frac{Q_{sur}}{T_0} + \frac{Q_R}{T_R}$$

Note that the entropy change of the reservoir is included in the universe entropy. It is important to note that Q_R in the above two equations is to be assigned the proper sign. That is, it is $- Q_R$ for the hot reservoir (source) and $+ Q_R$ for the cold reservoir (sink).

For non-flow process:

$$W_{rev} = mc_v(T_1 - T_2) - T_0(S_1 - S_2) - Q_R\left(1 - \frac{T_0}{T_R}\right)$$

$$W_{u,rev} = W_{rev} + p_0(V_1 - V_2)$$

$$= mc_v(T_1 - T_2) + p_0(V_1 - V_2) - T_0(S_1 - S_2) - Q_R\left(1 - \frac{T_0}{T_R}\right)$$

$$= (\phi_1 - \phi_2) - Q_R\left(1 - \frac{T_0}{T_R}\right)$$

For steady flow process:

$$W_{rev} = mc_p(T_1 - T_2) - T_0(S_1 - S_2) - Q_R\left(1 - \frac{T_0}{T_R}\right)$$

$$= W_{u,rev} - Q_R\left(1 - \frac{T_0}{T_R}\right)$$

7.11   Irreversibility

The irreversibility (also called anergy) of the non-flow work-producing devices is the difference between the reversible useful work $W_{u,\ rev}$ and the useful work W_u which is due to the presence of irreversibilities during the process.

Mathematically,

Irreversibility: $I = W_{u,rev} - W_u$ for non-flow work-producing devices.

For a closed system,

$$W_{u,rev} = (U_1 - U_2) + p_0(V_1 - V_2) - T_0(S_1 - S_2)$$

and

$$W_u = \delta Q - dU + p_0(V_1 - V_2)$$

$$= \delta Q - (U_2 - U_1) + p_0(V_1 - V_2)$$

$\therefore$

$$I = (U_1 - U_2) + p_0(V_1 - V_2) - T_0(S_1 - S_2) - \delta Q + (U_2 - U_1) - p_0(V_1 - V_2)$$
$$I = -T_0(S_1 - S_2) - \delta Q$$
$$I = T_0(S_2 - S_1) - T_0\frac{\delta Q}{T_0}$$
$$I = T_0\Delta S_{sys} + T_0\Delta S_{surr}$$
$$I = T_0(\Delta S_{sys} + \Delta S_{surr})$$
$$I = T_0\Delta S_{uni}$$

The irreversibility of the system is also defined as **the product of absolute temperature of the surroundings and universe entropy.**

For reversible process, the actual and reversible work terms are identical (i.e., $\Delta S_{uni} = 0$), and thus the irreversibility is zero. The irreversibility is a positive quantity for all actual (irreversible) processes.

$\therefore$

$$I \geq 0$$
$$I \begin{cases} > 0 & \text{for irreversible process} \\ = 0 & \text{for reversible process} \end{cases}$$

The irreversibility can be viewed as the wasted work or the lost opportunity to do work. So, the irreversibility is also called the lost work W_L.

$\therefore$ Lost work:

$$W_L = I = W_{u,rev} - W_u$$
$$= T_0\Delta S_{uni}$$

For a steady flow process,
Irreversibility:

$$I = W_{rev} - W \quad \text{for work - producing device (turbine)}$$
$$= T_0\Delta S_{uni}$$

Irreversibility: $I = W - W_{rev}$ for work - consuming device (compressor).

The exergy accounts for the irreversibility of a process is due to increase in entropy. Exergy is always destroyed when a process is irreversible. This destruction is proportional to the universe entropy (i.e., entropy change of the system + entropy change of its surroundings). The destroyed exergy has been called anergy. For a reversible process, exergy and energy are interchangeable terms, and there is no anergy.

7.12 Second-Law Efficiency

To measure the performance of any process, device, or system, we make use of the concept of efficiency. The thermal efficiency (also called first-law efficiency) widely used in thermodynamics is usually based on the concept of energy, in which no attempt is made to distinguish low-grade energy from high-grade energy. A simple example is the thermal efficiency of a heat engine, which is defined as the ratio of net work output to the amount of heat supplied (refer Fig. 7.11). In this definition, heat and work are given the same weightage. So, the first law efficiency does not give an accurate measure of thermodynamic performance. To overcome this deficiency, we defined a second-law efficiency η_{11} as the ratio of the actual work output to the maximum (reversible) work output of the work-producing devices.

Mathematically,

Second-law efficiency:

$$\eta_{II} = \frac{actual\ work : W}{\max imum\ work : W_{rev}}$$

$$\eta_{II} = \frac{W}{W_{rev}} \quad for\ work - producing\ devices$$

$$\eta_{II} = \frac{W}{W_{rev}} \times \frac{Q_1}{Q_1} = \frac{W}{Q_1 \times W_{rev}/Q_1}$$

$$\eta_{II} = \frac{\eta_{th}}{\eta_{rev}} = \frac{\eta_1}{\eta_{Carnot}}$$

where

$$\eta_{rev} = \eta_{Carnot} = \frac{W_{rev}}{Q_1}$$

and

$$\eta_{th} = \eta_1 = \frac{W}{Q_1}$$

Fig. 7.11 Heat engine

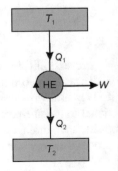

Fig. 7.12 Refrigerator

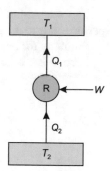

The second-law efficiency is also defined as the ratio of actual thermal efficiency to the reversible (Carnot) efficiency under the same condition.

It is important to note that the upper limit of the second-law efficiency is 100%, which corresponds to the ideal case with no exergy destruction.

For work-consuming devices

The second-law efficiency for work-consuming devices (such as compressors, refrigerators, and heat pumps) is the ratio of the minimum (reversible) work input to the actual work input (refer Fig. 7.12).

Mathematically,

Second-law efficiency:

$$\eta_{II} = \frac{minimum\ work\ input}{actual\ work\ input}$$
$$\eta_{II} = \frac{W_{rev}}{W}$$

For a refrigerator (refer Fig. 7.12),

$$\eta_{II} = \frac{W_{rev}}{W} \times \frac{Q_2}{Q_2}$$

$$\eta_{II} = \frac{Q_2}{W \times Q_2/W_{rev}}$$

$$\eta_{II} = \frac{(COP)_R}{(COP_{rev})_R}$$

where

$$(COP)_R = \frac{Q_2}{W}, \quad \text{actual coefficient of performance of the refrigerator}$$

and

$$(COP_{rev})_R = \frac{Q_2}{W_{rev}}, \quad \text{maximum coefficient of performance of the refrigerator}$$

The second-law efficiency is also defined as the ratio of actual coefficient of performance to the maximum (reversible) coefficient of performance.

Similarly for heat pump,

Second-law efficiency:

$$\eta_{II} = \frac{W_{rev}}{W} = \frac{(COP)_{HP}}{(COP_{rev})_{HP}}$$

7.13   Helmholtz Function: F

The Helmholtz function is defined as the difference between the internal energy and product of absolute temperature and entropy. It is denoted by F.

Mathematically, it is defined as.

Helmhotz function:

$$F = U - TS \tag{7.14}$$

The Helmholtz function F is also called Helmholtz free energy.

For a system undergoing a change of state from state 1 to state 2, Eq. (7.14) gives

$$
\begin{aligned}
F_2 - F_1 &= (U_2 - T_2 S_2) - (U_1 - T_1 S_1) \\
&= U_2 - U_1 - (T_2 S_2 - T_1 S_1)
\end{aligned}
\tag{7.15}
$$

If the system is closed, the first law gives

$$Q_{1-2} = (U_2 - U_1) + W_{1-2}$$

or

$$U_2 - U_1 = Q_{1-2} - W_{1-2} \tag{7.16}$$

Substituting the value of $(U_2 - U_1)$ from Eqs. (7.16) in (7.15), we get

$$F_2 - F_1 = Q_{1-2} - W_{1-2} - (T_2 S_2 - T_1 S_1)$$

If the process is isothermal, *i.e.*, $T_1 = T_2 = T$

$$F_2 - F_1 = Q_{1-2} - W_{1-2} - T(S_2 - S_1) \tag{7.17}$$

We know that for reversible and irreversible process.

$$dS \geq \frac{\delta Q}{T}$$

or

$$T dS \geq \delta Q.$$

For process 1–2,

$$T(S_2 - S_1) \geq Q_{1-2} \quad \because T = C.$$

or

$$Q_{1-2} \leq T(S_2 - S_1)$$

Hence, Eq. (7.17) becomes as

$$F_2 - F_1 \leq -W_{1-2}$$

The equality can be written as

$$F_2 - F_1 + T \Delta S_{irr} = -W_{1-2} \qquad (7.18)$$

and thus, during a reversible isothermal process, for which $\Lambda S_{irr} - 0$, the amount of work done by the system W_{max} is equal to the decrease in the value of the Helmholtz free energy. Furthermore, for an isothermal process conducted at constant volume, which necessarily, does not perform p–V work, Eq. (7.18) gives

$$F_2 - F_1 + T \Delta S_{irr} = 0$$

For an increment of such a process,

$$dF + T dS_{irr} = 0$$
$$or \quad dF = -T dS_{irr}$$

As dS_{irr} is always positive during a spontaneous process, it is thus seen that the Helmholtz function F decreases during spontaneous process, and as $dS_{irr} = 0$ for a reversible process, equilibrium requires that $dF = 0$.

Thus in a closed system held at constant temperature and volume, the Helmholtz free energy can only decrease or remain constant, and equilibrium is attained in such a system when Helmholtz function F achieves its minimum value. The Helmholtz function thus provides a criterion for equilibrium in a system at constant temperature and constant volume.

7.14   Gibbs Function: G

The Gibbs function is defined as the difference between the enthalpy and product of absolute temperature and entropy. It is denoted by G.

Mathematically, it is defined as.

Gibbs function:

$$G = H - TS \qquad (7.19)$$

The Gibbs function G is also called Gibbs free energy.

For a system undergoing a change of state from state 1 to state 2, Eq. (7.19) gives

$$G_2 - G_1 = (H_2 - T_2 S_2) - (H_1 - T_1 S_1)$$
$$= H_2 - H_1 - (T_2 S_2 - T_1 S_1)$$

where

$$H_2 = U_2 + p_2 V_2, \text{ and}$$
$$H_1 = U_1 + p_1 V_1 \qquad (7.20)$$
$$\therefore \quad G_2 - G_1 = (U_2 - U_1) + (p_2 V_2 - p_1 V_1) - (T_2 S_2 - T_1 S_1)$$

If the system is closed, the first law gives

$$Q_{1-2} = (U_2 - U_1) + W_{1-2} \qquad (7.21)$$

Substituting the value of $(U_2 - U_1)$ from (7.21) in Eq (7.20), we get

$$G_2 - G_1 = Q_{1-2} - W_{1-2} + (p_2 V_2 - p_1 V_1) - (T_2 S_2 - T_1 S_1)$$

If the process is isothermal, $i.e.$, $T_1 = T_2 = T$, the temperature of the heat reservoir which supplies or withdraws heat from the system, and also if $p_1 = p_2 = p$, the constant pressure at which the system undergoes a change in volume, then

$$G_2 - G_1 = Q_{1-2} - W_{1-2} + p(V_2 - V_1) - T(S_2 - S_1) \qquad (7.22)$$

In the expression for the first law the work W_{1-2} is the total work done on or by the system during.

the process. Thus if the system performs electrical or chemical work in addition to the work of expansion against the external pressure, these work terms are included in W_{1-2}. Thus W_{1-2} can be expressed as

$$W_{1-2} = W + p(V_2 - V_1) \qquad (7.23)$$

where $p\,(V_2 - V_1) =$ displacement work done by the change in volume at the constant pressure p, and $W =$ sum of all of the non-p–V forms of work done.

Substituting the value of W_{1-2} from Eq. (7.23) to Eq. (7.22), we get

$$
\begin{aligned}
G_2 - G_1 &= Q_{1-2} - W - p(V_2 - V_1) + p(V_2 - V_1) - T(S_2 - S_1) \\
G_2 - G_1 &= Q_{1-2} - W - T(S_2 - S_1)
\end{aligned}
\tag{7.24}
$$

We know that for reversible and irreversible process

$$
dS \geq \frac{\delta Q}{T}
$$

or

$$
T\,dS \geq \delta Q
$$

For process 1–2,

$$
T(S_2 - S_1) \geq Q_{1-2}
$$

or

$$
Q_{1-2} < T(S_2 - S_1)
$$

Hence, Eq. (7.24) becomes as

$$
G_2 - G_1 \leq -W
$$

The equality can be written as

$$
G_0 - G_1 + T\Delta S_i = -W
$$

Thus, for an isothermal, isobaric process, during which no form of work other than p–V work (i.e., displacement work) is performed, i.e., $W = 0$

$$
\therefore \quad G_2 - G_1 + T\Delta S_{im} = 0
$$

Such a process can only occur spontaneously (with a consequent increase in entropy) if the Gibbs free energy decreases. As $dS_{irr} = 0$ is a criterion for thermodynamic equilibrium, then an increment of an isothermal-isobaric process (i.e., $T = c$, $p = c$) occurring at equilibrium requires that.

$dG = 0$.

Thus, for a system undergoing a process at constant temperature and constant pressure, the Gibbs free energy can only decrease or remain constant, and the attainment of equilibrium in the system coincides with the system having the minimum value of G consistent with the values of T and p.

Note: The Helmholtz function $F = U - TS$ and Gibbs function $G = H - TS$ are composite properties of the system only, whereas availability function for closed and open systems.

$\phi = U - T_n S$ and $\Psi = H - T_0 S$ are also composite properties but they are composite properties of the system and surroundings.

Problem 7.1 A heat engine receives heat 320 kJ from a source at 700 K and rejects the waste heat to the sink at 305 K. If the work output of the engine is 98 kJ, determine the reversible work and the irreversibility for this process.

Solution Given data:

$$Q_1 = 320 \, \text{kJ}$$
$$T_1 = 700 \, \text{K}$$
$$T_2 = 305 \, \text{K}$$

Work output of the heat engine:

$$W = 98 \, \text{kJ}.$$

From Fig. 7.13b,
Reversible efficiency:

$$\eta_{rev} = 1 - \frac{T_2}{T_1}$$

also

Fig. 7.13 Schematic diagrams for Problem 7.1

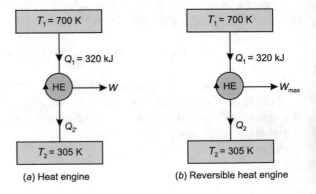

(a) Heat engine   (b) Reversible heat engine

$$\therefore \quad \eta_{rev} = \frac{W_{\max}}{Q_1}$$

or $W_{\max} = \; = \; = \mathbf{180.57\ kJ.}$

Irreversibility: $I = W_{\max} - W = 180.57 - 98 = \mathbf{82.57\ kJ.}$

Problem 7.2 A thermal reservoir at 1000 K can supply heat at steady rate of 2000 kJ/s. Determine the availability of heat if $T_0 = 15\ °C$. What is the unavailable energy of the heat supply?

Solution Given data (refer Fig. 7.14):

$$T_1 = 1000\ K$$
$$Q_1 = 2000\ kJ/s$$
$$T_0 = 15°C = 288K$$

The availability is nothing but the amount of work that a reversible heat engine operating between the source temperature T_1 and surroundings temperature T_0 can produce.

$$\therefore \quad \eta_{rev} = 1 - \frac{T_0}{T_1}$$

also

$$\eta_{rev} = \frac{W_{\max}}{Q_1}$$

$$\therefore \quad \frac{W_{\max}}{Q_1} = 1 - \frac{T_0}{T_1}$$

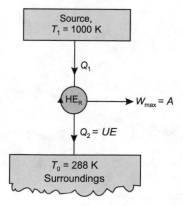

Fig. 7.14 Schematic for Problem 7.2

or

$$W_{\max} = Q_1\left(1 - \frac{T_0}{T_1}\right)$$
$$= 2000\left(1 - \frac{288}{1000}\right)$$
$$= 1424 \text{ kJ/s}$$

Therefore, the availability

$$A = W_{\max} = \mathbf{1424kJ/s}$$

By energy balance, we get

$$Q_1 = W_{\max} + UE$$

or Unavailable energy:

$$UE = Q_1 - W_{\max} = 2000 - 1424 = \mathbf{576kJ/s}$$

Also, unavailable energy:

$$UE = T_0\frac{Q_1}{T_1} = T_0\Delta S_1 = 288 \times \frac{2000}{1000} = \mathbf{576kJ/s}$$

Problem 7.3 A metal piece of 1 kg mass with constant specific heat of 0.4 kJ/kgK is cooled from 200 °C to 100 °C by transferring heat to the surroundings air at 25 °C. Determine the reversible work and irreversibility for this process.

Solution Given data:
Mass of metal:

$$m = 1\,\text{kg}.$$

Specific heat:

$$c = 0.4 \text{ kJ/kgK}.$$

Initial temperature of metal:

$$T_1 = 200°C = 478\,\text{K}.$$

Final temperature of metal:

$$T_2 = 100°C = 373\,\text{K}.$$

Surroundings temperature:

$$T_0 = 25°C = 298 \, \text{K}.$$

Heat transferred from the metal to the surroundings:

$$Q = mc(T_1 - T_2) = 1 \times 0.4 \times (473 - 373) = 40 \, \text{kJ}$$

That is, 40 kJ of heat is removed from the metal.

Now, the reversible work that can be done by a heat engine supplied with 40 kJ of heat and rejecting waste heat to the surrounding at $T_0 = 298$ K is

$$W_{rev} = Q - T_0(S_1 - S_2)$$

where

$(S_1 - S_2) = $ entropy change associated with the cooling of the metal piece

$$- mc \log_e \frac{T_1}{T_2} = 1 \times 0.4 \log_e \frac{473}{373} = 0.095 \, \text{kJ/K}$$

$$\therefore \qquad W_{rev} = 40 - 298 \times 0.095 = \textbf{11.69 kJ}$$

Irreversibility: $\qquad I = W_{rev} - W$

In this case, the system is rigid, the actual work W is zero.

$$\therefore \quad I = W_{rev} = \textbf{11.69kJ}$$

Note that the reversible work and irreversibility are identical for processes that involve no actual work.

Alternative method.

Irreversibility: $I = T_0 \, \Delta S_{uni}$.

where

$$\Delta S_{uni} = \Delta S_{sys} + \Delta S_{surr}$$

$$= mc \log_e \frac{T_2}{T_1} + \frac{Q_{surr}}{T_0} = 1 \times 0.4 \log_e \frac{373}{473} + \frac{40}{298}$$

$$= -0.095 + 0.134228 = 0.039228 \, \text{kJ/K}$$

$$\therefore \quad I = 298 \times 0.039228 = \textbf{11.69kJ}$$

Problem 7.4 Fuel and air are burned at atmospheric pressure in a furnance and a temperature of 1500 °C is attained. Determine the availability of heat in the furance gas per kg. Assume that for the furance gas $c_p = 1.045$ kJ/kgK. Take $T_0 = 15$ °C.

Solution Given data:

Temperature of gas in a furance:

$$T_1 = 1500°C = 1773 \text{ K}.$$

Presume of gas in a furance:

$$p_1 = p_0.$$

Specific heat of the gas:

$$c_p = 1.045 \text{ kJ/kgK}.$$

Temperature of surrounding:

$$T_0 = 15°C = 288 \text{ K}.$$

Availability of heat in the furnace gas per kg,

$$a = (h_1 - h_0) - T_0(S_1 - S_0)$$

$$= c_p(T_1 - T_0) - T_0\left[c_p \log_e \frac{T_1}{T_0} - R \log_e \frac{p_1}{p_0}\right]$$

$$= c_p(T_1 - T_0) - T_0 c_p \log_e \frac{T_1}{T_0} \quad | \because p_1 = p_0$$

$$= 1.045(1773 - 288) - 288 \times 1.045 \times \log_e \frac{1773}{288}$$

$$= 1551.82 - 546.98 = \mathbf{1004.84 kJ/kg}$$

Problem 7.5 In a boiler, water is vapourated at 260 °C, while the combustion gas is cooled from 1300 to 320 °C. The surroundings are at 25 °C. Determine the loss in available energy due to the above heat transfer per kg of water vapourated. Take the latent heat of vapourization of water at 260 °C = 1662.5 kJ/kg, and specific heat at constant pressure for combustion gas is 1.08 kJ/kgK.

Solution Given data:

For water	For combustion gas.
$T_1 = 260 °C = 533 \text{ K}$,	$T_{g1} = 1300 °C = 1573 \text{ K}$,
$m_w = 1 \text{ kg}$,	$T_{g2} = 320 °C = 593 \text{ K}$,
Latent heat: $L = 1662.5 \text{ kJ/kg}$.	$c_{pg} = 1.08 \text{ kJ/kgK}$.
For surroundings: $T_0 = 25 °C = 298$.	

According to energy balance equation,

latent heat gained by water $=$ heat lost by combustion gas

$$m_w L = m_g c_{pg} (T_{g1} - T_{g2})$$
$$1 \times 1662.5 = m_g \times 1.08(1573 - 593)$$

or

$$m_g = 1.57 \text{ kg}.$$

Entropy change of the water,

$$\Delta S_w = \frac{mL}{T_1} = \frac{1 \times 1662.5}{533} = 3.12 \text{ kJ/K}.$$

Entropy change of the combustion gas,

$$\Delta S_g = m_g c_{pg} \log_e \frac{T_{g2}}{T_{g1}}$$
$$= 1.57 \times 1.08 \times \log_e \frac{593}{1573} = -1.65 \text{ kJ/K}$$

Total entropy change:

$$\Delta S = \Delta S_w + \Delta S_g$$
$$= 3.12 - 1.65 = 1.47 \, kJ/K$$

Loss of available energy: $I = T_0 \, \Delta S = 298 \times 1.47 = \textbf{438.06 kJ.}$

Problem 7.6 Determine the availability per unit mass of air under the following states in non-flow system.

(a) $p_1 = 400$ kPa, $T_1 = 100$ °C, (b) $p_1 = 400$ kPa, $T_1 = 25$ °C,
(c) $p_1 = 400$ kPa, $T_1 = 10$ °C, (d) $p_1 = 100$ kPa, $T_1 = 100$ °C,
(e) $p_1 = 100$ kPa, $T_1 = 25$ °C, (f) $p_1 = 100$ kPa, $T_1 = 10$ °C,
(g) $p_1 = 50$ kPa, $T_1 = 100$ °C, (h) $p_1 = 50$ kPa, $T_1 = 25$ °C,
(i) $p_1 = 50$ kPa, $T_1 = 10$ °C.

The atmospheric pressure and temperature are 100 kPa and 25 °C, respectively. Neglect the changes in KE and PE. Take $c_p = 1.005$ kJ/kgK, $c_v = 0.718$ kJ/kgK and $R = 0.287$ kJ/kgK.

Solution Given data for atmospheric condition:
$p_0 = 100$ kPa,
$T_0 = 25$ °C $= (25 + 273)$ K $= 298$ K.

(a) $p_1 = 400$ kPa,
$T_1 = 100$ °C $= (100 + 273)$ K $= 373$ K.
Availability: $A = (U_1 - U_0) + p_0 (V_1 - V_0) - T_0 (S_1 - S_0)$.
Since changes in KE and PE are neglected.
For unit mass,

$$
\begin{aligned}
a &= \frac{A}{m} \\
&= (u_1 - u_0) + p_0(v_1 - v_0) - T_0(s_1 - s_0) \\
&= c_v(T_1 - T_0) + p_0 v_1 - p_0 v_0 - T_0\left[c_p \log_e \frac{T_1}{T_0} - R \log_e \frac{p_1}{p_0}\right] \\
&= c_v(T_1 - T_0) + \frac{p_0 p_1 v_1}{p_1} - p_0 v_0 - T_0\left[c_p \log_e \frac{T_1}{T_0} - R \log_e \frac{p_1}{p_0}\right] \\
&= c_v(T_1 - T_0) + \frac{p_0}{p_1} R T_1 - R T_0 - T_0\left[c_p \log_e \frac{T_1}{T_0} - R \log_e \frac{p_1}{p_0}\right] \\
&= 0.718(373 - 298) + \frac{100}{400} \times 0.287 \times 373 - 0.287 \times 298 \\
&\quad - 298\left[1.005 \log_e \frac{373}{298} - 0.287 \log_e \frac{400}{100}\right] \\
&= 51.85 + 26.76 - 85.52 - 298(0.2256 - 0.3978) \\
&= 51.85 + 26.76 - 85.52 + 51.31 = \mathbf{44.4\,kJ/kg}
\end{aligned}
$$

(b) $p_1 = 400$ kPa,
$T_1 = 25$ °C $= 298$ K $= T_0$.
Availability per unit mass,

$$
\begin{aligned}
a &= c_v(T_1 - T_0) + \frac{p_0 R T_1}{p_1} - R T_0 - T_0\left[c_p \log_e \frac{T_1}{T_0} - R \log_e \frac{p_1}{p_0}\right] \\
&= 0 + \frac{100}{400} \times 0.287 \times 298 - 0.287 \times 298 - 298\left[0 - 0.287 \log_e \frac{400}{100}\right] \\
&= 21.38 - 85.52 + 118.56 = \mathbf{54.42kJ/kg}
\end{aligned}
$$

(c) $p_1 = 400$ kPa,
$T_1 = 10$ °C $= 283$ K.
Availability per unit mass,

$$a = c_v(T_1 - T_0) + \frac{p_0 R T_1}{p_1} - R T_0 - T_0\left[c_p \log_e \frac{T_1}{T_0} - R \log_e \frac{p_1}{p_0}\right]$$

$$= 0.718(283 - 298) + \frac{100 \times 0.287 \times 283}{400} - 0.283 \times 298$$

$$- 298\left[1.005 \log_e \frac{283}{298} - 0.287 \log_e \frac{400}{100}\right]$$

$$= -10.77 + 20.30 - 84.33 - 298[-0.0519 - 0.3978]$$

$$= -10.77 + 20.30 - 84.33 + 134.01 = \mathbf{59.21\,kJ/kg}$$

(d) $p_1 = 100$ kPa $= p_0$.
$T_1 = 100\,°C = 373$ K.
Availability per unit mass,

$$a = c_v(T_1 - T_0) + \frac{p_0 R T_1}{p_1} - R T_0 - T_0\left[c_p \log_e \frac{T_1}{T_0} - R \log_e \frac{p_1}{p_0}\right]$$

$$= c_v(T_1 - T_0) + R T_1 - R T_0 - T_0 c_p \log_e \frac{T_1}{T_0}$$

$$= c_v(T_1 - T_0) + R(T_1 - T_0) - T_0 c_p \log_e \frac{T_1}{T_0}$$

$$= 0.718(373 - 298) + 0.287(373 - 298) - 298 \times 1.005 \log_e \frac{373}{298}$$

$$= 51.85 + 21.52 - 67.23 = \mathbf{6.14kJ/kg}$$

(e) $p_1 = 100$ kPa $= p_0$.
$T_1 = 25\,°C = 298$ K $= T_0$.
Availability per unit mass,

$$a = c_v(T_1 - T_0) + \frac{p_0 R T_1}{p_1} - R T_0 - T_0\left[c_p \log_e \frac{T_1}{T_0} - R \log_e \frac{p_1}{p_0}\right]$$

As $p_1 = p_0$ and $T_1 = T_0$

$$\therefore \quad a = c_v(T_0 - T_0) + \frac{p_0 R T_0}{p_0} - R T_0 - T_0\left[c_p \log_e \frac{T_0}{T_0} - R \log_e \frac{p_0}{p_0}\right] = \mathbf{0}$$

(f) $p_1 = 100$ kPa $= p_0$.
$T_1 = 10\,°C = 283$ K.
Availability per unit mass,

$$a = c_v(T_1 - T_0) + \frac{p_0 R T_1}{p_1} - R T_0 - T_0\left[c_p \log_e \frac{T_1}{T_0} - R \log_e \frac{p_1}{p_0}\right]$$

$$= c_v(T_1 - T_0) + R T_1 - R T_0 - T_0 c_p \log_e \frac{T_1}{T_0}$$

$$= c_v(T_1 - T_0) + R(T_1 - T_0) - T_0 c_p \log_e \frac{T_1}{T_0}$$

$$= (c_v + R)(T_1 - T_0) - T_0 c_p \log_e \frac{T_1}{T_0}$$

$$= c_p(T_1 - T_2) - T_0 c_p \log_e \frac{T_1}{T_0}$$

$$= 1.005(283 - 298) - 298 \times 1.005 \log_e \frac{283}{298}$$

$$= -15.07 + 15.46 = \mathbf{0.39 kJ/kg}$$

(g) $p_1 = 50$ kPa $= p_0$.
$T_1 = 100\ °C = 373$ K.
Availability per unit mass,

$$a = c_v(T_1 - T_0) + \frac{p_0 R T_1}{p_1} - R T_0 - T_0\left[c_p \log_e \frac{T_1}{T_0} - R \log_e \frac{p_1}{p_0}\right]$$

$$= 0.718(373 - 298) + \frac{100}{50} \times 0.287 \times 373 - 0.287 \times 298$$

$$- 298\left[1.005 \log_e \frac{373}{298} - 0.287 \log_e \frac{50}{100}\right]$$

$$= 53.85 + 214.1 - 85.52 - 298[0.2256 + 0.1989]$$

$$= \mathbf{55.929\,kJ/kg}$$

(h) $p_1 = 50$ kPa $= p_0$.
$T_1 = 25\ °C = 298$ K $= T_0$.
Availability per unit mass,

$$a = c_v(T_1 - T_0) + \frac{p_0 R T_1}{p_1} - R T_0 - T_0\left[c_p \log_e \frac{T_1}{T_0} - R \log_e \frac{p_1}{p_0}\right]$$

As $T_1 = T_0$.
$\therefore$

$$a = 0 + \frac{p_0 R T_0}{p_1} - R T_0 - T_0\left[0 - R \log_e \frac{p_1}{p_0}\right]$$

$$= \left(\frac{p_0}{p_1} - 1\right) R T_0 + R T_0 \log_e \frac{p_1}{p_0}$$

$$= \left(\frac{100}{50} - 1\right) 0.287 \times 298 + 0.287 \times 298 \log_e \frac{50}{100}$$

$$= 85.526 - 59.282 = \mathbf{26.244 kJ/kg}$$

(i) $p_1 = 50$ kPa.

$T_1 = 10\,°C = 283$ K.

Availability per unit mass,

$$a = c_v(T_1 - T_0) + \frac{p_0 R T_1}{p_1} - RT_0 - T_0\left[c_p \log_e \frac{T_1}{T_0} - R \log_e \frac{p_1}{p_0}\right]$$

$$= 0.718(283 - 298) + \frac{50}{100} \times \times 0.287 \times 283 - 0.287 \times 298$$

$$- 298\left[1.005 \log_e \frac{283}{298} - 0.287 \log_e \frac{50}{100}\right]$$

$$= -10.77 + 162.44 - 85.52 - 298[-0.0519 + 0.1989]$$

$$= \mathbf{22.34\,kJ/kg}$$

Problem 7.7 Exhaust gases leave an IC engine at 800 °C and 100 kPa, after having done 1050 kJ of work per kg of gas in the engine (c_p of gas $= 1.1$ kJ/kgK).

The surroundings are at 30 °C, 100 kPa. Determine (a) the available energy per kg of gas is lost by throwing away the exhaust gases. (b) the ratio of the lost available energy to the engine work.

Solution Given data:

At exhaust condition:

$$T_1 = 800\,°C = 1073\mathrm{K}$$
$$p_1 = 100\,\mathrm{kPa}$$

Work done by the engine:

$$W = 1050\,\mathrm{kJ/kg}$$
$$c_{pg} = 1.1\,\mathrm{kJ/kgk}$$

At surrounding condition:

$$T_0 = 30°C = 303\,\mathrm{K}$$
$$p_0 = 100\,\mathrm{kPa}$$

(a)   The loss of available energy per kg of the exhaust gases:

The availability of the exhaust gases per kg:

$$a = (h_1 - h_0) - T_0(s_1 - s_0)$$

$$= c_{pg}(T_1 - T_0) - T_0 c_{pg} \log_e \frac{T_1}{T_0}$$

$$= 1.1(1073 - 303) - 303 \times 1.1 \log_e \frac{1073}{303}$$

$$= 847 - 421.45 = 425.55\,\mathrm{kJ/kg}$$

Thus, 425.55 kJ/kg of work potential of exhaust gases is wasted to the surrounding.
∴Loss of available energy per kg = **425.55 kJ/kg.**

(b)   The ratio of the lost AE to engine work:

$$\frac{\text{lost of AE}}{\text{enginework}} = \frac{425.55}{1050} = \mathbf{0.4052}$$

Problem 7.8 Determine the decrease in availability when 30 kg of water at 77 °C mixed with 25 kg of water 50 °C at constant pressure. Take temperature of the surroundings at 25 °C.

Solution Given data:

State-1	State–2
Mass of water: $m_1 = 30$ kg	Mass of water: $m_2 = 25$ kg
Temperature: $T_1 = 77°C = 350$ K	Temperature: $T_2 = 50 °C = 323$ K

Temperature of the surrounding:

$$T_0 = 25°C = 298\,\text{K}.$$

We know that the specific heat of the water:

$$c_p w = 4.18\,\text{kJ/kgK}.$$

Availability at state-1,

$$A_1 = m_1 c_{pw}(T_1 - T_0) - T_0 m_1 c_{pw} \log_e \frac{T_1}{T_0}$$

$$= m_1 c_{pw}\left[(T_1 - T_0) - T_0 \log_e \frac{T_1}{T_0}\right]$$

$$= 30 \times 4.18\left[(350 - 298) - 298 \log_e \frac{350}{298}\right]$$

$$= 125.4[52 - 47.93] = 510.378\,\text{kJ}$$

Availability at state-2,

$$A_1 = m_2 c_{pw}(T_2 - T_0) - T_0 m_2 c_{pw} \log_e \frac{T_2}{T_0}$$

$$= m_2 c_{pw}\left[(T_2 - T_0) - T_0 \log_e \frac{T_2}{T_0}\right]$$

$$= 25 \times 4.18\left[(323 - 298) - 298 \log_e \frac{323}{298}\right]$$

$$= 104.5[25 - 24] = 104.5 \text{ kJ}$$

∴ Total availability before mixing: $A = A_1 + A_2 = 510.378 + 104.5 = 614.878$ kJ.
Let T_3 is the temperature after mixing. Applying the energy balance equation,

$$H_1 + H_2 = H_3$$
$$m_1 c_{pw} T_1 + m_2 c_{pw} T_2$$
$$= m_3 c_{pw} T_3$$
$$m_1 T_1 + m_2 T_2$$
$$= m_3 T_3$$

or

$$T_3 = \frac{m_1 T_1 + m_2 T_2}{m_3}$$
$$= \frac{m_1 T_1 + m_2 T_2}{m_1 + m_2} (\because m_3 = m_1 + m_2 = 30 + 25 = 55 \text{ kg})$$
$$= \frac{30 \times 350 + 25 \times 323}{30 + 25} = 337.72 \text{ K}$$

Availability at state-3 (i.e., after mixing):

$$A_3 = m_3 c_{pw} (T_3 - T_0) - T_0 m_3 c_{pw} \log_e \frac{T_3}{T_0}$$
$$= m_3 c_{pw} \left[(T_3 - T_0) - T_0 \log_e \frac{T_3}{T_0} \right]$$
$$= 55 \times 4.18 \left[(337.72 - 298) - 298 \log_e \frac{337.72}{298} \right]$$
$$= 229.9[39.72 - 37.28] = 560.956 \text{ kJ}$$

Decrease in availability due to mixing = Total availability before mixing
$$- \text{ availability after mixing}$$
$$= 614.878 - 560.956 = 53.922 \text{ kJ}$$

Problem 7.9 A mass of 2 kg of air in a vessel expands from 300 kPa, 70 °C to 100 kPa 40 °C, while receiving 1.2 kJ of heat from a reservoir at 120 °C. The environment is at 98 kPa, 27 °C. Determine the maximum work and the work done on the atmosphere.

Solution Given data:

At initial state,   At final state,

$m = 2$ kg, $p_2 = 100$ kPa,

$p_1 = 300$ kPa, $T_2 = 40\ °C = 313$ K.

$T_1 = 70\ °C = 343$ K.

For a reservoir:

$$Q_R = -1.2\,\text{kJ}$$
$$T_R = 120°C = 393\,\text{K}$$

For environment:

$$P_0 = 98\,\text{kPa}$$
$$T_0 = 27°C = 300\,\text{K}$$

We know that for air:

$$c_p = 1.005\,\text{kJ/kgK}$$
$$c_v = 0.718\,\text{kJ/kgK}$$
$$R = 0.287\,\text{kJ/kgK}$$

The maximum work: W_{max}

$$W_{\text{max}} = mc_v(T_1 - T_2) - T_0(S_1 - S_2) - Q_R\left(1 - \frac{T_0}{T_R}\right)$$

$$= mc_v(T_1 - T_2) - T_0\left(mc_p \log_e \frac{T_1}{T_2} - mR \log_e \frac{p_1}{p_2}\right) - Q_R\left(1 - \frac{T_0}{T_R}\right)$$

$$= 2 \times 0.718(343 - 313) - 300\left(2 \times 1.005 \log_e \frac{343}{313} - 2 \times 0.287 \log_e \frac{300}{100}\right)$$

$$- (-1.2)\left[1 - \frac{300}{393}\right]$$

$$= 43.08 - 300[0.1839 - 0.6306] + 0.284$$

$$= \mathbf{177.37\,kJ}$$

The work done on the atmosphere,

$$W_{atm} = p_0(V_2 - V_1)$$

$$= p_0\left(\frac{p_2 V_2}{p_2} - \frac{p_1 V_1}{p_1}\right) = p_0\left(\frac{mRT_2}{p_2} - \frac{mRT_1}{p_1}\right)$$

$$= mRp_0\left(\frac{T_2}{p_2} - \frac{T_1}{p_1}\right) = 2 \times 0.287 \times 98\left(\frac{313}{100} - \frac{343}{300}\right)$$

$$= \mathbf{111.94\ kJ}$$

Problem 7.10 A rigid tank of volume $1\,\text{m}^3$ contains air at 1400 kPa and 448 K. The air is cooled to the atmospheric temperature by heat transfer to the atmosphere. The atmosphere is at 100 kPa, 298 K. Determine the availability in the initial and final states and the irreversibility of this process.

Solution Given data (refer Fig. 7.15):
   At initial state 1

$$V_1 = 1\,\text{m}^3$$
$$p_1 = 1400\,\text{kPa}$$
$$T_1 = 448\,\text{K}$$

At final state 2

$$T_2 = 298\,\text{K}$$
$$V_2 = V_1 = 1\,\text{m}^3, \text{ since the tank is rigid}$$

For constant volume process,

$$\frac{p_2}{T_2} = \frac{p_1}{T_1}.$$

or

$$p_2 = p_1 \frac{T_2}{T_1} = 1400 \times \frac{298}{448} = 931.25\text{kPa}.$$

Atmosphere condition,   We know that for air,
$T_0 = 298\,\text{K},$ $c_p = 1.005\,\text{kJ/kgK},$
$p_0 = 100\,\text{kPa}.$ $c_v = 0.718\,\text{kJ/kgK}$
   $R = 0.287\,\text{kJ/kgK}.$

Heat transferred to the atmosphere per unit mass of air:

$$q = c_v(T_1 - T_2) = 0.718(448 - 298) = 107.7\,\text{kJ/kg}$$

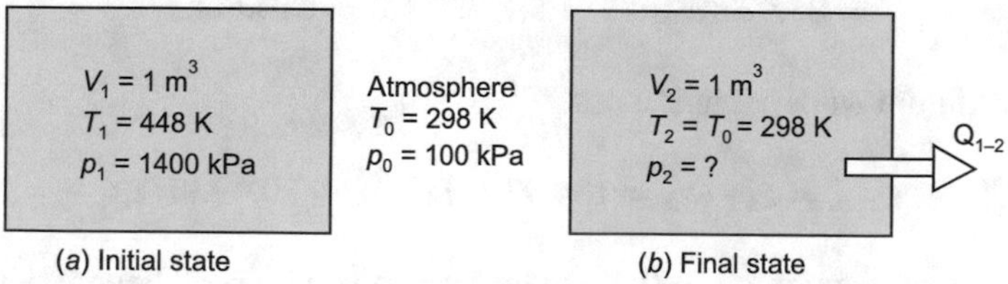

(a) Initial state   (b) Final state

Fig. 7.15 Schematic for Problem 7.10

The availability in the initial state per unit mass:

$$a_1 = (u_1 - u_0) + p_0(v_1 - v_0) - T_0(s_1 - s_0)$$

$$= c_v(T_1 - T_0) + \frac{p_0 p_1 v_1}{p_1} - p_0 v_0 - T_0\left[c_p \log_e \frac{T_1}{T_0} - R \log_e \frac{p_1}{p_0}\right]$$

$$= c_v(T_1 - T_0) + \frac{p_0}{p_1} RT_1 - RT_0 - T_0\left[c_p \log_e \frac{T_1}{T_0} - R \log_e \frac{p_1}{p_0}\right]$$

$$= 0.718(448 - 298) + \frac{100}{1400} \times 0.287 \times 448 - 0.287 \times 298 - 298$$

$$\left[1.005 \log_e \frac{448}{298} - 0.287 \log_e \frac{1400}{100}\right]$$

$$= 107.7 + 9.18 - 85.52 - 298[0.4097 - 0.7574]$$

$$= 107.7 + 9.18 - 85.52 + 103.61 = 134.97 kJ/kg$$

The availability in the final state per unit mass:

$$a_2 = (u_2 - u_0) + p_0(v_2 - v_0) - T_0(s_2 - s_0)$$

$$= c_v(T_2 - T_0) + \frac{p_0 p_2 v_2}{p_2} - p_0 v_0 - T_0\left(c_p \log_e \frac{T_2}{T_0} - R \log_e \frac{p_2}{p_0}\right)$$

Since.

$$T_2 = T_0.$$

$\therefore$

$$a_2 = 0 + \frac{p_0 RT_2}{p_2} - RT_0 - T_0\left(-R \log_e \frac{p_2}{p_0}\right)$$

$$= \frac{p_0 RT_0}{p_2} - RT_0 + RT_0 \log_e \frac{p_2}{p_0}$$

$$= RT_0\left(\frac{p_0}{p_2} - 1 + \log_e \frac{p_2}{p_0}\right)$$

$$= 0.287 \times 298\left(\frac{100}{931.25} - 1 + \log_e \frac{931.25}{100}\right)$$

$$= 85.526(0.1073 - 1 + 2.2313) = \mathbf{114.48 kJ/kg}$$

The maximum work per unit mass,

$$w_{max} = a_1 - a_2 = 134.97 - 114.48 = 20.49 \text{ kJ/kg}$$

The actual work output $w = 0$, since the system is with rigid boundaries.
$\therefore$ Irreversibility per unit mass:

$$i = w_{max} - w = 20.49 - 0 = \textbf{20.49 kJ/kg}$$

The mass of air in the rigid tank is calculated by using equation of state at initial state,
i.e.,

$$p_1 V_1 = m R T_1$$
$$1400 \times 1 = m \times 0.287 \times 448$$

or

$$m = 10.88 \text{kg}.$$

Irreversibility of the process:

$$I = mi = 10.88 \times 20.49 = \textbf{222.93 kJ}.$$

OR.
Irreversibility of the process:

$$I = T_0 \Delta S_{uni} = T_0 \left[\Delta S_{sys} + \Delta S_{sur} \right]$$
$$= T_0 \left[(S_2 - S_1) + \frac{Q}{T_0} \right]$$
$$= T_0 \left[mc_v \log_e \frac{T_2}{T_1} + mR \log_e \frac{V_2}{V_1} + \frac{mq}{T_0} \right]$$

Since

$$V_1 = V_2.$$

∴

$$I_0 = T_0 \left[mc_v \log_e \frac{T_2}{T_1} + 0 + \frac{mq}{T_0} \right] = mc_v T_0 \log_e \frac{T_2}{T_1} + mq$$
$$= m \left[c_v T_0 \log_e \frac{T_2}{T_1} + q \right] = 10.88 \left[0.718 \times 298 \log_e \frac{298}{448} + 107.7 \right]$$
$$= 10.88[-87.23 + 107.7] = \textbf{222.93 kJ}$$

Problem 7.11 5 kg of air at 550 K and 4 bar is enclosed in a closed system. Determine the availability of the system if the surroundings are at 1 bar 290 K. If the air is cooled at constant pressure to the atmospheric temperature, determine the change in availability.

Solution Given data:

For closed system: We know that for air:

$m = 5$ kg $c_v = 0.718$ kJ/kgK

$T_1 = 550$ K $c_p = 1.005$ kJ/kgK

$p_1 = 4$ bar $= 400$ kPa $R = 0.287$ kJ/kgK

For surroundings:

$p_0 = 1$ bar $= 100$ kPa,

$T_0 = 290$ K.

(a)   Availability of the system:

$$A = (U_1 - U_0) + p_0(V_1 - V_0) - T_0(S_1 - S_0)$$

$$= mc_v(T_1 - T_0) + \frac{p_0 p_1 V_1}{p_1} - p_0 V_0 - T_0\left[mc_p \log_e \frac{T_1}{T_0} - mR \log_e \frac{p_1}{p_0}\right]$$

$$= mc_v(T_1 - T_0) + \frac{p_0}{p_1} mRT_1 - mRT_0 - T_0 m\left[c_p \log_e \frac{T_1}{T_0} - R \log_e \frac{p_1}{p_0}\right]$$

$$= m\left[c_v(T_1 - T_0) + R\left(\frac{p_0}{p_1} T_1 - T_0\right) - T_0\left(c_p \log_e \frac{T_1}{T_0} - R \log_e \frac{p_1}{p_0}\right)\right]$$

$$= 5\left[\begin{array}{c} 0.718(550 - 290) + 0.287\left(\dfrac{100}{400} \times 550 - 290\right) - 290 \\[2mm] \left(1.005 \log_e \dfrac{550}{290} - 0.287 \log_e \dfrac{400}{100}\right) \end{array}\right]$$

$$= 5[186.68 + 0.287(137.5 - 290) - 290(0.6432 - 0.3978)]$$

$$= 5[186.68 - 43.76 - 71.16] = 358.8 \text{ kJ}$$

(b)   The change in availability: $\phi_1 - \phi_2$. If the air is cooled at constant pressure to the atmospheric temperature,

i.e.,

$$p_2 = p_1$$
$$T_2 = T_0$$

Change in availability $= (U_1 - U_2) + p_0(V_1 - V_2) - T_0(S_1 - S_2)$

$$= mc_v(T_1 - T_2) + p_0\left(\frac{p_1 V_1}{p_1} - \frac{p_2 V_2}{p_2}\right) - T_0(S_1 - S_2)$$

$$= mc_v(T_1 - T_2) + p_0\left(\frac{mRT_1}{p_1} - \frac{mRT_2}{p_1}\right) - T_0(S_1 - S_2)$$

$$= mc_v(T_1 - T_2) + \frac{p_0}{p_1} mR(T_1 - T_2)$$

$$- T_0\left(mc_p \log_e \frac{T_1}{T_2} - mR \log_e \frac{p_1}{p_2}\right)$$

$$= mc_v(T_1 - T_0) + \frac{mRp_0}{p_1}(T_1 - T_0) - T_0 m\left(c_p \log_e \frac{T_1}{T_0} - 0\right)$$

$$[\because T_2 = T_0 \text{ and } p_2 = p_1]$$

$$= m\left[c_v(T_1 - T_0) + \frac{Rp_0}{p_1}(T_1 - T_0) - T_0 c_p \log_e \frac{T_1}{T_0}\right]$$

$$= 5\left[0.718(550 - 290) + \frac{0.287 \times 100}{400}(550 - 290)\right.$$

$$\left. -290 \times 1.005 \log_e \frac{550}{290}\right]$$

$$= 5[186.68 + 18.65 - 186.53] = 94\,\text{kJ}$$

Problem 7.12 A hot spring produces water at a temperature of 56 °C. The water flows into a large lake, with a mean temperature of 14 °C, at a rate of 0.1 m³ of water per minute. What is the rate of working of an ideal heat engine which uses all the available energy?

Solution Given data:

Temperature of hot water:

$$T_1 = 56°C = 329\,\text{K}.$$

Temperature of lake:

$$T_2 = 14°C = 287\,\text{K}$$

$$= T_0, \text{ temperature of surroundings.}$$

$$\text{Rate of flow} = 0.1\,m^3/\min \frac{0.1}{60} m^3/s.$$

also

$$\text{Rate of flow} = mv$$

$$\therefore \quad \frac{0.1}{60} = \frac{m}{\rho} \quad \because \text{ specific volume: } v = \frac{1}{\rho}$$

or

$$m = \frac{0.1}{60}\rho$$

$$= \frac{0.1}{60} \times 1000 = 1.67\,\text{kg/s} \quad | \because \rho = 1000\,\text{kg/m}^3 \text{ for water}$$

The heat transferred from hot spring water to the lake water:

$$Q = mc(T_1 - T_0)$$

$$= 1.67 \times 4.18(329 - 287) = 293.18 \text{ kW}$$

That is, 293.18 kW of heat is removed from the hot water.

Now, the reversible work that can be done by a heat engine supplied with 293.18 kW of heat and rejecting waste heat to the lake $T_0 = 287$ K is

$$
\begin{aligned}
W_{\text{rev}} &= Q - T_0(S_1 - S_0) \\
&= Q - T_0 mc \log_e \frac{T_1}{T_0} \\
&= 293.18 - 287 \times 1.6 \times 4.18 \log_e \frac{329}{287} \\
&= 293.18 - 273.62 = \mathbf{19.56kW}
\end{aligned}
$$

Problem 7.13 Air expands in a turbine adiabatically from 500 kPa, 400 K, and 150 m/s to 100 kPa, 300 K, and 70 m/s. The atmosphere is at 100 kPa, 290 K. Determine per unit mass of air.

(a)   the actual work output,
(b)   the maximum work output, and
(c)   the irreversibility.

Solution Given data (refer Fig. 7.16):
    At inlet condition:

$$
\begin{aligned}
p_1 &= 500 \text{ kPa} \\
T_1 &= 400 \text{ K} \\
V_1 &= 150 \text{ m/s}
\end{aligned}
$$

At exit condition:

$$
\begin{aligned}
p_2 &= 100 \text{ kPa} \\
T_2 &= 300 \text{ K} \\
V_2 &= 70 \text{ m/s}
\end{aligned}
$$

Fig. 7.16 Schematic for Problem 7.13

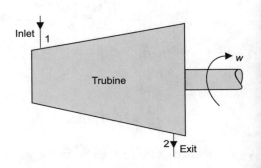

Inlet

1

w

Trubine

2

Exit

At atmosphere condition:

$$p_0 = 100\,\text{kPa}$$
$$T_0 = 290\,\text{K}$$

(a)   The actual work output per unit mass: w

The actual work is calculated by application of steady flow energy equation:
According to steady flow energy equation per unit mass,

$$\left(h_1 + \frac{V_1^2}{2} + gz_1\right) + q = \left(h_2 + \frac{V_2^2}{2} + gz_2\right) + w$$

Here $q = 0$, adiabatic flow and neglect the change in potential energy, i.e.,

$$Z_1 = Z_2.$$

$\therefore$

$$
\begin{aligned}
w &= (h_1 - h_2) + \frac{V_1^2}{2} - \frac{V_2^2}{2} \\
&= c_p(T_1 - T_2) + \frac{V_1^2}{2000} - \frac{V_2^2}{2000} \\
&= 1.005(400 - 300) + \frac{(150)^2}{2000} - \frac{(70)^2}{2000} \\
&= 100.5 + 11.25 - 2.45 = \textbf{109.3kJ/kg}
\end{aligned}
$$

(b)   The maximum work output per unit mass: $w_{\max}$

$$
\begin{aligned}
w_{\max} &= (h_1 - h_2) + \frac{V_1^2 - V_2^2}{2000} - T_0(s_1 - s_2) \\
&= w - T_0(s_1 - s_2) \\
&= w - T_0\left[c_p \log_e \frac{T_1}{T_2} - R \log_e \frac{p_1}{p_2}\right] \\
&= 109.3 - 290\left[1.005 \log_e\left(\frac{400}{300}\right) - 0.287 \log_e\left(\frac{500}{100}\right)\right] \\
&= 109.3 - 290[0.289 - 0.462] \\
&= 109.3 + 50.17 = \textbf{159.47kJ/kg}
\end{aligned}
$$

(c)   The irreversibility per unit mass: i

$$i = w_{\max} - w$$

$$= 159.47 - 109.3 = \mathbf{50.17 kJ/kg}$$

OR

$$
\begin{aligned}
i &= T_0 \Delta s_{\text{uni}} = T_0 \left[\Delta s_{\text{sys}} + \Delta s_{\text{surr}} \right) \\
&= T_0 \Delta s_{\text{sys}} \\
&= T_0 (s_2 - s_1) \\
&= T_0 \left[c_p \log_e \frac{T_2}{T_1} - R \log_e \frac{p_2}{p_1} \right] \\
&= 290 \left[1.005 \log_e \left(\frac{300}{400} \right) - 0.287 \log_e \left(\frac{100}{500} \right) \right] \\
&= 290[-0.289 + 0.462] = \mathbf{50.17 kJ/kg}
\end{aligned}
$$

Problem 7.14 An adiabatic gas turbine receives gas at 7 bar and 1000 °C and discharges at 1.5 Bar and 665 °C. Determine the Second Law and Isentropic Efficiencies of the Turbine. Take for a Gas,
$c_p = 1.09$ kJ/kgK and $c_v = 0.838$ kJ/kgK. The temperature of the surroundings is 25 °C.

Solution Given data:
At in let condition:

$$
\begin{aligned}
p_1 &= 7 \, \text{bar} = 700 \, \text{kPa} \\
T_1 &= 1000°C = 1273 \, \text{K}
\end{aligned}
$$

At exit condition:

$$
\begin{aligned}
p_2 &= 1.5 \, \text{bar} = 150 \, \text{kPa} \\
T_2 &= 665°C = 938 \, \text{K}
\end{aligned}
$$

For a gas:

$$
\begin{aligned}
c_p &= 1.09 \, \text{kJ/kgK} \\
c_v &= 0.838 \, \text{kJ/kgK}
\end{aligned}
$$

∴

$$R = c_p - c_v = 1.09 - 0.838 = 0.252 \, \text{kJ/kgK}.$$

and

$$\gamma = \frac{c_p}{c_v} = \frac{1.09}{0.838} = 1.3$$

Surroundings temperature:

$$T_0 = 25°C = 298 \, K$$

The maximum work output per unit mass,

$$
\begin{aligned}
w_{max} &= (h_1 - h_2) - T_0(s_1 - s_2) \\
&= c_p(T_1 - T_2) - T_0\left(c_p \log_e \frac{T_1}{T_2} - R \log_e \frac{p_1}{p_2}\right) \\
&= 1.09(1273 - 938) - 298\left(1.09 \log_e \frac{1273}{938} - 0.252 \log_e \frac{700}{150}\right) \\
&= 365.15 - 298(0.3328 - 0.3882) \\
&= 365.15 + 16.51 = 381.66 \, kJ/kg
\end{aligned}
$$

The actual work output per unit mass is calculated by using steady flow energy equation, and.
we have

$$h_1 + \frac{V_1^2}{2} + gz_1 + q = h_2 + \frac{V_2^2}{2} + gz_2 + w$$

since

$$q = 0 \, \text{adiabatic flow}$$

Neglect the change in KE and PE.
∴

$$h_1 = h_2 + w$$

or

$$
\begin{aligned}
w = h_1 - h_2 &= c_p(T_1 - T_2) \\
&= 1.09(1273 - 938) \\
&= 365.15 \, kJ/kg
\end{aligned}
$$

The second-law efficiency:

$$
\begin{aligned}
\eta_{II} &= \frac{\text{actual work output}}{\text{maximum work output}} \\
&= \frac{w}{w_{max}} = \frac{365.15}{381.66} \\
&= 0.9567 = 95.67\%
\end{aligned}
$$

Fig. 7.17 Schematic for
Problem 7.14

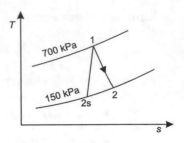

Process 1–2, actual adiabatic irreversible process (refer Fig. 7.17).
Process 1-2 s, adiabatic frictionless process, i.e., adiabatic isentropic.
For process 1-2 s

$$\frac{T_{2s}}{T_1} = \left(\frac{p_2}{p_1}\right)^{\frac{\gamma-1}{\gamma}}$$

$$\frac{T_{2s}}{1273} = \left(\frac{150}{700}\right)^{\frac{1.3-1}{1.3}}$$

$$\frac{T_{2s}}{1273} = 0.7$$

or

$$T_2 s = 891.1 \text{K}$$

The isentropic efficiency: η_{isen}

$$
\begin{aligned}
\eta_{isen} &= \frac{\text{actual enthalpy drop}}{\text{isentropicenthalpydrop}} \\
&= \frac{h_1 - h_2}{h_1 - h_{2s}} = \frac{c_p(T_1 - T_2)}{c_p(T_1 - T_{2s})} \\
&= \frac{T_1 - T_2}{T_1 - T_{2s}} = \frac{1273 - 938}{1273 - 891.1} = 0.8772 = \mathbf{87.72\%}
\end{aligned}
$$

Problem 7.15 Air enters a compressor in steady flow at 140 kPa, 17 °C, and 70 m/s
and exits it at 350 kPa, 127 °C, and 110 m/s. The environment is at 100 kPa, 7 °C.
Determine per unit mass of air:

(a)   The actual work required,
(b)   The minimum work required, and
(c)   The irreversibility.

Solution Given data (refer Fig. 7.18):
    At inlet condition:

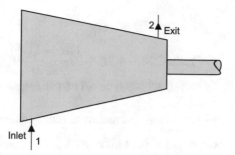

Fig. 7.18 Schematic for Problem 7.15

$$p_1 = 140 \, \text{kPa}$$
$$T_1 = 17°C = 290 \, \text{K}$$
$$V_1 = 70 \, \text{m/s}$$

At exit condition:

$$p_2 = 350 \, \text{kPa}$$
$$T_2 = 127°C = 400 \, \text{K}$$
$$V_2 = 110 \, \text{m/s}$$

At environment condition:

$$T_0 = 7°C = 280 \, \text{K}$$
$$p_0 = 100 \, \text{kPa}$$

(a)   The actual work required per unit mass: w

The actual work is calculated by the application of steady flow energy equation per unit mass.

$$\left(h_1 + \frac{V_1^2}{2} + gz_1 \right) + q = \left(h_2 + \frac{V_2^2}{2} + gz_2 \right) + w$$

Here $q = 0$, adiabatic flow and neglect the change in potential energy, i.e.,

$$z_1 = z_2$$

$\therefore$

$$w = (h_1 - h_2) + \frac{V_1^2 - V_2^2}{2}$$

$$= c_p(T_1 - T_2) + \frac{V_1^2 - V_2^2}{2000}$$

$$= 1.005(290 - 400) + \frac{(70^2 - 110^2)}{2000}$$

$$= -110.55 - 3.6 = \mathbf{-114.15 kJ/kg}$$

The $-ve$ sign indicates 114.15 kJ/kg actual work required to drive the compressor.

(b)   The minimum work required per unit mass: w_{min}

$$w_{min} = (h_1 - h_2) + \frac{(V_1^2 - V_2^2)}{2000} - T_0(s_1 - s_2)$$

$$= w - T_0(s_1 - s_2)$$

$$= w - T_0 \left[c_p \log_e \frac{T_1}{T_2} - R \log_e \frac{p_1}{p_2} \right]$$

$$= -114.15 - 280 \left[1.005 \log_e \left(\frac{290}{400} \right) - 0.287 \log_e \left(\frac{140}{350} \right) \right]$$

$$= -114.15 - 280[-0.3232 + 0.2629]$$

$$= -114.15 + 16.88 = -97.27 \text{ kJ/kg}$$

The $-ve$ sign indicates 97.27 kJ/kg minimum work required to drive the compressor.

(c)   The irreversibility per unit mass: i

$$i = w_{min} - w$$

$$= -97.27 - (-114.15)$$

$$= -97.27 + 114.15 = 16.88 \text{ kJ/kg}$$

OR

$$i = T_0 \Delta s_{uni}$$

$$i = T_0[\Delta s_{svs} + \Delta s_{surr}]$$

$$= T_0 \Delta s_{sys} \quad | \because \Delta s_{uni} = 0$$

$$= T_0(s_2 - s_1)$$

$$= T_0 \left[c_p \log_e \frac{T_2}{T_1} - R \log_e \frac{p_2}{p_1} \right]$$

$$= 280 \left[1.005 \log_e \left(\frac{400}{290} \right) - R \log_e \left(\frac{350}{140} \right) \right]$$

$$= 280[0.3232 - 0.2629] = \mathbf{16.88 kJ/kg}$$

Problem 7.16 2.5 kg of air at 6 bar, 90 °C expands adiabatically in a closed system until its volume is doubled and its temperature becomes equal to that of the surroundings which are at 1 bar, 5 °C. Determine.

(a) The maximum work,
(b) The change in availability, and
(c) The irreversibility.

Solution Given data:

$$m = 2.5 \text{ kg} \qquad \text{Working fluids air:}$$
$$p_1 = 6 \text{ bar} = 600 \text{ kPa} \qquad c_v = 0.718 \text{ kJ/kgK}$$
$$T_1 = 90 \text{ °C} = 363 \text{ K} \qquad R = 0.287 \text{ kJ/kgK}$$

$$V_2 = 2V_1$$

or

$$\frac{V_1}{V_2} = \frac{1}{2}$$

$$T_2 = T_0 = 5\,°C = 278 \text{ K}$$
$$p_0 = 1\text{bar} = 100\text{kPa}$$

Applying the equation of state at the initial state 1, we have.

$$p_1 V_1 = m R T_1$$

Here,

$$p_1 \text{ is in kPa}$$
$$V_1 \text{ is in m}^3$$
$$R \text{ is in kJ/kgK}$$

and

$$T \text{ is in K}$$

$\therefore$

$$600 \times V_1 = 2.5 \times 0.287 \times 363$$

or

$$V_1 = 0.4340 \, \text{m}^3$$

and

$$V_2 = 2V_1 = 2 \times 0.4340 = 0.868 \, \text{m}^3$$

(a) The maximum work: $W_{\max}$

$$
\begin{aligned}
W_{\max} &= (U_1 - U_2) - T_0(S_1 - S_2) \\
&= mc_v(T_1 - T_2) - T_0\left[m\left(c_v \log_e \frac{T_1}{T_2} + R \log_e \frac{V_1}{V_2}\right)\right] \\
&= m\left[c_v(T_1 - T_2) - T_0\left(c_v \log_e \frac{T_1}{T_2} + R \log_e \frac{V_1}{V_2}\right)\right] \\
&= 2.5\left[0.718(363 - 278) - 278\left(0.718 \log_e \frac{363}{278} + 0.287 \log_e \frac{1}{2}\right)\right] \\
&= 2.5[61.03 - 278(0.1915 - 0.1989)] \\
&= 2.5[61.03 + 2.057] = \textbf{157.71kJ}
\end{aligned}
$$

(b) The change in availability: $\phi_1 - \phi_2$

$$
\begin{aligned}
\text{Change availability} &= \phi_1 - \phi_2 \\
&= (U_1 - U_2) - T_0(S_1 - S_2) + p_0(V_1 - V_2) \\
&= W_{\max} + p_0(V_1 - V_2) \\
&= 157.71 + 100(0.4340 - 0.868) \\
&= 157.71 - 43.4 = 114.31 \text{kJ}
\end{aligned}
$$

(c) The irreversibility: I

$$I = W_{\max} - W$$

where W = actual work. It is determined by the application of the first law of thermodynamics.

$$Q = (U_2 - U_1) + W$$

or

$$
\begin{aligned}
W &= U_1 - U_2 \quad \because Q = 0, \text{ adiabatic process} \\
&= mc_v(T_1 - T_2) \\
&= 2.5 \times 0.718(363 - 278) = 152.57 \text{ kJ}
\end{aligned}
$$

$\therefore$

$$I = 157.71 - 152.57 = \textbf{5.14kJ}$$

OR

$$
\begin{aligned}
I &= T_0 \Delta S_{\text{uni}} = T_0 \left[\Delta S_{sys} + \Delta S_{\text{surr}} \right] \\
&= T_0 \Delta S_{sys} \quad \because \Delta S_{\text{surr}} = 0 \\
&= T_0 (S_2 - S_1) \\
&= T_0 \left[mc_v \log_e \frac{T_2}{T_1} + mR \log_e \frac{V_2}{V_1} \right] \\
&= mT_0 \left[c_v \log_e \frac{T_2}{T_1} + R \log_e \frac{V_2}{V_1} \right] \\
&= 2.5 \times 278 \left[0.718 \log_e \frac{278}{363} + 0.287 \log_e 2 \right] \\
&= 695[-0.1915 + 0.1989] = \textbf{5.14kJ}
\end{aligned}
$$

Problem 7.17 An IC engine contains gases at 2500 °C, 58 bar. Expansion takes place through a volume ratio of 9 according to $pV^{1.38} = C$. The environment is at 20 °C, 100 kPa. Determine per unit mass.

(a)   the heat transfer,
(b)   the actual work output,
(c)   the maximum work output,
(d)   the change in availability, and
(e)   the irreversibility.

Take $c_v = 0.82$ kJ/kgK and $R = 0.26$ kJ/kgK.

Solution Given data:
    At initial state-1,

$$T_1 = 2500°C = 2773 \text{ K}$$
$$p_1 = 58 \text{ bar} = 5800 \text{ kPa}$$

At finial state-2,

$$\frac{V_2}{V_1} = 9$$

Expansion takes place according to the law of $pV^{1.38} = C$.
Here

$$n = 1.38, \text{ polytropic index}$$

At environment condition:

$$T_0 = 20°C = 293\,K$$
$$p_0 = 100\,kPa$$

For gases:

$$c_v = 0.82\ kJ/kgK$$
$$R = 0.26\ kJ/kgK$$

also

$$R = c_p - c_v$$

∴

$$0.26 = c_p - 0.82$$

or

$$c_p = 1.08\ kJ/kgK$$

Adiabatic index:

$$\gamma = \frac{c_p}{c_v} = \frac{1.08}{0.82} = 1.317$$

For polytropic expansion,

$$\frac{T_2}{T_1} = \left(\frac{V_1}{V_2}\right)^{n-1}$$

$$\frac{T_2}{2773} = \left(\frac{1}{9}\right)^{1.38-1}$$

or

$$T_2 = 1203.2\ K$$

and

$$\frac{p_2}{p_1} = \left(\frac{V_1}{V_2}\right)^{n}$$

$$\frac{p_2}{5800} = \left(\frac{1}{9}\right)^{1.38}$$

$$p_2 = 279.62 \text{kPa}$$

(a) The heat transfer per unit mass: q_{1-2}

$$q_{1-2} = c_v \left(\frac{\gamma - n}{1 - n} \right) [T_2 - T_1]$$

$$= 0.82 \left(\frac{1.317 - 1.38}{1 - 1.38} \right) [1203.2 - 2773]$$

$$= \mathbf{-213.41 kJ/kg}$$

(b) The actual work output per unit mass: w_{1-2}.

According to the first law of thermodynamics for process,

$$q_{1-2} = (u_2 - u_1) + w_{1-2}$$

or

$$\begin{aligned} w_{1-2} &= q_{1-2} - (u_2 - u_1) \\ &= q_{1-2} - c_v(T_2 - T_1) \\ &= -213.14 - 0.82(1203.2 - 2773) \\ &= -213.14 + 1287.23 = \mathbf{1074.09 kJ/kg} \end{aligned}$$

OR.

For non-flow polytropic process,

$$w_{1-2} = \frac{R(T_1 - T_2)}{n - 1} = \frac{0.26(2773 - 1203.2)}{1.38 - 1} = \mathbf{1074.09 kJ/kg}$$

(c) The maximum work output per unit mass: $w_{\max}$

$$\begin{aligned} w_{\max} &= (u_1 - u_2) - T_0(s_1 - s_2) \\ &= c_v(T_1 - T_2) - T_0 \left[c_v \log_e \frac{T_1}{T_2} + R \log_e \frac{V_1}{V_2} \right] \\ &= 0.82(2773 - 1203.2) - 293 \left[0.82 \log_e \frac{2773}{1203.2} + 0.26 \log_e \frac{1}{9} \right] \\ &= 1287.23 - 293[0.6846 - 0.5712] \\ &= 1287.23 - 33.22 = \mathbf{1254.01 kJ/kg} \end{aligned}$$

(d) The change in availability per unit mass: $\phi_1 - \phi_2$

Change in availability $= (u_1 - u_2) - T_0(s_1 - s_2) + p_0(v_1 - v_2)$

$$= w_{max} + p_0(v_1 - v_2)$$

$$= 1254.01 + p_0\left(\frac{p_1 v_1}{p_1} - \frac{p_2 v_2}{p_2}\right)$$

$$= 1254.01 + p_0\left(\frac{RT_1}{p_1} - \frac{RT_2}{p_2}\right)$$

$$= 1254.01 + p_0 R\left(\frac{T_1}{p_1} - \frac{T_2}{p_2}\right)$$

$$= 1254.01 + 100 \times 0.26\left(\frac{2773}{5800} - \frac{1203.9}{279.62}\right)$$

$$= 1254.01 + 26(0.478 - 4.303)$$

$$= 1254.01 - 99.45 = 1154.36 \text{ kJ/kg}$$

$$= \text{loss of availability}$$

(e)   The irreversibility per unit mass: i

$$i = w_{max} - w_{1-2} = 1254.01 - 1074.09 = \mathbf{179.92 kJ/kg}$$

OR

$$i = T_0 \Delta s_{uni} = T_0\left[\Delta s_{sys} + \Delta s_{surr}\right]$$

where

$$\Delta s_{sys} = s_2 - s_1$$

$$= c_v \log_e \frac{T_2}{T_1} + R \log_e \frac{V_2}{V_1}$$

$$= 0.82 \log_e \frac{1203.2}{2773} + 0.26 \log_e 9$$

$$= -0.6846 + 0.5712 = -0.1134 \text{ kJ/kgK}$$

also

$$\Delta s_{sys} = c_v\left(\frac{\gamma - n}{n - 1}\right) \log_e \frac{T_2}{T_1}$$

$$= 0.82\left(\frac{1.317 - 1.38}{1 - 1.38}\right) \log_e \frac{1203.2}{2773}$$

$$= -0.1134 \text{ kJ/kgK}$$

and

$$\Delta s_{surr} = \frac{q_{1-2}}{T_0} = \frac{213.14}{293} \quad + \text{ve sign takes, surroundings heat gain}$$

$$= 0.7274 \text{ kJ/kgK}$$

$$\therefore \quad i = 293[-0.1134 + 0.7274] = \mathbf{179.92 kJ/kgK}$$

Problem 7.18 A rigid tank contains air at 1.5 bar and 60 °C. The pressure of air is raised to 2.5 bar by transfer of heat from a constant temperature reservoir at 400 °C. The temperature of surrounding is 27 °C. Determine per kg of air, the loss of available energy due to heat transfer.

Solution Given data (refer Fig. 7.19):
    At initial state in a rigid tank,

$$p_1 = 1.5 \, bar$$
$$T_1 = 60°C = 333 \text{ K}$$

At final state in a rigid tank,

$$p_2 = 2.5 \text{ bar}$$

As the tank is rigid, volume of the tank remains constant, i.e., $V = \text{constant}$.
$$\therefore$$

$$\frac{T_1}{p_1} = \frac{T_2}{p_2}$$
$$\frac{333}{1.5} = \frac{T_2}{2.5}$$

or

$$T_2 = 555 \text{ K}$$

Temperature of reservoir:

$$T_R = 400°C = 673K$$

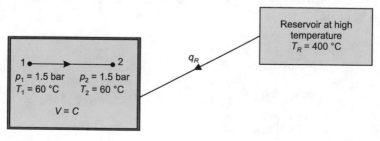

Fig. 7.19 Schematic for Problem 7.18

Temperature of surrounding:

$$T_0 = 27°C = 300K$$

Heat transfer to the tank per kg of air:

$$q_{1-2} = c_v(T_2 - T_1)$$
$$= 0.718(555 - 333) = 159.39 \text{ kJ/kg}$$
$$= -q_R, \text{ heat transfer from the reservoir to the tank}$$

The maximum work output per kg of air by transfer of heat from a constant temperature reservoir T_R,

$$w_{max} = c_v(T_1 - T_2) - T_0(s_1 - s_2) - q_R\left(1 - \frac{T_0}{T_R}\right)$$

where $q_R = -159.39$ kJ/kg, heat transfer from the reservoir.
$\therefore$

$$w_{max} = 0.718(333 - 555) - 300\left[c_v \log_e \frac{T_1}{T_2}\right] + 159.39\left(1 - \frac{300}{673}\right)$$
$$= -159.39 - 300 \times 0.718 \times \log_e \frac{333}{500} + 159.39 \times 0.554$$
$$= -159.39 + 110.03 + 88.33 = 38.97 \text{ kJ/kg}$$

The actual work output per kg of air,

$$w = 0 \quad \because \text{ Tank is rigid, } V = C$$

$\therefore$ Loss of available energy per kg of air due to heat transfer.

$$= w_{max} - w = \mathbf{38.93\,kJ/kg}$$

Also
Loss of available energy per kg of air due to heat transfer = Irreversibility of the tank per kg of air.
$\therefore$

$$i = T_0 \Delta s_{uni}$$
$$= T_0\left[\Delta s_{sys} + \frac{q_R}{T_R}\right] = T_0\left[(s_2 - s_1) - \frac{159.39}{673}\right]$$
$$= 300\left[c_v \log_e \frac{T_2}{T_1} - 0.236\right]$$

$$= 300\left[0.718\log_e\frac{555}{333} - 0.236\right]$$
$$= 300[0.3667 - 0.2368] = \mathbf{38.97kJ/kg}$$

Problem 7.19 An adiabatic cylinder of volume 10 m³ is divided into two compartments A and B, each of volume 6 m³ and 4 m³, respectively, by a thin sliding partition. Initially, the compartment B is filled with air at 6 bar and 600 K, while there is a vacuum in the compartment A. Suddenly the partition is removed, the air in compartment B expands and fills both the compartments. Calculate the loss in available energy. Assume atmosphere is at 1 bar and 300 K.

Solution Given data (refer Fig. 7.20):
Volume of the compartment A:

$$V_A = 6\,\text{m}^3$$

Volume of the compartment B:

$$V_B = 4\,\text{m}^3$$

Pressure of the compartment B:

$$p_B = 6\,\text{bar} = 600\,\text{kPa}$$

Temperature of the compartment B:

$$T_B = 600\,\text{K}$$

At atmosphere conditions:

$$p_0 = 1\,\text{bar} = 100\,\text{kPa}$$
$$T_0 = 300\,\text{K}$$

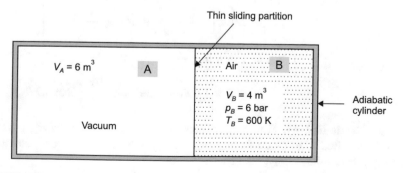

Fig. 7.20 Schematic for Problem 7.19

Applying the equation of state in compartment B,

$$p_B V_B = m_B R T_B$$
$$600 \times 4 = m_B \times 0.287 \times 600$$

or

$$m_B = 13.93\,\text{kg}$$

When the partition is removed, the air in compartment B expands and fills both the compartments,

∴. Volume of the air after removing the partition:

$$V = V_A + V_B = 6 + 4 = 10\,\text{m}^3$$

According to the first law of thermodynamic, $\delta Q = dU + \delta W$
where

$$\delta Q = 0, \text{adiabatic cylinder}$$
$$\delta W = 0, \text{free expansion}$$

∴

$$dU = 0$$

$$m c_v (T - T_B) = 0$$

or

$$T = T_B$$

Loss in available energy $= T_0 \Delta S_{\text{uni}} = T_0 \left[\Delta S_{\text{sys}} + \Delta S_{\text{surr}} \right]$

$$= T_0 \left[m \left(c_v \log_e \frac{T}{T_B} + R \log_e \frac{V}{V_B} \right) + 0 \right]$$

$$= T_0 m R \log_e \frac{V}{V_B}$$

$$= 300 \times 13.93 \times 0.287 \log_e \frac{10}{4} = \mathbf{1098.97kJ}$$

1. **High-Grade Energy.** The energy which can be converted entirely into useful work is called high-grade energy, for example, Mechanical work, electrical energy, hydraulic energy, tidal energy.

2. **Low-Grade Energy.** If a small part of energy is converted into mechanical, it is called low grade energy, for example, Heat energy.

3. **Available and Unavailable Energy.** The part of heat energy that is converted into useful work is called available energy. The remaining part of heat energy that is going to the sink (i.e., waste) is called unavailable energy.

4. **Dead State.** It is that state at which a system is in thermodynamic equilibrium with its surroundings.

5. **Availability.** The availability of the system is defined as the maximum useful work that can be obtained by reducing the system from its initial state to the dead state (i.e., atmospheric condition). It is denoted by letter A. The availability is also called exergy.

6. **Availability for Heat Engine Cycle.**

$$A = W_{max} = Q_1\left(1 - \frac{T_0}{T_1}\right)$$

where Q_1 = heat supplied to the system from the heat source at temperature T_1. T_1 = atmosphere temperature.

7. **Useful Work.** The difference between the actual work W and the surroundings work W_{surr} is called the useful work W_u.

Mathematically, $W_u = W - W_{surr}$.
For reversible system,
Maximum useful work: $W_{u,max} = W_{rev} - W_{surr}$.
For a steady flow process,
Surroundings work: $W_{surr} = 0$.
∴ Useful work: $W_u = W$.
and maximum useful work:
$W_{u,max} = W_{rev}$.

8. **Reversible Work in a Steady Flow Process:**

$$W_{rev} = m\left[h_1 - T_0 s_1 + \frac{V_1^2}{2} + g z_1\right] - m\left[h_2 - T_0 s_2 + \frac{V_2^2}{2} + g z_2\right]$$

For unit mass,
$$w_{rev} = \left(h_1 - T_0 s_1 + \frac{V_1^2}{2} + g z_1\right) - \left(h_2 - T_0 s_2 + \frac{V_2^2}{2} + g z_2\right)$$

$$= (h_1 - h_2) - T_0(s_1 - s_2) + \frac{(V_1^2 - V_2^2)}{2} + g(z_1 - z_2)$$

Maximum useful work: $w_{u,max} = w_{rev}$.
where surroundings work: $w_{surr} = 0$ for a steady flow process.

9. **Availability Function for a Steady Flow Process:** ψ

$$\psi = H - T_0S + \frac{mV^2}{2} + mgz$$

If kinetic and potential energies are neglected, then

$$\psi = H - T_0S$$

For unit mass, $\psi = h - T_0s$.

10.   **Reversible Work for a Closed System:**

$$W_{\text{rev}} = (U_1 - U_2) - T_0(S_1 - S_2) + \frac{m}{2}\left(V_1^2 - V_2^2\right) + mg(z_1 - z_2)$$

If changes in kinetic and potential energies are neglected, then.

$$W_{\text{rev}} = (U_1 - U_2) - T_0(S_1 - S_2)$$

For unit mass,

$$w_{\text{rev}} = (u_1 - u_2) - T_0(s_1 - s_2)$$

Maximum useful work:

$$w_{u,\text{max}} = w_{\text{rev}} - p_0(v_2 - v_1)$$
$$= w_{\text{rev}} + p_0(v_1 - v_2)$$
$$w_{u,\text{max}} = (u_1 - u_2) - T_0(s_1 - s_2) + p_0(v_1 - v_2)$$

11.   **Availability Function for a Closed System:** ϕ

$$\phi = U + p_0V - T_0S$$

12.   **Irreversibility.** The irreversibility of the non-flow work-producing devices is the difference between the reversible useful work $W_{u,\text{rev}}$ and the useful work W_u which is due to the presence of irreversibilities during the process. It is also called anergy.

Mathematically,
Irreversibility:

$$I = W_{u,\text{rev}} - W_u$$

for work-producing device.
Also irreversibility:

$$I = T_0 \Delta S_{\text{uni}} .$$

The irreversibility of the system is also defined as the product of absolute temperature of the surroundings and universe entropy.

For a steady flow process,

Irreversibility:

$$I = W_{rev} - W \text{ for work - producing device (turbine)}$$
$$= T_0 \Delta S_{uni}$$

for work-producing device (turbine).

13. **Second-Law Efficiency: η_{II}.** The second-law efficiency is defined as the ratio of the actual work output to the maximum (reversible) work output of the work-producing devices.

Mathematically,

$$\eta_{II} = \frac{W}{W_{rev}}$$

For a heat engine;

$$\eta_{II} = \frac{W}{W_{rev}} = \frac{\eta_1}{\eta_{Carnot}}.$$

For a work $-$ consuming device : $\eta_{II} = \dfrac{W_{rev}}{W}$ $for\, compressor$

$$\eta_{II} = \frac{W_{rev}}{W} = \frac{(COP)_R}{(COP_{rev})_R} \, for\, a\, refrigerator$$

$$\eta_{II} = \frac{W_{rev}}{W} = \frac{(COP)_{HP}}{(COP_{rev})_{HP}}, for\, a\, heat\, pump.$$

14. **Helmholtz Function: F.** The Helmholtz function is defined as the difference between the internal energy and product of absolute temperature and entropy. It is denoted by F.

Mathematically, it is defined as.
Helmholtz function:$F = U - TS$.
It is also called Helmholtz free energy.

15. **Gibbs Function: G.** The Gibbs function is defined as the difference between the enthalpy and product of absolute temperature and entropy.

It is denoted by G.
Mathematically, it is defined as.
Gibbs function:$G = H - TS$.

It is also called Gibbs free energy.

1. What is difference between high-grade energy and low-grade energy? Give examples.
2. Explain the concept of available energy and unavailable energy.
3. What is the availability of the system?
4. What do you understand by the dead state?
5. Define the following terms:

   (i) Available energy,
   (ii) Unavailable energy, and
   (iii) Availability

6. What do you understand by exergy and anergy?
7. If energy is conserved. Why is energy crisis?
8. What is the available energy, when one litre petrol is ignited on floor?
9. What do you understand by quality of energy?
10. What is the unavailable and the irreversibility in the following cases?

    (i) The reversible heat engine.
    (ii) The irreversible heat engine.

11. The availability is conserved as mass and energy. Yes or No, explain.
12. How does reversible work differ from useful work for non-flow process?
13. What do you understand by energy and exergy?
14. Under what conditions does the reversible work equals the irreversibility for a process?
15. For getting maximum work output of a heat engine, what should be the final state?
16. Is the exergy of a system different in different environments?
17. Explain the difference between the maximum useful work and useful work as concerned to the availability of a closed system.
18. Explain why the work done by the system on the atmosphere is not taken into account in a cycle process?
19. What is irreversibility of the process, when the actual useful work for that process equals the reversible work?
20. Derive an expression for reversible work in a steady flow process under a given environment.
21. Derive an expression for reversible work done by a closed system under a given environment.
22. What are the availability function for a (i) non-flow process (ii) steady flow process?
23. What is the second-law efficiency? How does it differ from the first-law efficiency?
24. Prove that the second-law efficiency in the following cases:

    (i) $\eta_{II} = \frac{\eta_1}{\eta_{rev}}$ for heat engine,

(ii) $\eta_{II} = \frac{(COP)_R}{(COP_{rev})_R}$ for refrigerator.

25. What is the difference between the availability function for non-flow system and Gibbs function. When will these two be equal?

26. Define the following terms:

(i) Helmholtz function, and

(ii) Gibbs function.

Assignment–2

1. A constant temperature source is maintained at 727 °C and the surroundings are at 17 °C. If 4000 kJ heat is transferred from the source reversibly, determine available and unavailable energy. [**Ans.** 2840 kJ, 1160 kJ].

2. A thermal energy reservoir at 1200 K can supply heat at a rate of 100,000 kJ/h. If the environment is at 20 °C, determine the availability of this supply heat. [**Ans.** 21 kW].

3. A turbine expands air adiabatically from 6 bar, 400 K, 200 m/s to 1 bar, 290 K, 50 m/s. Determine the actual work output and the optimum work output possible for the same operating conditions. Also, find the irreversibility of the process. Take the surrounding atmosphere to be at 1 bar and 280 K. [**Ans.** 129.24 kJ/kg, 182.78 kJ/kg, 53.54 kJ/kg].

4. Determine the closed system availability of nitrogen gas at 10 MPa and 320 K, when the environment is at 101 kPa and 25 °C. [**Ans.** 313.55 kJ/kg].

5. Determine the availability of a nitrogen gas stream at 10 MPa, 320 K, and 10 m/s, if the environment is at 101 kPa and 298 K.[**Ans.** 457.26 kJ/kg].

6. Determine the availability per unit mass of air at pressure of 6 bar and temperature of 500 K. Assuming atmosphere at 1 bar and 300 K. Neglect the change in kinetic energy and potential energy. [**Ans.** 81.69 kJ/kg].

7. Nitrogen gas at 500 kPa and 30 °C is throttled through as well-insulated throttle valve to a pressure of 130 kPa. The surrounding atmosphere is at 20 °C. Determine the irreversibility of this process. [**Ans.** 117.19 kJ/kg].

8. A heat engine receives heat from a source at 1000 K at the rate of 500 kJ/s, and rejects waste heat to a sink at 350 K. The surrounding environment is at 300 K. If the measured power output of the heat engine is 200 kW, determine (a) the availability, (b) the rate of irreversibility, and (c) the second-law efficiency of the heat engine. [**Ans.** 325 kJ/s, 125 kW, 61.53%].

9. A heat engine that receives heat from a furnace at 630 °C and rejects waste heat to a river has a thermal efficiency of 30 per cent and the second-law efficiency of 45 per cent. Determine the temperature of the river water. [**Ans.** 24.85 °C].

10. A heat engine operating between thermal reservoirs at 775 K and 310 K has a thermal efficiency of 36 per cent. Determine the second-law efficiency of the engine. [**Ans.** 60%].

11. How much of the 1200 kJ of thermal energy at 750 K can be converted to useful work when the environment is at 300 K?[**Ans.** 720 kJ].

12. Determine the decrease in available energy when 25 kg of water at 95 °C is mixed with 35 kg of water at 35 °C, the pressure being taken as constant and the temperature of the surroundings being 15 °C. Take c_p of water is 4.2 kJ/kgK [**Ans.** 281.81 kJ].

13. A rigid tank of volume 2.5 m^3 contains air at 200 kPa and 300 K. The air is heated by supplying heat from a reservoir at 600 K until the temperature reaches 500 K. The surrounding atmosphere is at 100 kPa and 300 K. Determine the maximum useful work and the irreversibility associated with the process. [**Ans.** 221.56 kJ, 221.56 kJ].

14. Air expands through a turbine from 500 kPa, 520 °C to 100 kPa, 300 °C. During expansion 10 kJ/kg of heat is lost to the surroundings which is at 98 kPa, 20 °C. Determine per unit mass of air (i) the decrease in availability, (ii) the maximum work, and (iii) the irreversibility. Take for air, $c_p = 1.005$ kJ/kgK, $R = 0.287$ kJ/kgK. [**Ans.** (i) 260.7 kJ/kg (ii) 260.7 kJ/kg (iii) 49.6 kJ/kg].

15. A piston-cylinder device contains 2 kg of air at 1 MPa and 400 °C. The air expands to a final state of 300 kPa and 150 °C, doing work. Heat losses from the system to the surroundings are estimated to be 75 kJ during the process. If the surrounding environment is at 100 kPa and 300 K, determine (i) the availability of the air at initial and final states, (ii) the reversible work, (iii) the irreversibility, and (iv) the second-law efficiency for the process. Treat the air to be a perfect gas. [**Ans.** (i) 311.12 kJ, 67.4 kJ (ii) 243.72 kJ (iii) 2.526 kJ (iv) 90%].

16. An ideal gas is flowing through an insulated pipe at the rate of 3 kg/s. There is a 10% pressure drop from inlet to exit of the pipe. What is the rate exergy loss because of the pressure drop due to friction? Take $R = 0.287$ kJ/kgK and $T_0 = 27$ °C. [**Ans.** 26.94 kW].

17. Air enters a compressor at 100 kPa, 30 °C, which is also the state of the environment. It leaves at 350 kPa, 141 °C, and 90 m/s. Neglecting inlet velocity and change in potential energy, determine (i) whether the compression is adiabatic or polytropic, (ii) the polytropic index if compression is not adiabatic? (iii) the isothermal efficiency, (iv) the minimum work input and the irreversibility, and (v) the second law efficiency. Take c_p of air is 1.0035 kJ/kgK. [**Ans.** (i) Polytropic (ii) 1.33 (iii) 97.7% (iv) –101.8 kJ/kg, 13.9 kJ/kg (v) 88%].

18. An insulated rigid tank contains 0.9 kg of air at 150 kPa and 20 °C. A paddle wheel inside the tank is now rotated by an external power source until the temperature in the tank rises to 55 °C. If the surrounding air is at $T_0 = 20$ °C, determine (i) the reversible work, and (ii) the irreversibility for this process.[**Ans.** (i) 1.25 J (ii) 21.36 kJ].

19. Air is compressed by a compressor from 95 kPa and 27 °C to 600 kPa and 277 °C at a rate of 3.6 kg/min. Neglecting the changes in kinetic and potential energies and assuming the surroundings is at 25 °C. Determine the reversible power input for this process. [**Ans.** 13.52 kW].

20. Argon gas enters an adiabatic compressor at 120 kPa and 30 °C with a velocity of 20 m/s and exits at 1.2 MPa, 530 °C, and 80 m/s. The inlet area of the

compressor is 130 cm^2. Assuming the surroundings to be at 25 °C, determine the reversible power input and exergy destroyed.[**Ans.** 126 kW, 4.12 kW].

21. Steam is throttled from 9 MPa and 500 °C to a pressure of 7 MPa. Determine the decrease in exergy of the steam during this process. Assume the surroundings to be at 25 °C. [**Ans.** 32.3 kJ/kg].

22. Air enters a nozzle steadily at 300 kPa and 87 °C with a velocity of 50 m/s and exits at 95 kPa and 300 m/s. The heat loss from the nozzle to the surrounding medium at 17 °C is estimated to be 4 kJ/kg. Determine (a) the exit temperature and (b) the exergy destroyed during this process.[**Ans.** (a) 39.5 °C, (b) 58.4 kJ/kg].

23. Steam enters an adiabatic turbine at 6 MPa, 600 °C, and 80 m/s and leaves at 50 kPa, 100 °C, and 140 m/s. If the power output of the turbine is 5 MW, determine (a) the reversible power output and (b) the second-law efficiency of the turbine. Assume the surroundings to be. at 25 °C.[**Ans.** (a) 5.84 MW, (b) 85.61 per cent].

24. The cylinder of an internal combustion engine contains gases at 2500 °C, 60 bar at TDC (top dead center). Expansion takes place through a volume ratio of 9 and the index of expansion is 1.38 on p–v plane. The surroundings are at 20 °C, 1.1 bar. For unit mass, determine (i) the loss of availability involved, and (ii) the loss of available energy of this process. Treat the gases as perfect having $R = 0.26$ kJ/kgK and $c_v = 0.82$ kJ/kgK. [**Ans.** (i) 1148.28 kJ/kg (ii) 180 kJ/kg].

Hints:

(i) Loss of availability a per unit mass:

$$\phi_1 - \phi_2 = c_v(T_1 - T_2) - T_0(s_1 - s_2) + p_0(v_1 - v_2)$$
$$= c_v(T_1 - T_2) - T_0\left(c_v \log_e \frac{T_1}{T_2} + R \log_e \frac{v_1}{v_2}\right) + p_0(v_1 - v_2)$$

(ii) Loss of available energy per unit mass:

$$i = w_{max} - w$$

OR

$$i = T_0 \Delta s_{uni}$$
$$= T_0\left(\Delta s_{sys} + \Delta s_{sur}\right) = T_0\left[(s_2 - s_1) + \frac{q}{T_0}\right]$$

where

$$q = c_v\left(\frac{\gamma - n}{1 - n}\right)(T_1 - T_2).$$

Appendix

See Tables A.1, A.2, A.3, A.4, A.5, A.6, A.7, A.8, A.9, A.10, A.11, A.12, A.13, and A.14.

© The Author(s) 2022
S. Kumar, *Thermal Engineering Volume 1*,
https://doi.org/10.1007/978-3-030-67274-4

Table A.1 Properties of saturated steam (temperature based)

Sat. temp. °C T_{sat}	Sat. pressure kPa p_{sat}	Specific volume m³/kg		Specific Internal Energy kJ/kg		Specific enthalpy kJ/kg			Specific Entropy kJ/kgK		
		Sat. liquid v_f	Sat. vapor v_g	Sat. liquid u_f	Sat. vapor u_g	Sat. liquid h_f	Evap. h_{fg}	Sat. vapor h_g	Sat. liquid s_f	Evap. s_{fg}	Sat. vapor s_g
0.01	0.6113	0.0010002	206.2	0.00	3237.3	0.01	2501.3	2501.4	0.000	9.156	9.156
1	0.6567	0.0010002	192.6	4.12	2376.7	4.2	2499.0	2503.2	0.015	9.115	9.130
2	0.6556	0.0010001	179.9	8.4	2378.1	8.4	2496.7	2505.0	0.031	9.073	9.104
3	0.7577	0.0010001	168.1	12.6	2379.5	12.6	2494.3	2506.9	0.046	9.032	9.077
4	0.8131	0.0010001	157.2	16.8	2380.9	16.8	2491.9	2508.7	0.061	8.990	9.051
5	0.8721	0.0010001	147.1	21.0	2382.3	21.0	2489.6	2510.6	0.076	8.950	9.026
6	0.9349	0.0010001	137.7	25.2	2383.6	25.2	2487.2	2512.4	0.091	8.909	9.000
7	1.002	0.0010002	129.0	29.4	2385.0	29.4	2484.8	2514.2	0.106	8.869	8.975
8	1.072	0.0010002	120.9	33.6	2386.4	33.6	2482.5	2516.1	0.121	8.829	8.950
9	1.148	0.0010003	113.4	37.8	2387.8	37.8	2480.1	2517.9	0.136	8.789	8.925
10	1.228	0.0010004	106.4	42.0	2389.2	42.0	2477.7	2519.7	0.151	8.750	8.901
11	1.312	0.0010004	99.86	46.2	2390.5	46.2	2475.4	2521.6	0.166	8.711	8.877
12	1.402	0.0010005	93.78	50.4	2391.9	50.4	2473.0	2523.4	0.181	8.672	8.852
13	1.497	0.001007	88.12	54.6	2393.3	54.6	2470.7	2525.3	0.195	8.632	8.828
14	1.598	0.0010008	82.85	58.8	2394.7	58.8	2468.3	2527.1	0.210	8.595	8.805
15	1.705	0.0010009	77.93	63.0	2396.1	63.0	2465.9	2528.9	0.224	8.5.57	8.781
16	1.818	0.001001	73.33	67.2	2397.4	67.2	2463.6	2530.8	0.239	8.519	8.758
17	1.938	0.001001	69.04	71.4	2398.8	71.4	2461.2	2532.6	0.253	8.482	8.735

(continued)

Table A.1 (continued)

Sat. temp. °C	Sat. pressure kPa	Specific volume m³/kg		Specific Internal Energy kJ/kg		Specific enthalpy kJ/kg			Specific Entropy kJ/kgK		
T_{sat}	p_{sat}	Sat. liquid v_f	Sat. vapor v_g	Sat. liquid u_f	Sat. vapor u_g	Sat. liquid h_f	Evap. h_{fg}	Sat. vapor h_g	Sat. liquid s_f	Evap. s_{fg}	Sat. vapor s_g
18	2.064	0.001001	65.04	75.6	2400.2	75.6	2458.8	2534.4	0.268	8.444	8.712
19	2.198	0.001002	61.29	79.8	2401.6	79.8	2456.5	2536.3	0.282	8.407	8.690
20	2.339	0.001002	57.79	84.0	2402.9	84.0	2454.1	2538.1	0.297	8.371	8.667
21	2.487	0.001002	54.51	88.1	2404.3	88.1	2451.8	2539.9	0.311	8.334	8.645
22	2.645	0.001002	51.45	92.3	2405.7	92.3	2449.4	2541.7	0.325	8.298	8.623
23	2.810	0.001002	48.57	96.5	2407.0	96.5	2447.0	2543.5	0.339	8.262	8.601
24	2.985	0.001003	45.88	100.7	2408.4	100.7	2444.7	2545.4	0.353	8.226	8.579
25	3.169	0.001003	43.36	104.9	2409.8	104.9	2442.3	2547.2	0.367	8.191	8.558
26	3.363	0.001003	40.99	109.1	2411.1	109.1	2439.9	2549.0	0.382	8.155	8.537
27	3.567	0.001004	38.77	113.2	2412.5	113.2	2437.6	2550.8	0.396	8.120	8.516
28	3.782	0.001004	36.69	117.4	2413.9	117.4	2435.2	2552.6	0.409	8.086	8.495
29	4.008	0.001004	34.73	121.6	2415.2	121.6	2432.8	2554.5	0.423	8.051	8.474
30	4.246	0.001004	32.89	125.8	2416.6	125.8	2430.5	2556.3	0.437	8.016	8.453
31	4.496	0.001005	31.17	130.0	2418.0	130.0	2428.1	2558.1	0.451	7.982	8.433
32	4.759	0.001005	29.54	134.2	2419.3	134.2	2425.7	2559.9	0.464	7.948	8.413
33	5.034	0.001005	28.01	138.3	2420.7	138.3	2423.4	2561.7	0.478	7.915	8.393
34	5.324	0.001006	26.57	142.5	2422.0	142.5	2421.0	2563.5	0.492	7.881	8.373

(continued)

Table A.1 (continued)

Sat. temp. °C	Sat. pressure kPa	Specific volume m³/kg		Specific Internal Energy kJ/kg		Specific enthalpy kJ/kg			Specific Entropy kJ/kgK		
T_{sat}	p_{sat}	Sat. liquid v_f	Sat. vapor v_g	Sat. liquid u_f	Sat. vapor u_g	Sat. liquid h_f	Evap. h_{fg}	Sat. vapor h_g	Sat. liquid s_f	Evap. s_{fg}	Sat. vapor s_g
35	5.628	0.001006	25.22	146.7	2423.4	146.7	2418.6	2565.3	0.505	7.848	8.353
36	5.947	0.001006	23.94	150.9	2424.7	150.9	2416.2	2567.1	0.519	7.815	8.334
37	6.281	0.001007	22.74	155.0	2426.1	155.0	2413.9	2568.9	0.532	7.782	8.314
38	6.632	0.001007	21.60	159.2	2427.4	159.2	2411.5	2570.7	0.546	7.749	8.295
39	6.999	0.001007	20.53	163.4	2428.8	163.4	2409.1	2572.5	0.559	7.717	8.276
40	7.384	0.001008	19.52	167.6	2430.1	167.6	2406.7	2574.3	0.573	7.685	8.257
41	7.786	0.001008	18.57	171.7	2431.5	171.7	2404.3	2576.0	0.586	7.652	8.238
42	8.208	0.001009	17.67	175.9	2432.8	175.9	2401.9	2577.8	0.599	7.621	8.220
43	8.649	0.001009	16.82	180.1	2434.2	180.1	2399.5	2579.6	0.612	7.589	8.201
44	9.111	0.001010	16.02	184.3	2435.5	184.3	2397.2	2581.5	0.626	7.557	8.183
45	9.593	0.001010	15.26	188.4	2436.8	188.4	2394.8	2583.2	0.639	7.526	8.165
46	10.10	0.001010	14.54	192.6	2438.2	192.6	2392.4	2585.0	6.652	7.495	8.147
47	10.62	0.001011	13.86	196.8	2439.5	196.8	2390.0	2586.8	0.665	7.464	8.129
48	11.48	0.001011	13.22	201.0	2440.8	201.0	2387.6	2588.6	0.678	7.433	8.111
49	11.75	0.001012	12.61	205.1	2442.2	205.1	2385.2	2590.3	0.691	7.403	8.094
50	12.65	0.001012	12.03	209.3	2443.5	209.3	2382.7	2592.1	0.704	7.373	8.076
52	13.63	0.001013	10.97	217.7	2446.1	217.7	2377.9	2595.6	0.730	7.312	8.042

(continued)

Table A.1 (continued)

Sat. temp. °C	Sat. pressure kPa	Specific volume m³/kg		Specific Internal Energy kJ/kg		Specific enthalpy kJ/kg			Specific Entropy kJ/kgK		
T_{sat}	P_{sat}	Sat. liquid v_f	Sat. vapor v_g	Sat. liquid u_f	Sat. vapor u_g	Sat. liquid h_f	Evap. h_{fg}	Sat. vapor h_g	Sat. liquid s_f	Evap. s_{fg}	Sat. vapor s_g
54	15.02	0.001014	10.01	226.0	2448.8	226.0	2373.1	2599.1	0.755	7.253	8.008
55	15.76	0.001015	9.568	230.2	2450.1	230.2	2370.7	2600.9	0.768	7.223	7.991
56	16.53	0.001015	9.149	234.4	2451.4	234.4	2368.2	2602.6	0.781	7.194	7.975
58	18.17	0.001016	8.372	242.8	2454.0	242.8	2363.4	2606.2	0.806	7.136	7.942
60	19.94	0.001017	7.671	.251.1	2456.6	251.1	2358.5	2609.6	0.831	7.078	7.909
62	21.86	0.001018	7.037	259.5	2459.3	259.5	2353.6	2613.1	0.856	7.022	7.878
64	23.03	0.001019	6.463	267.9	2461.8	267.9	2348.7	2616.5	0.881	6.965	7.846
65	25.03	0.001020	6.197	272.0	2463.1	272.1	2346.2	2618.3	0.894	6.937	7.831
66	26.17	0.001020	5.943	276.2	2464.4	276.2	2343.7	2619.9	0.906	6.910	7.816
68	28.59	0.001022	5.471	284.6	2467.0	284.6	2338.8	2623.4	0.930	6.855	7.785
70	31.49	0.001023	5.042	293.0	2469.6	293.0	2333.8	2626.8	0.955	6.800	7.755
72	34.00	0.001024	4.650	301.4	2472.1	301.4	2329.8	2630.2	0.979	6.746	7.725
75	38.58	0.001026	4.131	313.9	2475.9	313.9	2321.4	2635.3	1.015	1.667	7.682
80	47.39	0.001029	3.407	334.9	2482.2	334.9	2308.8	2643.7	1.075	6.537	7.612
85	57.83	0.001033	2.828	355.9	2488.4	355.9	2296.0	2651.9	1.134	6.410	7.544
90	70.14	0.001036	2.361	376.9	2494.5	376.9	2283.2	2660.1	5.192	6.287	7.479
95	84.65	0.001040	1.982	397.9	2500.6	397.9	2270.2	2668.1	1.250	6.166	7.416

(continued)

Table A.1 (continued)

Sat. temp. °C T_{sat}	Sat. pressure kPa p_{sat}	Specific volume m³/kg		Specific Internal Energy kJ/kg		Specific enthalpy kJ/kg			Specific Entropy kJ/kgK		
		Sat. liquid v_f	Sat. vapor v_g	Sat. liquid u_f	Sat. vapor u_g	Sat. liquid h_f	Evap. h_{fg}	Sat. vapor h_g	Sat. liquid s_f	Evap. s_{fg}	Sat. vapor s_g
100	101.35	0.001044	1.673	418.9	2506.5	419.0	2257.0	2676.0	1.307	6.048	7.355
105	1.2082×10^2	0.001048	1.4194	440.02	2512.4	440.15	2243.7	2683.8	1.3630	5.9328	7.2958
110	1.4327×10^2	0.001052	1.2102	461.14	2518.1	461.30	2230.2	2691.5	1.4185	5.8202	7.2387
115	1.6906×10^2	0.001056	1.0366	482.30	2523.7	482.48	2216.5	2699.0	1.4734	5.7100	7.1833
120	1.9853×10^2	0.001060	0.8919	503.50	2529.3	503.71	2202.6	2706.3	1.5276	5.6020	7.1296
125	2.321×10^2	0.001065	0.7706	524.74	2534.6	524.99	2188.5	2713.5	1.5813	5.4962	7.0775
130	2.701×10^2	0.001070	0.6685	546.02	2539.9	546.31	2174.2	2720.5	1.6344	5.3925	7.0269
135	3.130×10^2	0.001075	0.5822	567.35	2545.0	567.69	2159.6	2727.3	1.6870	5.2907	6.9777
140	3.613×10^2	0.001080	0.5089	588.74	2550.0	589.13	2144.7	2733.9	1.7391	5.1908	6.9299
145	4.154×10^2	0.001085	0.4463	610.18	2554.9	610.63	2129.6	2740.3	1.7907	5.0926	6.8379
150	4.758×10^2	0.001091	0.3928	631.68	2559.5	632.20	2114.3	2746.5	1.8418	4.9960	6.8379
155	5.431×10^2	0.001096	0.3468	653.24	2564.1	653.84	2098.6	2752.4	1.8925	4.9010	6.7935
160	6.178×10^2	0.001102	0.3071	674.87	2568.4	675.55	2082.6	2758.1	1.9427	4.8075	6.7502
165	7.005×10^2	0.001108	0.2727	696.56	2572.5	697.34	2066.2	2763.5	1.9925	4.7153	6.7078
170	7.917×10^2	0.001114	0.2428	718.33	2576.5	719.21	2049.5	2768.7	2.0419	4.6244	6.6663
175	8.920×10^2	0.001121	0.2168	740.17	2580.2	741.17	2032.4	2773.6	2.0909	4.5347	6.6256
180	10.021×10^2	0.001127	0.19405	762.09	2583.7	763.22	2015.0	2778.2	2.1396	4.4461	6.5857

(continued)

Table A.1 (continued)

Sat. temp. °C T_{sat}	Sat. pressure kPa p_{sat}	Specific volume m³/kg		Specific Internal Energy kJ/kg		Specific enthalpy kJ/kg			Specific Entropy kJ/kgK		
		Sat. liquid v_f	Sat. vapor v_g	Sat. liquid u_f	Sat. vapor u_g	Sat. liquid h_f	Evap. h_{fg}	Sat. vapor h_g	Sat. liquid s_f	Evap. s_{fg}	Sat. vapor s_g
185	11.227×10^2	0.001134	0.17409	784.10	2587.0	785.37	1997.1	2782.4	2.1879	4.3586	6.5465
190	12.544×10^2	0.001141	0.15654	806.19	2590.0	807.62	1978.8	2786.4	2.2359	4.2720	6.5079
195	13.078×10^2	0.001149	0.14105	828.37	2592.8	829.98	1960.0	2790.0	2.2835	4.1863	6.4698
200	15.538×10^2	0.001157	0.12736	850.65	2595.3	852.45	1940.7	2793.2	2.3309	4.1014	6.4323
205	17.230×10^2	0.001164	0.11521	873.04	2597.5	875.04	1921.0	2796.0	2.3780	4.0172	6.3952
210	19.062×10^2	0.001173	0.10441	895.53	2599.5	897.76	1900.7	2798.5	2.4248	3.9337	6.3585
215	21.04×10^2	0.001181	0.09479	918.14	2601.1	920.62	1879.9	2800.5	2.4714	3.8507	6.3221
220	23.18×10^2	0.001190	0.08619	940.87	2602.4	943.62	1858.5	2802.1	2.5178	3.7683	6.2861
225	25.48×10^2	0.001199	0.07849	963.73	2603.3	966.78	1836.5	2803.3	2.5639	3.6863	6.2503
230	27.95×10^2	0.001209	0.07158	986.74	2603.9	990.12	1813.8	2804.0	2.6099	3.6047	6.2146
235	30.60×10^2	0.001219	0.06537	1009.89	2604.1	1013.62	1790.5	2804.2	2.6558	3.5233	6.1791
240	33.44×10^2	0.001229	0.05976	1033.21	2604.0	1037.32	1766.5	2803.8	2.7015	3.4422	6.1437
245	36.48×10^2	0.001240	0.05471	1056.71	2603.4	1061.23	1741.7	2803.0	2.7472	3.3612	6.1083
250	39.73×10^2	0.001251	0.05013	1080.39	2602.4	1085.36	1716.2	2801.5	2.7927	3.2802	6.0730
255	43.19×10^2	0.001263	0.04598	1104.28	2600.9	1109.73	1689.8	2799.5	2.8383	3.1992	6.0375
260	46.88×10^2	0.001276	0.04221	1128.39	2599.0	1134.37	1662.5	2796.9	2.8838	3.1181	6.0019
265	50.81×10^2	0.001289	0.03877	1152.74	2596.6	1159.28	1634.4	2793.6	2.9294	3.0368	5.9662

(continued)

Table A.1 (continued)

T_{sat}	p_{sat}	Sat. liquid v_f	Sat. vapor v_g	Sat. liquid u_f	Sat. vapor u_g	Sat. liquid h_f	Evap. h_{fg}	Sat. vapor h_g	Sat. liquid s_f	Evap. s_{fg}	Sat. vapor s_g
Sat. temp. °C	Sat. pressure kPa	Specific volume m³/kg		Specific Internal Energy kJ/kg		Specific enthalpy kJ/kg			Specific Entropy kJ/kgK		
270	54.99×10^2	0.001302	0.03564	1177.36	2593.7	1184.51	1605.2	2789.7	2.9751	2.9551	5.9301
275	59.42×10^2	0.001317	0.03279	1202.25	2590.2	1210.07	1574.9	2785.0	3.0208	2.8730	5.8938
280	64.12×10^2	0.001332	0.03017	1227.46	2586.1	1235.99	1543.6	2779.6	3.0668	2.7903	5.8571
285	69.09×10^2	0.001348	0.02777	1253.00	2581.4	1262.31	1511.0	2773.3	3.1130	2.7070	5.8199
290	74.36×10^2	0.001366	0.02557	1278.92	2576.0	1289.07	1477.1	2766.2	3.1594	2.6227	5.7821
295	79.93×10^2	0.001384	0.02354	1305.2	2569.9	1316.3	1441.8	2758.1	3.2062	2.5375	5.7437
300	85.81×10^2	0.001404	0.02167	1332.0	2563.0	1344.0	1404.9	2749.0	3.2534	2.4511	5.7045
305	92.02×10^2	0.001425	0.019948	1359.3	2555.2	1372.4	1366.4	2738.7	3.3010	2.3633	5.6643
310	98.56×10^2	0.001447	0.018350	1387.1	2546.4	1401.3	1326.0	2727.3	3.3493	2.2737	5.6230

Table A.2 Properties of saturated steam (pressure based)

Sat. pressure bar p_{sat}	Sat. temp. °C T_{sat}	Specific volume m³/kg		Specific internal energy kJ/kg		Specific enthalpy kJ/kg			Specific entropy kJ/kgK		
		Sat. liquid v_f	Sat. vapor v_g	Sat. liquid u_f	Sat. vapor u_g	Sat. liquid h_f	Evap. h_{fg}	Sat. vapor h_g	Sat. liquid s_f	Evap. s_{fg}	Sat. vapor s_g
0.006113	0.01	0.0010002	206.140	0.00	2375.3	0.01	2501.3	2501.4	0.000	9.156	9.156
0.010	7.0	0.0010000	129.21	29.3	2385.0	29.3	2484.9	2514.2	0.106	8.870	8.976
0.105	13.0	0.0010007	87.98	54.7	2393.3	54.7	2470.6	2525.3	0.196	8.632	8.828
0.020	17.0	0.001001	67.00	73.5	2399.5	73.5	2460.0	2533.5-	0.261	8.463	8.724
0.025	21.1	0.001002	54.25	88.5	2404.4	88.5	2451.6	2540.1	0.312	8.331	8.643
0.030	24.1	0.001003	45.67	101.0	2408.5	101.0	2444.5	2545.5	0.355	8.223	8.578
0.035	26.7	0.001003	39.50	111.9	2412.1	111.9	2438.4	2550.3	0.391	8.132	8.523
0.040	29.0	0.001004	34.80	121.5	2415.2	121.5	2432.9	2554.4	0.423	8.052	8.475
0.045	31.0	0.001005	31.13	130.0	2417.9	130.0	2428.2	2558.2	0.451	7.982	8.433
0.050	32.9	0.001005	28.19	137.8	2420.5	137.8	2423.7	2561.5	0.476	7.919	8.395
0.055	34.6	0.001006	25.77	144.9	2422.8	144.9	2419.6	2565.5	0.500	7.861	8.361
0.060	36.2	0.001006	23.74	151.5	2425.0	151.5	2415.9	2567.4	0.521	7.809	8.330
0.065	37.6	0.001007	22.01	157.7	2426.9	157.7	2412.4	2570.1	0.541	7.761	8.302
0.070	39.0	0.001007	20.53	163.4	2428.8	163.4	2409.1	2572.5	0.559	7.717	8.276
0.075	40.3	0.001008	19.24	168.8	2430.5	168.8	2406.0	2574.8	0.576	7.675	8.251
0.080	41.5	0.001008	18.10	173.9	2432.2	173.9	2403.1	2577.0	0.593	7.636	8.229
0.085	42.7	0.001009	17.10	178.7	2433.7	178.7	2400.3	2579.0	0.608	7.599	8.207
0.090	43.8	0.001009	16.20	183.3	2435.2	183.3	2397.7	2581.0	0.622	7.565	8.187
0.095	44.8	0.001010	15.40	187.7	2436.6	187.7	2395.2	2582.9	0.636	7.532	8.168

(continued)

Table A.2 (continued)

Sat. pressure bar p_{sat}	Sat. temp. °C T_{sat}	Specific volume m³/kg		Specific internal energy kJ/kg		Specific enthalpy kJ/kg			Specific entropy kJ/kgK		
		Sat. liquid v_f	Sat. vapor v_g	Sat. liquid u_f	Sat. vapor u_g	Sat. liquid h_f	Evap. h_{fg}	Sat. vapor h_g	Sat. liquid s_f	Evap. s_{fg}	Sat. vapor s_g
0.10	45.8	0.001010	14.67	191.8	2437.9	191.8	2392.8	2584.7	0.649	7.501	8.150
0.11	47.7	0.001011	13.42	199.7	2440.4	199.7	2388.3	2588.0	0.674	7.453	8.117
0.12	49.4	0.001012	12.36	206.9	2442.7	206.9	2384.2	2591.1	0.696	7.390	8.086
0.13	51.0	0.001013	11.47	213.7	2.444.9	213.7	2380.2	2593.9	0.717	7.341	8.058
0.14	52.6	0.001013	10.69	220.0	2446.9	220.0	2376.6	2596.6	0.737	7.296	8.033
0.15	54.0	0.001014	10.02	225.9	2448.7	225.9	2373.2	2559.1	0.755	7.254	8.009
0.16	55.3	0.001015	9.43	231.6	2450.5	231.6	2369.9	2601.5	0.772	7.214	7.986
0.17	56.6	0.001015	8.91	236.9	2452.2	236.9	2366.8	2603.7	0.788	7.177	7.%5
0.18	57.8	0.001016	8.45	242.0	2453.8	242.0	2363.8	2505.8	0.804	7.141	7.945
0.19	59.0	0.001017	8.03	246.8	2455.3	246.8	2361.0	2607.8	0.818	7.108	7.926
0.20	60.1	0.001017	7.65	251.4	2456.7	251.4	2358.3	2609.7	0.832	7.077	7.909
0.22	62.1	0.001018	7.45	260.1	2459.4	260.1	2353.2	2613.3	0.858	7.018	7.876
0.24	64.1	0.001019	6.45	268.1	2461.9	268.1	2348.5	2616.6	0.882	6.964	7.846
0.25	65.0	0.001020	6.20	271.9	2463.1	271.9	2346.3	2618.2	0.893	6.938	7.831
0.26	65.9	0.001020	5.98	275.6	2464.2	275.6	2344.1	2619.7	0.904	6.914	7.818
0.28	67.5	0.001021	5.58	282.6	2466.4	282.6	2340.0	2622.6	0.925	6.868	7.793
0.30	69.1	0.001022	5.23	289.2	2468.4	289.2	2336.1	2625.3	0.944	6.825	7.768
0.32	70.6	0.001023	4.92	295.5	2470.3	295.5	2332.4	2627.9	0.962	6.784	7.746

(continued)

Table A.2 (continued)

Sat. pressure bar	Sat. temp. °C	Specific volume m³/kg		Specific internal energy kJ/kg		Specific enthalpy kJ/kg			Specific entropy kJ/kgK		
p_{sat}	T_{sat}	Sat. liquid v_f	Sat. vapor v_g	Sat. liquid u_f	Sat. vapor u_g	Sat. liquid h_f	Evap. h_{fg}	Sat. vapor h_g	Sat. liquid s_f	Evap. s_{fg}	Sat. vapor s_g
0.34	72.0	0.001024	4.65	301.4	2472.1	301.4	2328.8	2630.2	0.979	6.746	7.725
0.35	72.7	0.001024	4.53	304.2	2473.0	304.2	2327.2	2631.4	0.987	6.728	7.715
0.36	73.4	0.001025	4.41	307.0	2473.8	307.0	2325.5	2632.5	0.996	6.710	7.706
0.38	74.6	0.001026	4.19	312.4	2475.5	312.4	2322.3	2634.7	1.011	6.676	7.687
0.40	75.9	0.001027	3.99	317.6	2477.0	317.6	2319.2	2636.8	1.026	6.644	7.670
0.45	78.7	0.001028	3.58	329.6	2480.7	329.6	2312.0	2641.6	1.060	6.571	7.631
0.50	81.3	0.001030	3.24	340.5	2483.9	340.6	2305.4	2646.0	1.091	6.503	7.594
0.55	83.7	0.001032	2.96	350.5	2486.8	350.5	2299.3	2649.8	1.119	6.442	7.561
0.60	85.9	0.001033	2.73	359.8	2489.6	359.9	2293.6	2653.5	1.145	6.387	7.532
0.65	88.0	0.001035	2.53	368.5	2492.1	368.5	2288.3	2656.8	1.169	6.335	7.504
0.70	90.0	0.001036	2.37	376.7	2494.5	376.7	2283.3	2660.0	1.192	6.288	7.480
0.75	91.8	0.001037	2.22	384.3	2496.7	384.4	2278.6	2663.0	1.213	6.243	7.456
0.80	93.5	0.001039	2.087	391.6	2498.8	391.7	2274.1	2665.8	1.233	6.202	7.435
0.85	95.1	0.001040	1.972	398.5	2500.7	398.6	2269.8	2668.4	1.252	6.163	7.415
0.90	96.7	0.001041	1.869	405.1	2502.6	405.2	2265.7	2670.9	1.270	6.125	7.395
0.95	98.2	0.001042	1.777	411.3	25,044	411.4	2261.8	2673.2	1.287	6.090	7.377
1.00	99.6	0.001043	1.694	417.4	2506.1	417.5	2258.0	2675.5	1.303	6.057	7.360
1.0135	**100**	**0.001044**	**1.673**	**418.9**	**2506.5**	**419.0**	**2256.9**	**2676.0**	**1.307**	**6.048**	**7.355**

(continued)

Table A.2 (continued)

Sat. pressure bar p_{sat}	Sat. temp. °C T_{sat}	Specific volume m³/kg		Specific internal energy kJ/kg		Specific enthalpy kJ/kg			Specific entropy kJ/kgK		
		Sat. liquid v_f	Sat. vapor v_g	Sat. liquid u_f	Sat. vapor u_g	Sat. liquid h_f	Evap. h_{fg}	Sat. vapor h_g	Sat. liquid s_f	Evap. s_{fg}	Sat. vapor s_g
1.1	102.3	0.001045	1.549	428.7	2509.2	428.8	2250.9	2679.7	1.333	5.994	7.327
1.2	104.8	0.001047	1.428	439.2	2512.1	439.3	2244.2	2683.5	1.361	5.937	7.298
1.3	107.1	0.001049	1.325	449.0	2514.8	449.1	2238.0	2687.1	1.387	5.884	7.271
1.4	109.3	0.001051	1.237	458.2	2517.3	458.4	2232.0	2690.4	1.411	5.835	7.246
1.5	111.4	0.001053	1.159	466.9	2519.7	467.1	2226.5	2693.6	1.434	5.789	7.223
1.6	113.3	0.001054	1.091	475.2	2521.9	475.4	2221.1	2696.5	1.455	5.747	7.202
1.7	115.2	0.001056	1.031	483.0	2523.9	483.2	2211.2	2699.2	1.475	5.706	7.181
1.8	116.9	0.001058	0.977	490.5	2525.9	490.7	2211.2	2701.8	1.494	5.668	7.162
1.9	118.6	0.001059	0.929	497.6	2527.7	497.8	2206.5	2704.3	1.513	5.631	7.144
2.0	120.2	0.001061	0.886	504.5	2529.5	504.7	2201.9	2706.7	1.530	5.597	7.127
2.1	121.8	0.001062	0.846	511.1	2531.2	511.3	2197.6	2708.9	1347	5.564	7.111
2.2	123.3	0.001063	0.810	517.4	2532.8	517.6	2193.4	2711.0	1.563	5.532	7.095
2.3	124.7	0.001065	0.777	523.5	2534.3	523.7	2189.3	2713.1	1378	5302	7.080
2.4	126.1	0.001066	0.747	529.4	2535.8	529.6	2185.4	2715.0	1.593	5.473	7.066
2.5	127.4	0.001067	0.719	535.1	2537.2	535.4	2118.5	2716.9	1.607	5.446	7.053
2.6	128.7	0.001069	0.693	540.6	2538.6	540.9	2177.8	2718.7	1.621	5.419	7.040
2.7	130.0	0.001070	0.669	546.0	2539.9	546.3	2174.2	2720.5	1.634	5.393	7.027
2.8	131.2	0.001071	0.646	551.2	2541.2	551.4	2170.7	2722.1	1.647	5.368	7.015

(continued)

Table A.2 (continued)

Sat. pressure bar	Sat. temp. °C	Specific volume m³/kg		Specific internal energy kJ/kg		Specific enthalpy kJ/kg			Specific entropy kJ/kgK		
p_{sat}	T_{sat}	Sat. liquid v_f	Sat. vapor v_g	Sat. liquid u_f	Sat. vapor u_g	Sat. liquid h_f	Evap. h_{fg}	Sat. vapor h_g	Sat. liquid s_f	Evap. s_{fg}	Sat. vapor s_g
2.9	132.4	0.001072	0.625	556.2	2542.4	556.5	2167.3	2723.8	1.660	5.343	7.003
3.0	133.5	0.001073	0.606	561.1	2543.6	561.6	2163.8	2725.3	1.672	5.320	6.992
3.1	134.7	0.001074	0.588	565.9	2544.7	566.3	2160.6	2726.8	1.684	5.297	6.981
3.2	135.8	0.001075	0.570	570.6	2545.8	571.0	2157.3	2728.3	1.695	5.275	6.970
3.3	136.8	0.001076	0.554	575.2	2546.9	575.5	2154.2	2729.7	1.706	5.254	6.960
3.4	137.9	0.001078	0.539	579.6	2547.9	580.0	2151.1	2731.1	1.717	5.233	6.950
3.5	138.9	0.001079	0.524	583.9	2548.9	584.3	2148.1	2732.4	1.728	5.213	6.941
3.6	140.0	0.001080	0.511	588.2	2549.9	588.6	2145.1	2733.7	1.738	5.193	6.931
3.7	140.8	0.001081	0.498	592.4	2550.9	592.8	2142.2	2735.0	1.748	5.174	6.922
3.8	141.8	0.001082	0.485	596.4	2551.8	596.8	2139.4	2736.2	1.758	5.155	6.913
3.9	142.7	0.001083	0.474	600.4	2552.7	600.8	2136.6	2737.4	1.767	5.137	6.904
4.0	143.6	0.001084	0.463	604.3	2553.5	604.7	2133.8	2738.5	1.777	5.119	6.896
4.1	144.5	0.001085	0.452	608.1	2554.4	608.6	2131.1	2739.7	1.786	5.102	6.888
4.2	145.4	0.001286	0.442	611.9	2555.2	612.4	2128.4	2740.8	1.795	5.085	6.880
4.3	146.3	0.001086	0.432	615.6	2556.1	616.1	2125.8	2741.9	1.804	5.068	6.872
4.4	147.1	0.001087	0.423	619.2	2556.9	619.7	2123.2	2742.9	1.812	5.052	6.684
4.5	147.9	0.001088	0.414	622.8	2557.6	623.3	2120.6	2743.9	1.821	5.036	6.857
4.6	148.7	0.001089	0.406	626.3	2558.4	626.8	2118.2	2744.9	1.829	5.020	6.849

(continued)

Table A.2 (continued)

Sat. pressure bar p_{sat}	Sat. temp. °C T_{sat}	Specific volume m³/kg		Specific internal energy kJ/kg		Specific enthalpy kJ/kg			Specific entropy kJ/kgK		
		Sat. liquid v_f	Sat. vapor v_g	Sat. liquid u_f	Sat. vapor u_g	Sat. liquid h_f	Evap. h_{fg}	Sat. vapor h_g	Sat. liquid s_f	Evap. s_{fg}	Sat. vapor s_g
4.7	149.5	0.001090	0.397	629.7	2559.1	63,012	2115.7	2745.9	1.837	5.005	6.842
4.8	150.3	0.001091	0.390	633.1	2559.8	633.6	2113.2	2746.8	1.845	4.990	6.835
4.9	151.1	0.001092	0.382	636.4	2560.6	636.9	2110.8	2747.8	1.853	4.975	6.828
5.0	151.9	0.001093	0.375	639.7	2561.2	640.2	2108.5	2748.7	1.861	4.961	6.821
5.2	153.3	0.001094	0.361	646.1	2562.6	646.7	2103.8	2750.5	1.876	4.932	6.808
5.4	154.8	0.001096	0.349	652.3	2563.9	652.9	2099.3	2752.1	1.890	4.905	6.795
5.5	155.5	0.001096	0.343	655.3	2564.5	655.9	2097.0	2752.9	1.897	4.892	6.789
5.6	156.2	0.001097	0.337	658.3	2565.1	658.9	2094.8	2753.8	1.904	4.879	6.783
5.8	157.5	0.001099	0.326	664.2	2566.3	664.8	2090.5	2755.3	1.918	4.853	6.771
6.0	158.9	0.001101	0.316	669.9	2567.4	670.6	2086.3	2756.8	1.931	4.829	6.760
6.2	160.1	0.001002	0.306	675.5	2568.5	676.2	2082.1	2758.3	1.944	4.805	6.749
6.4	161.4	0.001104	0.297	680.9	2569.6	681.6	2078.0	2759.6	1.956	4.782	6.738
6.5	162.0	0.001104	0.293	683.6	2570.1	684.3	2076.0	2760.3	1.963	4.770	6.733
6.6	162.6	0.001105	0.288	686.2	2570.6	686.9	2074.0	2761.0	1.969	4.759	6.728
6.8	163.8	0.001107	0.281	691.4	2571.6	692.1	2070.1	2762.2	1.981	4.737	6.718
7.0	165.0	0.001108	0.273	696.4	2572.5	697.2	2066.3	2763.5	1.992	4.716	6.708
7.2	166.1	0.001109	0.266	701.4	2573.4	702.2	2062.5	2764.7	2.004	4.695	6.699
7.4	167.2	0.001111	0.259	706.2	2574.3	707.1	2058.8	2765.9	2.014	4.675	6.689

(continued)

Table A.2 (continued)

Sat. pressure bar P_{sat}	Sat. temp. °C T_{sat}	Specific volume m³/kg		Specific internal energy kJ/kg		Specific enthalpy kJ/kg			Specific entropy kJ/kgK		
		Sat. liquid v_f	Sat. vapor v_g	Sat. liquid u_f	Sat. vapor u_g	Sat. liquid h_f	Evap. h_{fg}	Sat. vapor h_g	Sat. liquid s_f	Evap. s_{fg}	Sat. vapor s_g
7.5	167.8	0.001112	0.256	708.6	2574.7	709.5	2057.0	2766.5	2.020	4.665	6.685
7.6	168.3	0.001112	0.252	711.0	2575.2	711.8	2055.2	2767.0	2.025	4.655	6-680
7.8	169.4	0.001113	0.246	715.7	2576.0	716.5	2051.6	2768.1	2.036	4.635	6.671
8.0	170.4	0.001115	0.240	720.2	2575.8	721.1	2048.0	2769.1	2.046	4.617	6.663
8.2	171.5	0.001116	0.235	724.7	2577.6	725.6	2044.6	2770.2	2.056	4.598	6.654
8.4	172.5	0.001117	0.230	729.1	2578.3	730.1	2041.1	2771.2	2.066	4.580	6.646
8.5	173.0	0.001118	0.227	731.3	2578.7	732.2	2039.4	2771.6	2.071	4.571	6.642
8.6	173.5	0.001119	0.225	733.4	2579.1	734.4	2037.7	2772.1	2.076	4.562	6.638
8.8	174.4	0.001120	0.220	737.7	2579.8	738.6	2034.4	2773.0	2.085	4.545	6.630
9.0	175.4	0.001121	0.215	741.8	2580.5	742.8	2031.1	2773.9	2.095	4.528	6.623
9.2	176.3	0.001122	0.211	745.9	2581.1	747.0	2027.8	2774.8	2.104	4.511	6.615
9.4	177.2	0.001124	0.206	750.0	2581.8	751.0	2024.7	2775.7	2.113	4.495	6.608
9.5	177.7	0.001124	0.204	751.9	2582.1	753.0	2023.1	2776.1	2.117	4.487	6.604
9.6	178.1	0.001125	0.202	753.9	2582.4	755.0	2021.5	2776.5	2.122	4.479	6.601
9.8	179.0	0.001126	0.198	757.8	2583.1	758.9	2018.4	2777.3	2.130	4.463	6.593
10.0	179.9	0.001127	0.194	761.7	2583.6	762.8	2015.3	2778.1	2.139	4.448	6.587
10.5	182.0	0.001130	0.186	771.1	2585.1	772.2	2007.7	2779.9	2.159	4.411	6.570
11.0	184.1	0.001133	0.178	780.1	2586.4	781.3	2000.4	2781.7	2.179	4.374	6.553

(continued)

Table A.2 (continued)

Sat. pressure bar p_{sat}	Sat. temp. °C T_{sat}	Specific volume m³/kg		Specific internal energy kJ/kg		Specific enthalpy kJ/kg			Specific entropy kJ/kgK		
		Sat. liquid v_f	Sat. vapor v_g	Sat. liquid u_f	Sat. vapor u_g	Sat. liquid h_f	Evap. h_{fg}	Sat. vapor h_g	Sat. liquid s_f	Evap. s_{fg}	Sat. vapor s_g
11.5	186.1	0.001136	0.170	778.8	2587.6	790.1	1993.2	2783.3	2.198	4.340	6.538
12.0	188.0	0.001139	0163	797.3	2588.8	798.6	1986.2	2784.8	2.217	4.306	6.533
12.5	189.8	0.001141	0.157	805.4	2589.9	806.8	1979.4	2786.2	2.243	4.275	6.509
13.0	191.6	0.001144	0.151	813.4	2591.0	314.9	1972.7	2787.6	2.251	4.244	6.495
13.5	193.4	0.001146	0.146	821.2	2591.9	822.6	1966.2	2788.8	2.268	4.214	6.482
14.0	195.0	0.001149	0.141	828.7	2592.8	830.3	1959.7	2790.0	2.284	4.185	6.469
14.5	196.7	0.001151	0.136	836.0	2593.7	837.6	1953.5	27,911	2.300	4.157	6.457
15.0	198.3	0.001154	0.132	843.2	2594.5	844.9	1947.3	2792.2	2.315	4.130	6.445
15.5	199.9	0.001156	0.128	850.1	2595.3	851.9	1941.2	2793.1	2.330	4.103	6.433
16.0	201.4	0.001169	0.124	856.9	2596.0	858.8	1935.2	2794.0	2.344	4.078	6.422
16.5	202.9	0.001161	0.120	863.6	2596.6	865.5	1929.4	2794.9	2.358	4.053	6.411
17.0	204.3	0.001163	0.117	870.1	2597.3	872.1	1923.6	2795.7	2.372	4.028	6.400
17.5	205.8	0.001166	0.133	876.5	2597.8	878.5	1917.9	2796.4	2.385	4.005	6.390
18.0	207.2	0.001168	0.110	882.7	2598.4	884.8	1912.4	2797.2	2.398	3.981	6.379
18.5	208.5	0.001170	0.108	888.8	2598.9	891.0	1906.8	2797.8	2.411	3.958	6.369
19.0	209.8	0.001172	0.105	894.8	2599.4	897.0	1901.4	2798.4	2.423	3.936	3.359
19.5	211.1	0.001175	0.102	900.7	2599.9	903.0	1896.0	2799.0	2.435	3.915	6.350
20.0	212.4	0.001177	0.0996	906.4	2600.3	908.8	1890.7	2799.5	2.447	3.894	6.341

(continued)

Table A.2 (continued)

Sat. pressure bar	Sat. temp. °C	Specific volume m³/kg		Specific internal energy kJ/kg		Specific enthalpy kJ/kg			Specific entropy kJ/kgK		
p_{sat}	T_{sat}	Sat. liquid v_f	Sat. vapor v_g	Sat. liquid u_f	Sat. vapor u_g	Sat. liquid h_f	Evap. h_{fg}	Sat. vapor h_g	Sat. liquid s_f	Evap. s_{fg}	Sat. vapor s_g
21.0	214.9	0.001181	0.0950	917.7	2601.0	920.2	1880.3	2800.5	2.470	3.852	6.323
22.0	217.3	0.001185	0.0907	928.5	2601.7	931.1	1870.2	2801.3	2.493	3.813	6.306
23.6	219.6	0.001189	0.0869	939.0	2602.3	941.8	1860.2	2802.0	2.514	3.775	6.289
24.0	221.8	0.001193	0.0833	949.2	2602.8	952.1	1850.5	2802.6	2.535	3.738	6.273
25.0	224.0	0.001197	0.0801	959.1	2603.1	962.1	1841.0	2803.1	2.555	3.703	6.258
26.0	226.1	0.001201	0.0769	968.7	2603.5	971.9	1831.6	2803.5	2.574	3.669	6.243
27.0	228.1	0.001205	0.0741	978.1	2603.7	981.3	1822.4	2803.8	2.593	3.635	6.228
28.0	230.1	0.001209	0.0715	987.2	2603.9	990.6	1813.4	2804.0	2.611	3.603	6.214
29.0	232.0	0.001213	0.0690	996.1	2604.0	999.6	1804.5	2804.1	2.628	3.572	6.200
30.0	233.9	0.001217	0.0667	1004.8	2604.1	1008.4	1795.7	2804.2	2.646	3.541	6.187
31.0	235.7	0.001220	00,645	1013.3	2604.1	1017.0	1787.1	2804.1	2.662	3.512	6.174
32.0	237.5	0.001224	0.0625	1021.6	2604.1	1025.5	1778.6	2804.1	2.679	3.483	6161
33.0	239.2	0.001227	0.0606	1029.7	2604.0	1033.7	1770.2	2803.9	2.695	3.454	6.149
34.0	240.9	0.001231	0.0588	1037.6	2603.9	1041.8	1761.9	2803.7	2.710	3.427	6.137
35.0	242.6	0.001235	0.0571	1045.4	2503.7	1049.7	1753.7	2803.4	2.725	3.400	6.125
36.0	244.2	0.001238	0.0555	1053.1	2603.5	1057.5	1745.6	2803.1	2.740	3.374	6.114
37.0	245.8	0.001242	0.0539	1060.6	2603.3	1065.2	1737.6	2802.8	2.755	3.348	6.103
38.0	247.4	0.001245	0.0523	1068.0	2603.0	1072.7	1729.7	2802.4	2.769	3.323	6.092

(continued)

Table A.2 (continued)

Sat. pressure bar p_{sat}	Sat. temp. °C T_{sat}	Specific volume m³/kg		Specific internal energy kJ/kg		Specific enthalpy kJ/kg			Specific entropy kJ/kgK		
		Sat. liquid v_f	Sat. vapor v_g	Sat. liquid u_f	Sat. vapor u_g	Sat. liquid h_f	Evap. h_{fg}	Sat. vapor h_g	Sat. liquid s_f	Evap. s_{fg}	Sat. vapor s_g
39.0	248.9	0.001249	0.0511	1075.2	2602.6	1080.1	1721.8	2801.9	2.783	3.298	6.081
40.0	250.4	0.001252	0.0498	1082.3	2602.3	1087.3	1714.1	2801.4	2.796	3.274	6.070
42.0	253.3	0.001259	0.0473	1096.2	2601.5	1101.5	1698.8	2800.3	2.823	3.227	6.050
44.0	256.1	0.001266	0.0451	1109.7	2600.6	1115.2	1683.8	2799.0	2.849	3.181	6.030
45.0	257.5	0.001269	0.0441	1116.2	2600.1	1121.9	1676.4	2798.3	2.861	3.159	6.020
46.0	258.8	0.001273	0.0431	1122.7	2599.5	1128.6	1669.0	2797.6	2.873	3.137	6.010
48.0	261.4	0.001279	0.0412	1135.4	2598.4	1141.5	1654.5	2796.0	2.897	3.095	5.992
50.0	264.0	0.001286	0.0394	1147.8	2597.1	1154.2	1640.1	2794.3	2.920	3.053	5.973
52.0	266.5	0.001293	0.0378	1159.9	2595.8	1166.6	1626.0	2792.6	2.943	3.013	5.956
54.0	268.8	0.001299	0.0363	1171.7	2594.4	1178.7	1612.0	2790.7	2.965	2.974	5.939
55.0	270.0	0.001302	0.0357	1177.4	2593.7	1184.5	1605.1	2789.6	2.975	2.955	5.930
56.0	271.2	0.001306	0.0350	1183.2	2592.9	1190.5	1598.2	2788.6	2.986	2.936	5.922
58.0	273.4	0.001312	0.0337	1194.4	2591.3	1202.0	1584.5	2786.5	3.006	2.899	5.905
60.0	275.6	0.001319	0.0324	1205.4	2589.7	1213.3	1571.0	2784.3	3.027	2.862	5.889
62.0	277.8	0.001325	0.0313	1216.3	2588.0	1224.5	1557.6	2782.1	3.046	2.827	5.873
64.0	279.9	0.001332	0.0302	1226.9	2586.2	1235.4	1544.3	2779.7	3.066	2.792	5.858
65.0	280.8	0.001335	0.0297	1232.0	2585.3	1240.7	1537.9	2778.6	3.075	2.775	5.850
66.0	281.9	0.001338	0.0292	1237.3	2584.4	1246.1	1531.1	2777.2	3.085	2.758	5.843

(continued)

Table A.2 (continued)

Sat. pressure bar p_{sat}	Sat. temp. °C T_{sat}	Specific volume m³/kg		Specific internal energy kJ/kg		Specific enthalpy kJ/kg			Specific entropy kJ/kgK		
		Sat. liquid v_f	Sat. vapor v_g	Sat. liquid u_f	Sat. vapor u_g	Sat. liquid h_f	Evap. h_{fg}	Sat. vapor h_g	Sat. liquid s_f	Evap. s_{fg}	Sat. vapor s_g
68.0	283.9	0.001345	0.0283	1247.5	2582.5	1256.6	1518.1	2774.7	3.103	2.725	5.828
70.0	285.9	0.001351	0.0274	1257.5	2580.5	1267.0	1505.1	2772.1	3.121	2.692	5.813
72.0	287.8	0.001358	0.0265	1267.4	2578.5	1277.2	1492.2	2769.4	3.139	2.660	5.799
74.0	289.7	0.001364	0.0257	1277.2	2576.4	1287.3	1479.4	2766.7	3.156	2.629	5.785
75.0	290.6	0.001368	0.0253	1282.0	2575.3	1292.2	1473.1	2765.3	3.165	2.613	5.778
76.0	291.5	0.001371	0.0249	1286.8	2574.3	1297.2	1466.6	2763.8	3.174	2.597	5.771
78.0	293.3	0.001378	0.0242	1296.2	2572.1	1307.0	1453.9	2760.9	3.190	2.567	5.757
80.0	295.1	0.001384	0.0235	1305.6	2569.8	1316.6	1441.3	2758.0	3.207	2.536	5.743
85.0	299.3	0.001401	0.0219	1328.4	2564.0	1340.3	1410.0	2750.3	3.247	2.463	5.710
90.0	303.4	0.001418	0.0205	1350.5	2557.8	1363.2	1378.9	2742.1	3.286	2.391	5.677
95.0	307.3	0.001435	0.0192	1372.0	2551.2	1385.6	1348.0	2733.6	3.323	2.322	5.645
100.00	311.1	0.001452	0.0180	1393.0	2544.4	1407.6	1317.1	2724.7	3.360	2.254	5.614
105.0	314.7	0.001470	0.0170	1413.6	2537.7	1429.0	1286.4	2715.4	3.395	2.188	5.583
110.0	318.2	0.001489	0.0160	1433.7	2529.8	1450.1	1255.5	2705.6	3.430	2.123	5.553
115.0	321.5	0.001507	0.0151	1453.5	2522.2	1470.8	1224.6	2695.4	3.463	2.059	5.522
120.0	324.8	0.001527	0.0143	1473.0	2513.7	1491.3	1193.6	2684.9	3.496	1.996	5.492
125.0	327.9	0.001547	0.0135	1492.1	2505.1	1511.5	1162.2	2673.7	3.528	1.934	5.462
130.0	330.9	0.001567	0.0128	1511.1	2496.1	1531.5	1130.7	2662.2	3.561	1.871	5.432

(continued)

Table A.2 (continued)

Sat. pressure bar p_{sat}	Sat. temp. °C T_{sat}	Specific volume m³/kg		Specific internal energy kJ/kg		Specific enthalpy kJ/kg			Specific entropy kJ/kgK		
		Sat. liquid v_f	Sat. vapor v_g	Sat. liquid u_f	Sat. vapor u_g	Sat. liquid h_f	Evap. h_{fg}	Sat. vapor h_g	Sat. liquid s_f	Evap. s_{fg}	Sat. vapor s_g
135.0	333.9	0.001588	0.0121	1529.9	2486.6	1551.4	1098.8	2650.2	3.592	1.810	5.402
140.0	336.8	0.001611	0.0115	1548.6	2476.8	1571.1	1066.5	2637.6	3.623	1.749	5.372
145.0	339.5	0.001634	0.0109	1567.1	2466.4	1590.9	1033.5	2624.4	3.654	1.687	5.341
150.0	342.2	0.001658	0.0103	1585.6	2455.5	1610.5	1000.0	2610.5	3.685	1.625	5.310
155.0	344.9	0.001684	0.00981	1604.1	2443.9	1630.3	965.7	2596.0	3.715	1.563	5.278
160.0	347.4	0.001711	0.00931	1622.7	2431.7	1650.1	930.6	7.580.6	3.746	1.499	5.245
165.0	349.9	0.001740	0.00883	1641.4	2418.8	1670.1	894.3	2564.4	3.777	1.435	5.212
170.0	352.4	0.001770	0.00836	1660.2	2405.0	1690.3	856.9	2547.2	3.808	1.370	5.178
175.0	354.7	0.001804	0.00793	1679.4	2390.2	1711.0	817.8	2528.8	3.839	1.302	5.141
180.0	357.1	0.001840	0.00749	1698.9	2374.3	1732.0	777.1	2509.1	3.871	1.233	5.104
185.0	359.3	0.001880	0.00708	1719.1	2357.0	1753.9	733.9	2487.8	3.905	1.160	5.065
190.0	361.5	0.001924	0.00666	1739.9	2338.1	1776.5	688.0	2464.5	3.939	1.084	5.023
195.0	363.7	0.001976	0.00625	1762.0	2316.9	1800.6	638.2	2438.8	3.975	1.002	4.977
200.0	365.8	0.002036	0.00583	1785.6	2293.0	1826.3	583.4	2409.7	4.014	0.913	4.927
205.0	367.9	0.002110	0.00541	1811.8	2265.0	1855.0	520.8	2375.8	4.057	0.812	4.869
210.0	369.9	0.002207	0.00495	1842.1	2230.6	1888.4	446.2	2334.6	4.107	0.694	4.801
215.0	371.9	0.002358	0.00442	1882.3	2183.0	1933.0	344.9	2277.9	4.175	0.535	4,710
220.0	373.8	0.002742	0.00357	1961.9	2087.1	2022.2	143.4	2165.6	4.311	0.222	4.533

(continued)

Table A.2 (continued)

Sat. pressure bar	Sat. temp. °C	Specific volume m³/kg		Specific internal energy kJ/kg		Specific enthalpy kJ/kg			Specific entropy kJ/kgK		
p_{sat}	T_{sat}	Sat. liquid v_f	Sat. vapor v_g	Sat. liquid u_f	Sat. vapor u_g	Sat. liquid h_f	Evap. h_{fg}	Sat. vapor h_g	Sat. liquid s_f	Evap. s_{fg}	Sat. vapor s_g
221.2	374.15	0.003155	0.003155	2029.6	2029.6	2099.3	0	2099.3	4.4298	0	4.4298

Table A.3 Superheated steam table

T °C	p = 0.1 bar (45.8 °C)				p = 0.5 bar (81.3 °C)				p = 1 bar (99.6 °C)			
	v m³/kg	u kJ/kg	h kJ/kg	s kJ/kgK	v m³/kg	u kJ/kg	h kJ/kg	s kJ/kgK	v m³/kg	u kJ/kg	h kJ/kg	s kJ/kgK
Sat	14.674	2437.9	2584.7	8.1502	3.240	2483.9	2645.9	7.5939	1.6940	2506.1	2675.5	7.3594
50	14.869	2483.9	2592.6	8.1749								
100	17.196	2515.5	2687.5	8.4479	3.418	2511.6	2682.5	7.6947	1.6958	2506.7	2676.2	7.3614
150	19.512	2587.9	2783.0	8.6882	3.889	2585.6	2780.1	7.9401	1.9364	2582.8	2776.4	7.6134
200	21.825	2661.3	2879.5	8.9038	4.356	2659.9	2877.7	8.1580	2.172	2658.1	2875.3	7.8343
250	24.136	2736.0	2977.3	9.1002	4.820	2735.0	2976.0	8.3556	2.406	2733.7	2974.3	8.0333
300	26.445	2812.1	3075.5	9.2813	5.284	2811.3	3075.5	8.5373	2.639	2810.4	3074.3	8.2158
400	31.063	2968.9	3279.6	9.6077	6.209	2968.5	3278.9	8.8642	3.103	2967.9	3278.2	8.5435
500	35.679	3132.3	3489.1	9.8978	7.134	3132.0	3488.7	9.1546	3.565	3131.6	3488.1	8.8342
600	40.295	3302.5	3705.4	10.1608	8.057	3302.2	3705.1	9.4178	4.028	3301.9	3704.4	9.0976
700	44.911	3479.6	3928.7	10.4028	8.981	3479.4	3928.5	9.6599	4.490	3479.2	3928.2	9.3398
800	49.526	3663.8	4159.0	10.6281	9.904	3663.6	4158.0	9.8852	4.952	3663.5	4158.6	9.5652
900	54.141	3855.0	4396.4	10.8396	10.828	3854.9	4396.3	10.0967	5.414	3854.8	4396.1	9.7767
1000	58.757	4053.0	4640.6	11.0393	11.751	4052.9	4640.5	10.2964	5.875	4052.8	4640.3	9.9764
1100	63.372	4257.5	4891.2	11.2287	12.674	4257.4	4891.1	10.4859	6.337	4257.3	4891.0	10.1659
1200	67.987	4467.9	5147.8	11.4091	13.597	4467.8	5147.7	10.6662	6.799	4467.7	5147.6	10.3463
1300	72.602	4683.7	5409.7	11.5811	14.521	4683.6	5409.6	10.8382	7.260	4683.5	5409.5	10.5183

Table A.4 Superheated steam table

T °C	v m³/kg	u kJ/kg	h kJ/kg	s kJ/kgK	v m³/kg	u kJ/kg	h kJ/kg	s kJ/kgK	v m³/kg	u kJ/kg	h kJ/kg	S kJ/kgK
	p = 2 bar (120.2 °C)				p = 3 bar (133.5 °C)				p = 4 bar (143.6 °C)			
Sat	0.8857	2529.5	2706.7	7.1272	0.6058	2543.6	2725.3	6.9919	0.4625	2553.6	2738.6	6.8959
150	0.9596	2576.9	2768.8	7.2795	0.6339	2570.8	2761.0	7.0778	0.4708	2564.5	2752.8	6.9299
200	1.0803	2654.4	2870.5	7.5066	0.7163	2650.7	2865.6	7.3115	0.5342	2646.8	2860.5	7.1706
250	1.1988	2731.2	2971.0	7.7086	0.7964	2728.7	2967.6	7.5166	0.5951	2726.1	2964.2	7.3789
300	1.3162	2808.6	3071.8	7.8926	0.8753	2806.7	3069.3	7.7022	0.6548	2804.8	3066.8	7.5662
400	1.5493	2966.7	3276.6	8.2218	1.0315	2965.6	3275.0	8.0330	0.7726	2964.4	3273.4	7.8985
500	1.7814	3130.8	3487.1	8.5133	1.1867	3130.0	3486.0	8.3251	0.8893	3129.2	3484.9	8.1913
600	2.013	3301.4	3704.0	8.7770	1.3414	3300.8	3703.2	8.5892	1.0055	3300.2	3702.4	8.4558
700	2.244	3478.8	3927.6	9.0194	1.4957	3478.4	3927.1	8.8319	1.1215	3477.9	3926.5	8.6987
800	2.475	3663.1	4158.2	9.2449	1.6499	3662.9	4157.8	9.0576	1.2312	3662.4	4157.3	8.9244
900	2.705	3854.5	4395.8	9.4566	1.8041	3854.2	4395.4	9.2692	1.3529	3853.9	4395.1	9.1362
1000	2.937	4052.5	4640.0	9.5663	1.9581	4052.3	4639.7	9.4690	1.4685	4052.0	4639.4	9.3360
1100	3.168	4257.0	4890.7	9.8458	2.1121	4256.8	4890.4	9.6585	1.5840	4256.5	4890.2	9.5256
1200	3.399	4467.5	5147.5	10.0262	2.2661	4467.2	5147.1	9.8389	1.6996	4467.0	5146.8	9.7060
1300	3.630	4683.2	5409.3	10.1982	2.4201	4583.0	5409.0	10.0110	1.8151	4682.8	5408.8	9.8780

Table A.5 Superheated steam table

T °C	v m³/kg	u kJ/kg	h kJ/kg	s kJ/kgK	v m³/kg	u kJ/kg	h kJ/kg	s kJ/kgK	v m³/kg	u kJ/kg	h kJ/kg	s kJ/kgK
	p = 5 bar (151.9 °C)				p = 6 bar (158.9 °C)				p = 8 bar (170.4 °C)			
Sat	0.3749	2561.2	2748.7	6.8213	0.3157	2567.4	2756.8	6.7600	0.2404	2576.8	2769.1	6.6628
200	0.4249	2642.9	2855.4	7.0592	0.3520	2638.9	2850.1	6.9665	0.2608	2630.6	2839.3	6.8158
250	0.4744	2723.5	2960.7	7.2709	0.3938	2720.9	2957.2	7.1816	0.2931	2715.5	2950.0	7.0384
300	0.5226	2802.9	3064.2	7.4599	0.4344	2801.0	3061.6	7.3724	0.3241	2797.2	3056.5	7.2328
350	0.5701	2882.6	3167.7	7.6329	0.4742	2881.2	3165.7	7.5464	0.3544	2878.2	3161.7	7.4089
400	0.6173	2963.2	3271.9	7.7938	0.5137	2962.1	3270.3	7.7079	0.3843	2959.7	3267.1	7.5716
500	0.7109	3128.4	3483.9	8.0873	0.5920	3127.6	3482.8	8.0021	0.4433	3126.0	3480.6	7.8673
600	0.8041	3299.6	3701.7	8.3522	0.6697	3299.1	3700.9	8.2674	0.5018	3297.9	3699.4	8.1333
700	0.8969	3477.5	3925.9	8.5952	0.7472	3477.0	3925.3	8.5107	0.5601	3476.2	3924.2	8.3770
800	0.9896	3662.1	4156.9	8.8211	0.8215	3661.8	4156.5	8.7367	0.6181	3661.1	4155.6	8.6033
900	1.0822	3853.6	4394.7	9.0329	0.9017	3853.4	4394.4	8.9486	0.6761	3852.8	4393.7	8.8153
1000	1.1747	4051.8	4639.1	9.2328	0.9788	4051.5	4638.8	9.1485	0.7340	4051.0	4638.2	9.0153
1100	1.2672	4256.3	4889.9	9.4224	1.0559	4256.1	4889.6	9.3381	0.7919	4255.6	4889.1	9.2050
1200	1.3596	4466.8	5146.6	9.6029	1.1330	4466.5	5146.3	9.5185	0.8497	4466.1	5145.9	9.3855
1300	1.4521	4682.5	5408.6	9.7749	1.2101	4682.3	5408.3	9.6906	0.9076	4681.8	5407.9	9.5575

Table A.6 Superheated steam table

T °C	v m³/kg	u kJ/kg	h kJ/kg	s kJ/kgK	v m³/kg	u kJ/kg	h kJ/kg	s kJ/kgK	v m³/kg	u kJ/kg	h kJ/kg	s kJ/kgK
	p = 10 bar (179.9 °C)				p = 12 bar (188 °C)				p = 14 bar (195 °C)			
Sat	0.19444	2583.6	2778.1	6.5865	0.16333	2583.8	2784.8	6.5233	0.14084	2592.8	2790.0	6.4693
200	0.2060	2621.9	2827.9	6.6940	0.16930	2612.8	2815.9	6.5898	0.14302	2603.1	2803.3	6.4975
250	0.2327	2709.9	2942.6	6.9247	0.19234	2704.2	2935.0	6.8294	0.16350	2698.3	2927.2	6.7467
300	0.2579	2793.2	3051.2	7.1229	0.2138	2789.2	3045.8	7.0317	0.18228	2785.2	3040.4	6.9534
350	0.2825	2875.2	3157.7	7.3011	0.2345	2872.2	3153.6	7.2121	0.2003	2869.2	3149.5	7.1370
400	0.3066	2957.3	3263.9	7.4651	0.2548	2954.9	3260.7	7.3774	0.2178	2952.5	3257.5	7.3026
500	0.3541	3124.4	3478.5	7.7622	0.2946	3122.8	3476.3	7.6759	0.2521	3121.1	3474.1	7.6027
600	0.4011	3296.8	3697.9	8.0290	0.3339	3295.5	3696.3	7.9435	0.2860	3294.4	3694.8	7.8710
700	0.4478	3475.3	3923.1	8.2731	0.3729	3474.4	3922.0	8.1881	0.3195	3473.6	3920.8	8.1160
800	0.4943	3660.4	4154.7	8.4996	0.4118	3659.7	4153.8	8.4148	0.3528	3659.0	4153.0	8.3431
900	0.5407	3852.2	4392.9	8.7118	0.4505	3851.6	4392.2	8.6272	0.3851	3851.1	4391.5	8.5556
1000	0.5871	4050.5	4637.6	8.9119	0.4892	4050.0	4637.0	8.8274	0.4192	4049.5	4636.4	8.7559
1100	0.6335	4255.1	4888.6	9.1017	0.5278	4254.6	4888.0	9.0172	0.4524	4254.1	4887.5	8.9457
1200	0.6798	4465.6	5145.4	9.2822	0.5665	4465.1	5144.9	9.1977	0.4855	4464.7	5144.4	9.1262
1300	0.7261	4681.3	5407.4	9.4543	0.6051	4580.9	5407.0	9.3698	0.5186	4680.4	5406.5	9.2984

Table A.7 Superheated steam table

T °C	v m³/kg	u kJ/kg	h kJ/kg	s kJ/kgK	v m³/kg	u kJ/kg	h kJ/kg	s kJ/kgK	v m³/kg	u kJ/kg	h kJ/kg	s kJ/kgK
	$p = 16$ bar (201.4 °C)				$p = 18$ bar (207.2 °C)				$p = 20$ bar (212.4 °C)			
Sat	0.12380	2596.0	2794.0	6.4218	0.11042	2598.4	2797.1	6.3794	0.09963	2600.3	2799.5	6.3409
225	0.13287	2644.7	2857.3	6.5518	0.11673	2636.6	2846.7	6.4808	0.10377	2628.3	2835.8	6.4147
250	0.14184	2692.3	2919.2	6.6732	0.12497	2686.0	2911.0	6.6066	0.11144	2679.6	2902.5	6.5453
300	0.15862	2781.1	3034.8	6.8844	0.14021	2776.9	3029.2	6.8226	0.12547	2772.6	3023.5	6.7664
350	0.17456	2866.1	3145.4	7.0694	0.15457	2863.0	3141.2	7.0100	0.13857	2859.8	3137.0	6.9563
400	0.19005	2950.1	3254.2	7.2374	0.16847	2947.7	3250.9	7.1794	0.15120	2945.2	3247.6	7.1271
500	0.2203	3119.5	3472.0	7.5390	0.19550	3117.9	3469.8	7.4825	0.17568	3116.2	3467.6	7.4317
600	0.2500	3293.3	3693.2	7.8080	0.2220	3292.1	3691.7	7.7523	0.19960	3290.9	3690.1	7.7024
700	0.2794	3472.7	3919.7	8.0535	0.2482	3471.8	3971.8	7.9983	0.2232	3470.9	3917.4	7.9487
800	0.3086	3658.3	4152.1	8.2808	0.2742	3657.6	4151.2	8.2258	0.2467	3657.0	4150.3	8.1765
900	0.3377	3850.5	4390.8	8.4935	0.3001	3849.9	4390.1	8.4386	0.2700	3849.3	4389.4	8.3895
1000	0.3668	4049.0	4635.8	8.6938	0.3260	4048.5	4635.2	8.6391	0.2933	4048.0	4634.6	8.5901
1100	0.3958	4253.7	4887.0	8.8837	0.3518	4253.2	4886.4	8.8290	0.3166	4252.7	4885.9	8.7800
1200	0.4248	4464.2	5143.9	9.0643	0.3776	4463.7	5143.4	9.0096	0.3398	4463.3	5142.9	8.9607
1300	0.4538	4679.9	5406.0	9.2364	0.4034	4679.5	5405.6	9.1818	0.3631	4679.0	5405.1	9.1329

Table A.8 Superheated steam table

T °C	p = 25 bar (224 °C)				p = 30 bar (233.9 °C)				p = 35 bar (242.6 °C)			
	v m³/kg	u kJ/kg	h kJ/kg	s kJ/kgK	v m³/kg	u kJ/kg	h kJ/kg	s kJ/kgK	v m³/kg	u kJ/kg	h kJ/kg	s kJ/kgK
Sat	0.07998	2603.1	2803.1	6.2575	0.06668	2604.1	2804.2	6.1869	0.05707	2603.7	2803.4	6.1253
225	0.08027	2605.6	2806.3	6.2639								
250	0.08700	2662.6	2880.1	6.4085	0.07058	2644.0	2855.8	6.2872	0.05872	2623.7	2829.2	6.1749
300	0.09890	2761.6	3008.8	6.6438	0.08114	2750.1	2993.5	6.5390	0.06842	2738.0	2977.5	6.4461
350	0.10976	2851.9	3126.3	6.8403	0.09053	2843.7	3115.3	6.7428	0.07678	2835.3	3104.0	6.6579
400	0.12010	2939.1	3239.3	7.0148	0.09936	2932.8	3230.9	6.9212	0.08453	2926.4	3222.3	6.8405
450	0.13014	3025.5	3350.8	7.1746	0.10787	3020.4	3344.0	7.0834	0.09196	3015.3	3337.2	7.0052
500	0.13993	3112.1	3462.1	7.3234	0.11619	3108.0	3456.5	7.2338	0.09918	3103.0	3450.9	7.1572
600	0.15930	3288.0	3686.3	7.5960	0.13243	3285.0	3682.3	7.5085	0.11324	3282.1	3678.4	7.4339
700	0.17832	3468.7	3914.5	7.8435	0.14838	3466.5	3911.7	7.7571	0.12699	3464.3	3908.8	7.6837
800	0.19716	3655.3	4148.2	8.0720	0.16414	3653.5	4145.9	7.9862	0.14056	3651.8	4143.7	7.9134
900	0.21590	3847.9	4387.6	8.2853	0.17980	3846.5	4385.9	8.1999	0.15402	3845.0	4384.1	8.1276
1000	0.2346	4046.7	4633.1	8.4861	0.19541	4045.4	4631.6	8.4009	0.16743	4044.1	4630.1	8.3288
1100	0.2532	4251.5	4884.6	8.6762	0.21098	4250.3	4883.3	8.5912	0.18080	4249.2	4881.9	8.5192
1200	0.2718	4462.1	5141.7	8.8569	0.22652	4460.9	5140.5	8.7720	0.19415	4459.8	5139.3	8.7000
1300	0.2905	4677.8	5404.0	9.0291	0.24206	4676.6	5402.8	8.9442	0.20749	4675.5	5401.7	8.8723

Table A.9 Superheated steam table

T °C	p = 40 bar (250.4 °C)				p = 45 bar (257.5 °C)				p = 50 bar (264 °C)			
	v m³/kg	u kJ/kg	h kJ/kg	s kJ/kgK	v m³/kg	u kJ/kg	h kJ/kg	s kJ/kgK	v m³/kg	u kJ/kg	h kJ/kg	s kJ/kgK
Sat	0.04978	2602.3	2801.4	6.0701	0.04406	2600.1	2798.3	6.0198	0.03944	2597.1	2794.3	5.9734
275	0.05457	2667.9	2886.2	6.2285	0.04730	2650.3	2863.2	6.1401	0.04141	2631.3	2838.3	6.0544
300	0.05884	2725.3	2960.7	6.3615	0.05135	2712.0	2943.1	6.2828	0.04532	2698.0	2924.5	6.2084
350	0.06645	2826.7	3092.5	6.5821	0.05840	2817.8	3080.6	6.5131	0.05194	2808.7	3068.4	6.4493
400	0.07341	2919.9	3213.6	6.7690	0.06475	2913.3	3204.7	6.7047	0.05781	2906.6	3195.7	6.6459
450	0.08002	3010.2	3330.3	6.9363	0.07074	3005.0	3323.3	6.8746	0.06330	2999.7	3316.2	6.8186
500	0.08643	3099.5	3445.3	7.0901	0.07651	3095.3	3439.6	7.0301	0.06857	3091.0	3433.8	6.9759
600	0.09885	3279.1	3674.4	7.3688	0.08765	3276.0	3670.5	7.3110	0.07869	3273.0	3666.5	7.2589
700	0.11095	3462.1	3905.9	7.6198	0.09847	3459.9	3903.0	7.5631	0.08849	3457.6	3900.1	7.5122
800	0.12287	3650.0	4141.5	7.8502	0.10911	3648.3	4139.3	7.7942	0.09811	3646.6	4137.1	7.7440
900	0.13469	3843.6	4382.3	8.0647	0.11965	3842.2	4380.6	8.0091	0.10762	3840.7	4378.8	7.9593
1000	0.14645	4042.9	4628.7	8.2662	0.13013	4041.6	4627.2	8.2108	0.11707	4040.4	4625.7	8.1612
1100	0.15817	4248.0	4880.6	8.4567	0.14056	4246.8	4879.3	8.4015	0.12648	4245.6	4878.0	8.3520
1200	0.16987	4458.6	5138.1	8.6376	0.15098	4457.5	5136.9	8.5825	0.13587	4456.3	5135.7	8.5331
1300	0.18156	4674.3	5400.5	8.8100	0.16139	4673.1	5399.4	8.7549	0.14526	4672.0	5398.2	8.7055

Table A.10 Superheated steam table

T °C	$p = 60$ bar (275.6 °C)				$p = 70$ bar (285.9 °C)				$p = 80$ bar (295.1 °C)			
	$v\,\mathrm{m^3/kg}$	$u\,\mathrm{kJ/kg}$	$h\,\mathrm{kJ/kg}$	$s\,\mathrm{kJ/kgK}$	$v\,\mathrm{m^3/kg}$	$u\,\mathrm{kJ/kg}$	$h\,\mathrm{kJ/kg}$	$s\,\mathrm{kJ/kgK}$	$v\,\mathrm{m^3/kg}$	$u\,\mathrm{kJ/kg}$	$h\,\mathrm{kJ/kg}$	$s\,\mathrm{kJ/kgK}$
Sat	0.03244	2589.7	2784.3	5.8892	0.02737	2580.5	2772.1	5.8133	0.02352	2569.0	2758.0	5.7432
300	0.03616	2667.2	2884.2	6.0674	0.02947	2632.2	2838.4	5.9305	0.02426	2590.9	2785.0	5.7906
350	0.04223	2789.6	3043.0	6.3335	0.03524	2769.4	3016.0	6.2283	0.02995	2747.7	2987.3	6.1301
400	0.04739	2892.0	3177.2'	6.5408'	0.03993	2878.6	3158.1	6.4478	0.03432	2863.8	3138.3	6.3634
450	0.05214	2988.9	3301.8	6.7193	0.04416	2978.0	3287.1	6.6327	0.03817	2966.7	3272.0	6.5551
500	0.05665	3082.2	3422.2	6.8803	0.04814	3073.4	3410.3	6.7980	0.04175	3064.3	3398.3	6.7240
550	0.06101	3174.6	3540.6	7.0288	0.05195	3167.2	3530.9	6.9486	0.04516	3159.8	3521.0	6.8778
600	0.06525	3266.9	3658.4	7.1677	0.05565	3260.7	3650.3	7.0894	0.04845	3254.4	3642.0	7.0206
700	0.07352	3453.1	3894.2	7.4234	0.06283	3448.5	3888.3	7.3476	0.05481	3443.9	3882.4	7.2812
800	0.08160	3643.1	4132.7	7.6566	0.06981	3639.5	4128.2	7.5822	0.06097	3636.0	4123.8	7.5173
900	0.08958	3837.8	4375.3	7.8727	0.07669	3835.0	4371.8	7.7991	0.06702	3832.1	4368.3	7.7351
1000	0.09749	4037.8	4622.7	8.0751	0.08350	4035.3	4619.8	8.0020	0.07301	4032.8	4616.9	7.9384
1100	0.10536	4243.3	4875.4	8.2661	0.09027	4240.9	4872.8	8.1933	0.07896	4238.6	4870.3	8.1300
1200	0.11321	4454.0	5133.3	8.4474	0.09703	4451.7	5130.9	8.3747	0.08489	4449.5	5128.5	8.3115
1300	0.12106	4669.6	5396.0	8.6199	0.10377	4667.3	5393.7	8.5475	0.09080	4665.0	5391.5	8.4842

Table A.11 Superheated steam table

T °C	v m³/kg	u kJ/kg	h kJ/kg	s kJ/kg K	v m³/kg	u kJ/kg	h kJ/kg	s kJ/kgK	v m³/kg	u kJ/kg	h kJ/kg	s kJ/kg K
	$p = 90$ bar (303.4 °C)				$p = 100$ bar (311.1 °C)				$p = 125$ bar (327.9 °C)			
Sat	0.02048	2557.8	2742.1	5.6112	0.018026	2544.4	2724.7	5.6141	0.013495	2505.1	2673.8	5.4624
325	0.02327	2646.6	2856.0	5.8712	0.019861	2610.4	2809.1	5.7568				
350	0.02580	2724.4	2956.6	6.0361	0.02242	2699.2	2923.4	5.9443	0.016126	2624.6	2826.2	5.7118
400	0.02993	2848.4	3117.8	6.2854	0.02641	2832.4	3096.5	6.2120	0.02000	2789.3	3039.3	6.0417
450	0.03350	2955.2	3256.6	6.4844	0.02975	2943.4	3240.9	6.4190	0.02299	2912.5	3199.8	6.2719
500	0.03677	3055.2	3386.1	6.6576	0.03279	3045.8	3373.7	6.5966	0.02560	3021.7	3341.8	6.4618
550	0.03987	3152.2	3511.0	6.8142	0.03564	3144.1	3500.9	6.7561	0.02801	3125.0	3475.2	6.6290
600	0.04285	3248.1	3633.7	6.9589	0.03837	3241.7	3625.3	6.9029	0.03029	3225.4	3604.0	6.7810
650	0.04574	3343.6	3755.3	7.0943	0.04101	3338.2	3748.2	7.0398	0.03248	3324.4	3730.4	6.9218
700	0.04857	3439.3	3876.5	7.2221	0.04358	3434.7	3870.5	7.1687	0.03460	3422.9	3855.3	7.0536
800	0.05409	3632.5	4119.3	7.4596	0.04859	3628.9	4114.8	7.4077	0.03869	3620.0	4103.6	7.2965
900	0.05950	3829.2	4364.8	7.6783	0.05349	3826.3	4361.2	7.6272	0.04267	3819.1	4352.5	7.5182
1000	0.06485	4030.3	4614.0	7.8821	0.05832	4027.8	4611.0	7.8315	0.04658	4021.6	4603.8	7.7237
1100	0.07016	4236.3	4867.7	8.0740	0.06312	4234.0	4865.1	8.0237	0.05045	4228.2	4858.8	7.9165
1200	0.07544	4447.2	5126.2	8.2556	0.06789	4444.9	5123.8	8.2055	0.05430	4439.3	5118.0	8.0937
1300	0.08072	4662.7	5389.2	8.4284	0.07265	4660.5	5387.0	8.3783	0.05813	4654.8	5381.4	8.2717

Table A.12 Superheated steam table

T °C	v m³/kg	u kJ/kg	h kJ/kg	s kJ/kg K	v m³/kg	u kJ/kg	h kJ/kg	s kJ/kg K	v m³/kg	u kJ/kg	h kJ/kg	s kJ/kg K
	p = 150 bar (342.2 °C)				p = 175 bar (354.7 °C)				p = 200 bar (365.8 °C)			
Sat	0.010337	2455.5	2610.5	5.3098	0.007920	2390.2	2528.8	5.1419	0.005834	2293.0	2409.7	4.9269
350	0.011470	2520.4	2692.4	5.4421								
400	0.015649	2740.7	2975.5	5.8811	0.012447	2685.0	2902.9	5.7213	0.009942	2619.3	2818.1	5.5540
450	0.018445	2879.5	3156.2	6.1404	0.015174	2844.2	3109.7	6.0184	0.012695	2806.2	3060.1	5.9017
500	0.02080	2996.6	3308.6	6.3443	0.017358	2970.3	3274.1	6.2383	0.014768	2942.9	3238.2	6.1401
550	0.02293	3104.7	3448.6	6.5199	0.019288	3083.9	3421.4	6.4230	0.016555	3062.4	3393.5	6.3348
600	0.02491	3208.6	3582.3	6.6776	0.02106	3191.5	3560.1	6.5866	0.018178	3174.0	3537.6	6.5048
650	0.02680	3310.3	3712.3	6.8224	0.02274	3296.0	3693.9	6.7357	0.019693	3281.4	3675.3	6.6582
700	0.02861	3410.9	3840.1	6.9572	0.02434	3324.5	3824.6	6.8736	0.02113	3386.4	3809.0	6.7993
800	0.03210	3610.9	4092.4	7.2040	0.02738	3601.8	4081.1	7.1244	0.02385	3592.7	4069.7	7.0544
900	0.03546	3811.9	4343.8	7.4279	0.03031	3804.7	4335.1	7.3507	0.02645	3797.5	4326.4	7.2830
1000	0.03875	4015.4	4596.6	7.6348	0.03316	4009.3	4589.5	7.5589	0.02897	4003.1	4582.5	7.4925
1100	0.04200	4222.6	4852.6	7.8283	0.03597	4216.9	4846.4	7.7531	0.03145	4211.3	4840.2	7.6874
1200	0.04523	4433.8	5112.3	8.0108	0.03876	4428.3	5106.6	7.9360	0.03391	4422.8	5101.0	7.8707
1300	0.04845	4649.1	5376.0	8.1840	0.04154	4643.5	5370.5	8.1093	0.03636	4638.0	5365.1	8.0442

Table A.13 Superheated steam table

T °C	p = 250 bar				p = 300 bar				p = 350 bar			
	v m³/kg	u kJ/kg	h kJ/kg	s kJ/kg K	v m³/kg	u kJ/kg	h kJ/kg	s kJ/kgK	v m³/kg	u kJ/kg	h kJ/kg	s kJ/kg K
375	0.001973	1798.7	1848.0	4.0320	0.0017892	1737.8	1791.5	3.9305	0.0017003	1702.9	1762.4	3.8722
400	0.006004	2430.1	2580.2	5.1418	0.002790	2067.4	2151.1	4.4728	0.002100	1914.1	1987.6	4.2126
425	0.007881	2609.2	2806.3	5.4723	0.005303	2455.1	2614.2	5.1504	0.003428	2253.4	2373.4	4.7747
450	0.009162	2720.7	2949.7	5.6744	0.006735	2619.3	2821.4	5.4424	0.004961	2498.7	2672.4	5.1962
500	0.011123	2884.3	3162.4	5.9592	0.008678	2820.7	3081.1	5.7905	0.006927	2751.9	2994.4	5.6282
550	0.012724	3017.5	3335.6	6.1765	0.010168	2970.3	3275.4	6.0342	0.008345	2921.0	3213.0	5.9026
600	0.014137	3137.9	3491.4	6.3602	0.011446	3100.5	3443.9	6.2331	0.009527	3062.0	3395.5	6.1179
650	0.015433	3251.6	3637.4	6.5229	0.012596	3221.0	3598.9	6.4058	0.010575	3189.8	3559.9	6.3010
700	0.016646	3361.3	3777.5	6.6707	0.013661	3335.8	3745.6	6.5606	0.011533	3309.8	3713.5	6.4631
800	0.018912	3574.3	4047.1	6.9345	0.015623	3555.5	4024.2	6.8332	0.013278	3536.7	4001.5	6.7450
900	0.021045	3783.0	4309.1	7.1680	0.017448	3768.5	4291.9	7.0718	0.014883	3754.0	4274.9	6.9386
1000	0.02310	3990.9	4568.5	7.3802	0.019196	3978.8	4554.7	7.2867	0.016410	3966.7	4541.1	7.2064
1100	0.02512	4200.2	4828.2	7.5765	0.020903	4189.2	4816.3	7.4845	0.017895	4178.3	4804.6	7.4037
1200	0.02711	4412.0	5089.9	7.7605	0.022589	4401.3	5079.0	7.6692	0.019360	4390.7	5068.3	7.5910
1300	0.02910	4626.9	5354.4	7.9342	0.024266	4616.0	5344.0	7.8432	0.020815	4605.1	5333.6	7.7653

Table A.14 Superheated steam table

T °C	v m³/kg	u kJ/kg	h kJ/kg	s kJ/kgK	v m³/kg	u kJ/kg	h kJ/kg	s kJ/kg K	v m³/kg	u kJ/kg	h kJ/kg	s kJ/kg K
	p = 400 bar				p = 500 bar				p = 600 bar			
375	0.0016407	1677.1	1742.8	3.8290	0.0015594	1638.6	1716.6	3.7639	0.0015028	1609.4	1699.5	3.7141
400	0.0019077	1854.6	1930.9	4.1135	0.0017309	1788.1	1874.6	4.0031	0.0016335	1745.4	1843.4	3.9318
425	0.002532	2096.9	2198.1	4.5029	0.002007	1959.7	2060.0	4.2734	0.0018165	1892.7	2001.7	4.1626
450	0.003693	2365.1	2512.8	4.9459	0.002486	2159.6	2284.0	4.5884	0.002085	2053.9	2179.0	4.4121
500	0.005622	2678.4	2903.3	5.4700	0.003892	2525.5	2720.1	5.1726	0.002956	2390.6	2567.9	4.9321
550	0.006984	2869.7	3149.1	5.7785	0.005118	2763.6	3019.5	5.5485	0.003956	2658.8	2896.2	5.3441
600	0.008094	3022.6	3346.4	6.0144	0.006112	2942.0	3247.6	5.8178	0.004834	2861.1	3151.2	5.6452
650	0.009063	3158.0	3520.6	6.2054	0.006966	3093.5	3441.8	6.0342	0.005595	3028.8	3364.5	5.8829
700	0.009941	3283.6	3681.2	6.3750	0.007727	3230.5	3616.8	6.2189	0.006272	3177.2	3553.5	6.0824
800	0.011523	3517.8	3978.7	6.6662	0.009076	3479.8	3933.6	6.5290	0.007459	3441.5	3889.1	6.4109
900	0.012962	3739.4	4257.9	6.9150	0.010283	3710.3	4224.4	6.7882	0.008508	3681.0	4191.5	6.6805
1000	0.014324	3954.6	4527.6	7.1356	0.011411	3930.5	4501.1	7.0146	0.009480	3906.4	4475.2	6.9121
1100	0.015642	4167.4	4793.1	7.3364	0.012496	4145.7	4770.5	7.2184	0.010409	4124.1	4748.6	7.1195
1200	0.016940	4380.1	5057.7	7.5224	0.013561	4359.1	5037.2	7.4058	0.011317	4338.2	5017.2	7.3083
1300	0.018229	4594.3	5323.5	7.6969	0.014616	4572.8	5303.6	7.5808	0.012215	4551.4	5284.3	7.4837

Bibliography

David R. Gaskell, Thermodynamics of material, 4th Edition, Taylor and Francis, New York.

V, Ganesan, Internal Combustion Engines, 3rd Edition, Tata McGraw Hill Publishing Company Ltd., New Delhi.

M. L. Mathur and R, P, Sharma, Internal Combustion Engines,7th Edition, Dhanpat Rai Publications, New Delhi.

Mark Zemansky, Richard Dittman, Heat and Thermodynamics, 7th Edition, Tata McGraw Hill Publishing Company Ltd., New Delhi.

Michel A. Saad, Thermodynamics for Engineers, 2nd Edition, Prentice-Hall, New Delhi.

P.K. Nag, Engineering Thermodynamics, 4th Edition, Tata McGraw Hill Publishing Company Ltd., New Delhi.

M.C. Potter and E.P. Scott, Thermal SCIENCE, Cengage Learning India Pvt. Ltd. 2008.

E. Rathakrishnan, Engineering Thermodynamics, 2nd Edition, PHI Learning Pvt. Ltd., New Delhi.

Shiv Kumar, Fluid Mechanics, 1st Edition, Ane books Pvt. Ltd., New Delhi.

Sonntag, Borgnakke, Van Wylen, Fundamentals of Thermodynamics, 5th Edition, *Wiley India* Private Limited, New Delhi.

Yunus A. Cengel and Michael A. Boles, Thermodynamics, 5th Edition, Tata McGraw Hill Publishing Company Ltd., New Delhi.

© The Author(s) 2022
S. Kumar, *Thermal Engineering Volume 1*,
https://doi.org/10.1007/978-3-030-67274-4

Index

© The Author(s) 2022
S. Kumar, *Thermal Engineering Volume 1*,
https://doi.org/10.1007/978-3-030-67274-4

Printed in the United States
by Baker & Taylor Publisher Services